Abraham P. Punnen
Editor

The Quadratic Unconstrained Binary Optimization Problem

Theory, Algorithms, and Applications

Springer

Editor
Abraham P. Punnen
Department of Mathematics
Simon Fraser University
Surrey, BC, Canada

ISBN 978-3-031-04519-6 ISBN 978-3-031-04520-2 (eBook)
https://doi.org/10.1007/978-3-031-04520-2

This Springer imprint is published by the registered company Springer Nature Switzerland AG
The registered company address is: Gewerbestrasse 11, 6330 Cham, Switzerland

The Quadratic Unconstrained Binary Optimization Problem

This book is dedicated to the memory of the millions of people who lost their lives due to the COVID-19 pandemic.

Preface

Studies on the quadratic unconstrained binary optimization problem (QUBO) can be traced back to the late 1950s and early 1960s, when there was much work on the more general class of pseudo-Boolean optimization problems. Since then, a large number of papers have been published on the topic from different points of view, including theory, algorithms, and applications. Moreover, QUBO has received renewed interest lately due to the development of quantum and quantum-inspired computers. The literature on QUBO has however been scattered across many disciplines and, up to now, no comprehensive volume had been published on the topic. This book is an effort to present important developments from the point of view of theory, algorithms, and applications of QUBO in consolidated form.

There are 11 chapters in this book, each focusing on a specific aspect of the problem. Chapter 1 provides an introduction to QUBO including some historical notes, sample applications, and general representations of the problem. Chapter 2 presents a thorough summary of various application areas, with a detailed reference list. It also covers some formulation techniques and sample experimental results. Chapter 3 highlights the computational complexity of the problem and contains a detailed discussion of polynomially solvable special cases. Chapter 4 deals with the Boolean quadric polytope and its applications in developing branch-and-cut algorithms for the problem. Chapter 5 deals with so-called autarkies and persistencies for QUBO. They can be used to fix variables a priori and to enhance exact and heuristic algorithms. Chapter 6 deals with exact algorithms. Various mixed-integer linear programming and semidefinite programming formulations of QUBO are discussed, and there is an overview of specially designed exact algorithms. The analysis of random QUBO instances is the topic of discussion in Chap. 7. Chapter 8 talks about approximation algorithms for QUBO, analyzed from a theoretical point of view, followed by Chap. 9, which provides a state-of-the-art account on metaheuristics for QUBO. The bipartite QUBO is the topic of discussion in Chap. 10, and Chap. 11 provides a summary of the currently available QUBO software, with pointers on how to access this software.

I believe researchers, students, and practitioners in various fields, including operations research, computer science, mathematics, and industrial engineering, will find this book useful as a state-of-the-art reference volume.

I would like to express my indebtedness to each of the authors who wrote various chapters of the book. Without their unrelenting support, this book would not have been possible. Each of the authors contributed significantly to the development of the field and, as one can see, citations to their works appear throughout the book. I am fortunate that such a distinguished and marvelous group of people collaborated on this project. The chapters were prepared at an unusually difficult time for everyone due to COVID-19, with universities closed, teaching completely moved online, worldwide travel bans, and, sadly, even seeing loved ones depart from the world.

Some chapter authors also participated in reading and providing feedback on various chapters. Many other distinguished scholars also shared their valuable time in providing feedback on the various chapters of the book. In particular, I am thankful to Fred Glover, Adam Letchford, Karthik Natarajan, Renata Sotirov, Brad Woods, Pooja Pandey, Snezana Minic, Gregory Gutin, Navpreet Kaur, Binay Bhattacharya, and Konstantin Makarychev for their feedback. I am thankful to Mikel Rodriguez for providing me with his drawing related to the "person-tracking" example. I am also thankful to Simon Fraser University for providing me with the facilities required to edit this book. The staff at Springer patiently supported the project, despite us missing one deadline after another. In particular, I would like to express my appreciation to Sujatha Chakkala, Katrin Petermann, and Rocio Torregrosa. In retrospect, perhaps the delays were a blessing in disguise: what a wonderful end product! Thank you again to all authors and everyone else involved. I believe that the book will remain an authoritative reference on QUBO for many years to come.

Surrey, BC, Canada, Abraham P. Punnen
31 December 2021

Contents

Contributors

Endre Boros MSIS Department and RUTCOR, Rutgers Business School, Rutgers University, Piscataway, NJ, USA

Eranda Çela Department of Discrete Mathematics, Graz University of Technology, Graz, Austria

Yu Du College of Business, University of Colorado at Denver, Denver, CO, USA

Fred Glover ECEE, College of Engineering and Applied Science, University of Colorado, Boulder, CO, USA

Jin-Kao Hao LERIA, Faculty of Sciences - University of Angers, Angers, France

Gary Kochenberger Entanglement, Inc., Westminster, CO, USA

Adam Letchford Department of Management Science, Lancaster University Management School, Lancaster, UK

Karthik Natarajan Engineering Systems and Design, Singapore University of Technology and Design, Singapore, Singapore

Abraham P. Punnen Department of Mathematics, Simon Fraser University Surrey, Surrey, BC, Canada

Renata Sotirov Department of Econometrics and Operations Research, Tilburg School of Economics and Management, Tilburg University, Tilburg, The Netherlands

Yang Wang School of Management, Northwestern Polytechnical University, Xi'an, China

Brad Woods 1QBit, Vancouver, BC, Canada

Acronyms

BQUBO	Bipartite quadratic unconstrained binary optimization problem
B-MaxCut	Maximum weight cut in bipartite graph
CMCS	Conditional Markov Chain Search
EMWCP	Euclidean maximum weight cut problem
FPTAS	Fully polynomial approximation scheme
ILP	Integer linear program
LP	Linear programming
MCCP	Maximum cardinality cut problem
MILP	Mixed-integer linear programming
MSSP	Maximum sum submatrix problem
MWBCP	Maximum weight biclique problem
MWClP	Maximum weight clique problem
MWCP	Maximum weight cut problem
MWSSP	Maximum weight stable set problem
QAP	Quadratic assignment problem
QCR	Quadratic convex reformulation
QUBO	Quadratic unconstrained binary optimization problem
PBF	Pseudo-Boolean function
psd	Positive semidefinite
RLT	Reformulation linearization technique
SCPT	Single machine scheduling with controllable processing times
SDP	Semidefinite programming
SEMWCP	Squared Euclidean maximum weight cut problem
SSP	Stable set problem
SWPT	Weighted shortest processing time rule
VLSI	Very large-scale system integration
VLSN	Very large-scale neighborhood

Chapter 1
Introduction to QUBO

Abraham P. Punnen

Abstract This chapter provides a general introduction to the quadratic unconstrained binary optimization problem (QUBO). Starting with a brief historical review, we present some basic definitions and notations, equivalent representations, examples of important combinatorial optimization problems that are equivalent to QUBO and some additional motivating examples. We also discuss some of the basic mathematical programming formulations of QUBO along with relevant pointers to the contents of other chapters of the book.

1.1 Introduction

Mathematical programming models play a vital role in the socio-economic developments of the modern-day society. Optimization frameworks such as linear programming, quadratic programming, combinatorial optimization, and mixed integer programming are effectively used in engineering design, finance, healthcare, economics, medicine, transportation, supply chains, environment, telecommunications among others. Perhaps the most fundamental among the various optimization modeling tools is linear programming. This model building framework, as we know it today, was originated around 1938 and evolved into an applicable modelling technique in the early 1950s with the discovery of the simplex method [14]. Realization of the benefits associated with having integrality restrictions on some or all of the decision variables of a linear program led to the development of integer programming and significant advancements in this area continues to emerge [42].

The literature on unconstrained nonlinear optimization on the other hand, can be traced back to early days of calculus [66]. However, the development of successful methods for solving constrained nonlinear optimization problems took a very long time. Special nonlinear programs of minimizing convex quadratic functions

A. P. Punnen (✉)
Department of Mathematics, Simon Fraser University, Surrey, BC, Canada
e-mail: apunnen@sfu.ca

A. P. Punnen (ed.), *The Quadratic Unconstrained Binary Optimization Problem*,
https://doi.org/10.1007/978-3-031-04520-2_1

over linear constraints can now be solved very efficiently. Study of constrained quadratic optimization problems started around the mid 1950s with applications in portfolio optimization [56, 57]. Systematic investigations on the quadratic programming problem with integrality restrictions on the decision variables took another decade to initiate [34, 53]. This book addresses a versatile unconstrained quadratic programming model called *quadratic unconstrained binary optimization problem* (QUBO) where the decision variables take values 0 or 1. The literature on QUBO is quite extensive and for a quick overview, we refer the reader to the survey papers [28, 45, 48].

Before getting into the technical details, let us start with some basic notations. All matrices are represented using bold capital letters and elements of a matrix are represented by the corresponding small letters along with accents, if any, and the location coordinates. For example the (i, j)th element of the matrix $\mathbf{A}$ is a_{ij}, of the matrix $\bar{\mathbf{B}}$ is $\bar{b}_{ij}$, and of the matrix $\mathbf{D}^k$ is d_{ij}^k. Similarly, vectors are represented by boldface small letters along with appropriate accents, as applicable, and elements of the vector is represented using the same letter (without boldface) along with its location coordinate and accents, if any. For example, the ith element of the vector $\mathbf{c}$ is c_i, of the vector $\mathbf{x}^k$ is x_i^k, and of the vector $\tilde{\mathbf{v}}$ is $\tilde{v}_i$. Exceptions to this rule will be stated explicitly and operators such as transpose etc. are not considered as accents in the case of vectors. The zero vector in any dimension is represented by $\mathbf{0}$. Additional notations will be introduced as need arises.

Let us now present a formal mathematical definition of QUBO. Let $\mathbf{Q}$ be an $n \times n$ matrix, $\mathbf{c}^T = (c_1, c_2, \ldots, c_n)$ be a row vector from $\mathbb{R}^n$, and $\mathbf{x}^T = (x_1, x_2, \ldots, x_n)$ be a row vector from $\{0, 1\}^n$. Then, QUBO is defined as the mathematical programming problem:

$$\begin{aligned}&\text{Maximize } \mathbf{x}^{\mathrm{T}}\mathbf{Q}\mathbf{x} + \mathbf{c}^T\mathbf{x}\\ &\text{Subject to}\\ &\qquad\qquad \mathbf{x} \in \{0, 1\}^n.\end{aligned}$$

Since $x_i^2 = x_i$ for any binary variable x_i, we can represent QUBO without the linear term by simply adding c_i to q_{ii} for $i = 1, 2, \ldots, n$. Alternatively, we can assume q_{ii} to be zero by replacing c_i with $c_i + q_{ii}$. In fact, QUBO is represented in many other alternative forms which are discussed in more detail later in this chapter. QUBO sometimes is presented as a minimization problem, different from the standard maximization form. The data $\mathbf{Q}, \mathbf{c}$ along with the orientation of optimization completely defines a problem instance. Thus, a maximization instance of QUBO can be represented by the triple $(\mathbf{Q}, \mathbf{c}, \max)$ and a minimization instance of QUBO can be represented by $(\mathbf{Q}, \mathbf{c}, \min)$. Note that, the instances $(\mathbf{Q}, \mathbf{c}, \max)$ and $(-\mathbf{Q}, -\mathbf{c}, \min)$ are equivalent. Unless otherwise specified, we assume that QUBO is presented in the maximization form and we normally represent such an instance by the ordered pair $(\mathbf{Q}, \mathbf{c})$.

Let us now look at an interpretation of QUBO from the point of view of matrix theory. A principal submatrix of $\mathbf{Q}$ is a square submatrix obtained by deleting rows and columns corresponding to an index set $S \subseteq \{1, 2, \ldots, n\}$. The value of a matrix is the sum of its elements. Without loss of generality assume $\mathbf{c} = \mathbf{0}$. Then QUBO is precisely the problem of computing a principal submatrix of $\mathbf{Q}$ with maximum value.

A graph theoretic interpretation of QUBO can be given as follows. Let $G = (V, E)$ be a graph such that $V = \{1, 2, \ldots, n\}$ and $(i, j) \in E$ if and only if $q_{ij} \neq 0$. The graph G is called the *support graph* of $\mathbf{Q}$. Let q_{ij} be the cost of edge $(i, j) \in E$ and $q_{ii} + c_i$ be the cost of the vertex $i \in V$. Let S be a subset of V and $G(S)$ be the subgraph of G induced by S. The cost of $G(S)$ is defined as the sum of the cost of its edges and vertices. Then the QUBO seeks a subset S of V such that the cost of the induced subgraph $G(S)$ of G is maximized.

Literature on QUBO, presented as a 0-1 quadratic program, can be traced back to 1960s, particularly with the work of Hammer and Rudeanu [34] on pseudo-Boolean functions and those of [13, 20, 21, 33, 80, 81]. Graph theoretic optimization problems such as the maximum weight stable set problem, the maximum weight clique problem, and the maximum cut problem have a hidden QUBO structure. As we will see later, these graph theoretic optimization problems are indeed equivalent to QUBO. In this sense, the history of QUBO is also linked to the origins of these graph theoretic structures.

A stable set of a graph $G = (V, E)$ is a subset S of V such that no two vertices of S are adjacent in G. Let d_i be a given weight associated with vertex $i \in V$. Then, the *maximum weight stable set problem* is to find a stable set S in G such that $\sum_{i \in S} d_i$ is maximized. It is easy to formulate the maximum weight stable set problem as a QUBO. Consider the binary decision variables x_i for $i = 1, 2, \ldots, n$ where $x_i = 1$ represents the event that vertex i is in the stable set. Then, for any stable set S and $(i, j) \in E$, if the distinct vertices $i, j \in S$ then $x_i x_j = 0$. This constraint can be forced using a quadratic term in the objective function. Define

$$q_{ij} = \begin{cases} d_i & \text{if } i = j \\ -M & \text{if } (i, j) \in E \\ 0 & \text{if } (i, j) \notin E, \end{cases}$$

where M is a large positive number. Also, choose $\mathbf{c}$ as the zero vector. Then, the maximum weight stable set problem can be solved as the QUBO instance $(\mathbf{Q}, \mathbf{c})$ where $\mathbf{Q}$ and $\mathbf{c}$ are as defined above. Interestingly, it is possible to formulate any instance $(\mathbf{Q}, \mathbf{c})$ of QUBO as a maximum weight stable set problem and this will be discussed in Sect. 1.4 along with the other equivalent forms of QUBO.

A *clique* in the graph G is a subset S of V such that the subgraph $G(S)$ of G induced by S is a complete graph. It is easy to see that S is a clique in G if and only if S is a stable set in the complement of G. Thus, in view of the discussion above, the maximum weight clique problem can also be formulated as a QUBO and it is also equivalent to QUBO.

Thus, from a historical perspective, cliques and stable sets play an indirect role in the evolution of QUBO. The literature on computing maximum weight clique and maximum weight stable set is quite extensive and we will not attempt a detailed review of historical developments of these structures. However, we briefly discuss below some of the earlier works that introduced these notions. Although clique (stable set) is a well established concept in graph theory at present, the terminology has its roots in social sciences [19, 22, 55]. Luce and Perry [55] in 1949 defined a clique as follows:

> A subset of the group forms a *clique* provided that it consists of three or more members each in the symmetric relation to each other member of the subset, and provided further that there can be found no element outside the subset that is in the symmetric relation to each of the elements of the subset.

The first algorithmic results on computing cliques was presented by Harrary and Ross [36]. Without using the name 'clique', the concept of complete subgraphs and independent sets (same as stable sets) were considered by Erdos and Szekeres [17] in 1935 while discussing Ramsey Theory in the context of graphs.

QUBO can also be written as a continuous optimization problem [70, 71]. Consider the box-constrained quadratic program

$$\begin{aligned}&\text{Maximize } \mathbf{x}^{\mathrm{T}}\mathbf{Q}\mathbf{x} + \mathbf{c}^{\mathrm{T}}\mathbf{x} - M\mathbf{x}^{\mathrm{T}}(\mathbf{e} - \mathbf{x})\\ &\text{Subject to}\\ &\qquad\qquad \mathbf{x} \in [0, 1]^n\end{aligned}$$

where M is a large positive number and $\mathbf{e}$ is the all-one vector in $\mathbb{R}^n$. Note that for large M, the objective function becomes convex and hence an optimal solution is attained at an extreme point of $[0, 1]^n$ [70, 71]. The collection of extreme points of $[0, 1]^n$ is precisely $\{0, 1\}^n$ establishing the validity of the above representation of QUBO.

QUBO is strongly NP-hard. A thorough complexity analysis of the problem and various polynomially solvable special cases are discussed in Chap. 3 and complexity results in connection with approximability is discussed in Chap. 8.

1.2 The Ising Model

Long before the QUBO become popular within the operations research community, the model was used in statistical mechanics in an alternative form [75]. This is popularly known as the Ising model where the variables take values from $\{-1, 1\}$. The Ising model was originally proposed by Ernst Ising and Wilhelm Lenz in the 1920s to understand the behaviour of magnetic materials. This model considers a magnetic material as a set of molecules. The Molecules have spins which 'align' or 'anti-align' when a magnetic field is introduced and have pairwise interaction with each other [3, 5]. Let $V = \{1, 2, \ldots, n\}$ be a collection of molecules and $x_i \in \{-1, 1\}$ represents the spin of molecule i. Let b_i be the strength of the magnetic field applied on molecule i and a_{ij} represents joint interaction field between neighbouring spins of i and j. For a given spin vector $\mathbf{x} = (x_1, x_2, \ldots, x_n)$, the corresponding energy value is defined as

$$E(\mathbf{x}) = \sum_{i=1}^{n}\sum_{j=1}^{n} a_{ij}x_ix_j + \sum_{i=1}^{n} b_ix_i.$$

At low temperature levels, the system tends to low energy states and hence the sign of a_{ij} indicates possible spin direction. Thus, when $a_{ij} > 0$ the system favors anti-aligned neighbouring spins, i.e. $x_ix_j = -1$ leaving $a_{ij}x_ix_j < 0$. Likewise, when $a_{ij} < 0$ the spins x_i and x_j are likely to align; i.e. $x_ix_j = 1$ leaving $a_{ij}x_ix_j < 0$. This simple hypothetical model has been validated through experimental considerations by physicists for decades [10, 40, 61–63]. This underlying model is the basis of some of the current quantum inspired computing machines [5] which essentially brings a system to its lowest energy level which in turn indirectly computes the minimum of the energy function (maximum of $-E(\mathbf{x})$) over variables that takes values -1 or 1.

The associated optimization problem, presented in the maximization problem, can be stated as

$$\begin{aligned} &\text{Maximize } \mathbf{y}^{\mathrm{T}}\mathbf{A}\mathbf{y} + \mathbf{b}^{\mathrm{T}}\mathbf{y} \\ &\text{Subject to} \\ &\qquad\qquad \mathbf{y} \in \{-1, 1\}^n \end{aligned}$$

where $\mathbf{A}$ is an $n \times n$ real valued matrix and $\mathbf{b}^{\mathrm{T}} = (b_1, b_2, \ldots, b_n)$ is a row vector from $\mathbb{R}^n$ and $\mathbf{y}$ is a column vector in $\{-1, 1\}^n$. We refer to this version of QUBO as the *Ising QUBO*.

The models QUBO and the Ising QUBO are equivalent from an optimality point of view. Note that the linear transformation $x_i = \frac{1}{2}(y_i + 1)$ for $i = 1, 2 \ldots, n$ reduces the QUBO $(\mathbf{Q}, \mathbf{c})$ to the Ising QUBO $(\mathbf{A}, \mathbf{b})$ where $\mathbf{A} = \frac{1}{4}\mathbf{Q}$, $b_i = \frac{1}{4}\left(2c_i + \sum_{j=1}^{n}(q_{ij} + q_{ji})\right)$ for $i = 1, 2, \ldots, n$ along with an additive constant $q_0 = \frac{1}{4}\left(\sum_{i=1}^{n}\sum_{j=1}^{n} q_{ij} + \sum_{i=1}^{n} c_i\right)$, which can be discarded. Similarly, an instance $(\mathbf{A}, \mathbf{b})$ of the Ising QUBO can be reduced to an instance $(\mathbf{Q}, \mathbf{c})$ of QUBO using the transformation $y_i = 2x_i - 1$ for $i = 1, 2 \ldots, n$ with $\mathbf{Q} = 4\mathbf{A}$, $\mathbf{c} = 2\mathbf{b} + 2\mathbf{A}^T\mathbf{e} + 2\mathbf{A}\mathbf{e}$ along with the additive constant $\mathbf{c}^T\mathbf{e} + \mathbf{e}^T\mathbf{A}\mathbf{e}$.

The original Ising model was in fact a special case of the Ising QUBO. There are various extensions and generalizations of this popular model. For a historical account of various developments related to this model, we refer to the survey papers [10, 40, 61–63]. Although the name 'Ising model' is widely accepted within the physics community, it is not without controversy. For example, Barry Simon in his book on *The Statistical Mechanics of Lattice Gases* [75] writes:

> Lattice models are caricatures invented to illuminate various aspects of elementary statistical mechanics, especially the phenomena of phase transitions and spontaneously broken symmetry. The simplest of all models is the Ising (or Lenz-Ising) model and this model

> was suggested to Ising by his thesis adviser, Lenz. Ising solved the one-dimensional model, ..., and on the basis of the fact that the one-dimensional model had no phase transition, he asserted that there was no phase transition in any dimension. As we shall see, this is false. It is ironic that on the basis of an elementary calculation and erroneous conclusion, Ising's name has become among the most commonly mentioned in the theoretical physics literature. But history has had its revenge. Ising's name, which is correctly pronounced "E-zing," is almost universally mispronounced "I-zing."

Such discussions are probably important in physics to put contributions to scientific developments in context. We continue to use the terminology QUBO and Ising QUBO to distinguish between the nature of the underlying variables and to recognize the linkages between the Ising model and Ising QUBO.

For QUBO, we have seen that the linear term can be absorbed into the $\mathbf{Q}$ matrix or the diagonal of $\mathbf{Q}$ can be extracted and added to the linear term without altering the optimal solutions set. It is possible to discard the diagonal elements of $\mathbf{A}$ from the Ising QUBO since these elements simply contribute a constant value to any feasible solution. Thus, it is customary to assume that diagonal entries of $\mathbf{A}$ are zeros. Further, it is possible to reformulate Ising QUBO without the linear term [18, 38]. Define the $(n+1) \times (n+1)$ matrix $\bar{\mathbf{A}}$ where

$$\bar{a}_{ij} = \begin{cases} a_{ij}, & \text{if } 1 \leq i, j \leq n \\ \frac{1}{2}b_i & \text{for } j = n+1, i = 1, 2, \ldots, n \\ \frac{1}{2}b_j & \text{for } i = n+1 j = 1, 2, \ldots, n \\ 0 & \text{for } i = j = n+1. \end{cases}$$

Now, consider the Ising QUBO,

$$\begin{aligned} \text{IQB:}\quad & \text{Maximize } \mathbf{y}^{\mathrm{T}}\bar{\mathbf{A}}\mathbf{y} \\ & \text{Subject to} \\ & \qquad\qquad \mathbf{y} \in \{-1, 1\}^{n+1}. \end{aligned}$$

Note that if $\mathbf{y}$ is a solution to the Ising QUBO IQB then $-\mathbf{y}$ is also a solution. Further, if both $\mathbf{y}$ and $-\mathbf{y}$ have the same objective function value in IQB. Thus, without loss of generality we can assume that $y_{n+1} = 1$ in an optimal solution to IQB. Now, it can be verified that IQB is equivalent to the Ising QUBO.

The validity of the Ising model is not restricted to physics. Its relevance has been established in various other fields such as finance, biology, psychology, sociology etc. For example, the Ising model can be adapted to study human behavior by interpreting similarity between the behaviour of molecules and the behaviour of human, as argued by Galam [24]. Specific applications of the Ising QUBO from this point of view can be found in [49, 77]. Applications of QUBO and the Ising QUBO are covered in detail in Chap. 8.

1.3 Representations of the Q Matrix

Let $f_{Q,c}(\mathbf{x}) = \mathbf{x}^T\mathbf{Q}\mathbf{x} + \mathbf{c}^T\mathbf{x}$ denote the objective function of the instance $(\mathbf{Q}, \mathbf{c})$ of QUBO. We remove the suffix Q, c from $f_{Q,c}$ when the explicit consideration of $\mathbf{Q}$ and $\mathbf{c}$ is unimportant. An instance $(\mathbf{Q}', \mathbf{c}')$ of QUBO is an *equivalent representation* of the instance $(\mathbf{Q}, \mathbf{c})$ if

1. $f_{Q,c}(\mathbf{x}) = f_{Q',c'}(\mathbf{x})$ for all $\mathbf{x} \in \{0, 1\}^n$
2. $\mathbf{Q}$ and $\mathbf{Q}'$ belongs to $\mathbb{R}^{n\times n}$, $\mathbf{c}$ and $\mathbf{c}'$ belongs to $\mathbb{R}^n$.

Equivalent representations, although preserving optimality, could generate instances with different structural properties that may be exploited to obtain computational advantages. Let us now look at some simple and commonly used equivalent representations of QUBO [65]. More involved equivalent representations and corresponding applications in computing strong lower bounds are discussed in Chaps. 6 and 7.

Remark 1.1 $(\mathbf{Q}^T, \mathbf{c})$ and $\left(\frac{1}{2}\left(\mathbf{Q} + \mathbf{Q}^T\right), \mathbf{c}\right)$ are equivalent representations of $(\mathbf{Q}, \mathbf{c})$.

The proof of this remark is straightforward. Note that $\frac{1}{2}\left(\mathbf{Q} + \mathbf{Q}^T\right)$ is a symmetric matrix. Thus, it is possible to assume without loss of generality that the matrix $\mathbf{Q}$ in the definition of QUBO is symmetric and this assumption is required for some of the algorithms to solve the problem and in computing certain types of lower bounds. For any matrix $\mathbf{Q} \in \mathbb{R}^{n\times n}$, $\mathsf{diagV}(\mathbf{Q})$ is the diagonal vector of $\mathbf{Q}$. i.e. $\mathsf{diagV}(\mathbf{Q})$ is the vector of size n and its ith element is q_{ii}.

Remark 1.2 ([69]) If $\mathbf{S}$ is a skew-symmetric matrix, $\mathbf{D}$ is a diagonal matrix, $\mathbf{Q}' = \mathbf{Q} + \mathbf{S} + \mathbf{D}$, and $\mathbf{c}' = \mathbf{c} - \mathsf{diagV}(\mathbf{D})$, then $(\mathbf{Q}', \mathbf{c}')$ is an equivalent representation of $(\mathbf{Q}, \mathbf{c})$.

The proof of Remark 1.2 follows from the fact that $\mathbf{x}^T\mathbf{S}\mathbf{x} = 0$ for any skew-symmetric matrix $\mathbf{S}$ and $x_i^2 = x_i$ for all binary variables x_i. Choosing $\mathbf{S}$ and $\mathbf{D}$ appropriately, we can get different standard representations of $\mathbf{Q}$. For example, choose $\mathbf{D}$ such that $d_{ii} = -q_{ii}$ for $i = 1, \ldots, n$ and choose $\mathbf{S}$ such that

$$s_{ij} = \begin{cases} q_{ji} & \text{if } j > i \\ -q_{ij} & \text{if } j < i \\ 0 & \text{otherwise.} \end{cases}$$

Then, the resulting matrix $\mathbf{Q}'$ is upper triangular with zeros on the diagonal. This is a common representation used in the QUBO literature and is also assumed in some of the chapters in this book. If we choose $\mathbf{S}$ as the zero matrix and $d_{ii} = M$ for $i = 1, \ldots, n$, where M is a sufficiently large nonnegative number, the resulting matrix $\mathbf{Q}'$ will be positive semidefinite. Consequently, without loss of generality one may assume $\mathbf{Q}$ to be positive semidefinite. When $\mathbf{Q}$ is symmetric and not positive semidefinite, choosing M to be the negative of the smallest eigenvalue of $\mathbf{Q}$ is sufficient to make $\mathbf{Q}'$ to be positive semidefinite [32]. This makes the continuous relaxation of the objective function of QUBO a convex quadratic function. In a similar way, choosing M as a sufficiently small negative number, we can get a representation of QUBO where the $\mathbf{Q}$ matrix is negative semidefinite and thereby making the continuous relaxation of the objective function of QUBO a concave quadratic function.

A QUBO $(\mathbf{Q}, \mathbf{c})$ of size n can be represented as a QUBO $(\mathbf{Q}', \mathbf{0})$ of size $n + 1$ such that the row sum and column sums of $\mathbf{Q}'$ are zeros. We call this *the zero sum representation*. Without loss of generality assume $\mathbf{c} = \mathbf{0}$. From the instance $(\mathbf{Q}, \mathbf{0})$ construct the matrix $\mathbf{Q}'$ as

$$q'_{ij} = \begin{cases} q_{ij} & \text{if } i, j \leq n \\ -\sum_{k=1}^{n} q_{ik} & \text{if } j = n+1, i \leq n \\ -\sum_{k=1}^{n} q_{kj} & \text{if } i = n+1, j \leq n \\ \sum_{k=1}^{n}\sum_{\ell=1}^{n} q_{k\ell} & \text{if } i = n+1, j = n+1 \end{cases}$$

It can be verified that the row and column sums of the matrix $\mathbf{Q}'$ are zeros. Consequently, the sum of all elements of Q' is also zero. If x_{n+1} is zero in an optimal solution for $(\mathbf{Q}', \mathbf{0})$ then $(\mathbf{Q}', \mathbf{0})$ is equivalent to $(\mathbf{Q}, \mathbf{0})$. Let $\bar{\mathbf{x}} = \mathbf{e} - \mathbf{x}$ where $\mathbf{e}$ is the all one vector in $\mathbb{R}^{n+1}$.

Lemma 1.1 *If row and column sums of* $\mathbf{Q}'$ *are zeros, then* $\mathbf{x}^T\mathbf{Q}'\mathbf{x} = \bar{\mathbf{x}}^T\mathbf{Q}'\bar{\mathbf{x}}$ *for all* $\mathbf{x} \in \{0, 1\}^{n+1}$.

Proof $\bar{\mathbf{x}}^T\mathbf{Q}'\bar{\mathbf{x}} = (\mathbf{e} - \mathbf{x})^T\mathbf{Q}'(\mathbf{e} - \mathbf{x}) = \mathbf{e}^T\mathbf{Q}'\mathbf{e} - \mathbf{x}^T\mathbf{Q}'\mathbf{e} - \mathbf{e}^T\mathbf{Q}'\mathbf{x} + \mathbf{x}^T\mathbf{Q}'\mathbf{x} = \mathbf{x}^T\mathbf{Q}'\mathbf{x}$. The last equality follows from the fact that row and column sums of $\mathbf{Q}'$ are zeros, which makes $\mathbf{e}^T\mathbf{Q}'\mathbf{e} = 0$, $\mathbf{x}^T\mathbf{Q}'\mathbf{e} = 0$ and $\mathbf{e}^T\mathbf{Q}'\mathbf{x} = 0$. □

From Lemma 1.1, if $\mathbf{x}$ is an optimal solution to $(\mathbf{Q}', \mathbf{0})$, then $\bar{\mathbf{x}}$ is also an optimal solution. Thus, $x_{n+1} = 0$ in one of these optimal solutions of $(\mathbf{Q}', \mathbf{0})$ and hence $(\mathbf{Q}', \mathbf{0})$ and $(\mathbf{Q}, \mathbf{0})$ are equivalent.

1.4 Some Equivalent Optimization Problems

We have seen that the problem of computing the maximum value principal minor of an $n \times n$ matrix, the maximum weight induced subgraph problem, and the Ising QUBO are alternative ways of representing a QUBO. In this section, we discuss some additional problems that are equivalent to QUBO and these equivalent forms are used in some of the chapters to follow to derive structural properties and to develop solution algorithms.

1.4.1 QUBO as a Bilinear Program

In this subsection, we assume that $\mathbf{Q}$ is symmetric and positive semidefinite. As discussed in the previous section, this assumption is without loss of generality. Then the continuous relaxation of QUBO is to

$$\begin{aligned} \text{QUBO(C):} \quad & \text{Maximize} \quad f(\mathbf{x}) = \mathbf{x}^{\mathrm{T}}\mathbf{Q}\mathbf{x} + \mathbf{c}^{\mathrm{T}}\mathbf{x} \\ & \text{Subject to:} \quad \mathbf{x} \in [0, 1]^n \end{aligned}$$

Since Q is symmetric and positive semidefinite, $f(\mathbf{x})$ is a convex function and hence there exists an optimal solution to QUBO(C) which at an extreme point of the hypercube $[0, 1]^n$. Now consider the *hypercube bilinear program*

$$\begin{aligned} \text{HBLP:} \quad & \text{Maximize} \quad \eta(\mathbf{x}, \mathbf{y}) = \mathbf{x}^{\mathrm{T}}\mathbf{Q}\mathbf{y} + \tfrac{1}{2}\mathbf{c}^{\mathrm{T}}\mathbf{x} + \tfrac{1}{2}\mathbf{c}^{\mathrm{T}}\mathbf{y} \\ & \text{Subject to:} \quad \mathbf{x} \in [0, 1]^n, \mathbf{y} \in [0, 1]^n \end{aligned}$$

Lemma 1.2 ([51]) *There exists an optimal solution* $(\mathbf{x}^*, \mathbf{y}^*)$ *to HBLP such that both* $\mathbf{x}^*$ *and* $\mathbf{y}^*$ *are extreme points of the hypercube* $[0, 1]^n$.

Proof Suppose that $(\mathbf{x}^0, \mathbf{y}^0)$ is an optimal solution to HBLP. Now fix $\mathbf{y} = \mathbf{y}^0$ in HBLP and let $\bar{\mathbf{x}}$ be an optimal extreme point solution of the resulting linear program. Then $(\bar{\mathbf{x}}, \mathbf{y}^0)$ is an optimal solution to HBLP. Next fix $\mathbf{x} = \bar{\mathbf{x}}$ in HBLP and let $\bar{\mathbf{y}}$ be an optimal extreme point solution of the resulting linear program. Then $(\bar{\mathbf{x}}, \bar{\mathbf{y}})$ is an optimal solution to HBLP and this completes the proof. □

Let us now prove a property of a symmetric positive semidefinite matrix which is used in the proof of the theorem that follows.

Lemma 1.3 *If* $\mathbf{Q}$ *is a symmetric positive semidefinite matrix such that* $\mathbf{x}^T\mathbf{Q}\mathbf{x} = 0$ *then* $\mathbf{Q}\mathbf{x} = \mathbf{0}$.

Proof Since $\mathbf{Q}$ is symmetric and positive semidefinite, there exists a matrix B such that $\mathbf{Q} = \mathbf{B}^T\mathbf{B}$. Then, $\mathbf{x}^T\mathbf{Q}\mathbf{x} = \mathbf{x}^{\mathrm{T}}\mathbf{B}^{\mathrm{T}}\mathbf{B}\mathbf{x} = (\mathbf{B}\mathbf{x})^T(\mathbf{B}\mathbf{x}) = ||\mathbf{B}\mathbf{x}||^2$. Thus, $\mathbf{x}^T\mathbf{Q}\mathbf{x} = 0$ implies $\mathbf{B}\mathbf{x} = \mathbf{0}$ and hence $\mathbf{Q}\mathbf{x} = \mathbf{B}^{\mathrm{T}}\mathbf{B}\mathbf{x} = \mathbf{0}$. □

We now show that QUBO is equivalent to HBLP.

Theorem 1.1 ([50, 51]) *If* $(\mathbf{x}^*, \mathbf{y}^*)$ *is an optimal extreme point solution of HBLP then both* $\mathbf{x}^*$ *and* $\mathbf{y}^*$ *are optimal solutions of QUBO(C). Conversely, if* $\mathbf{x}^*$ *is an optimal extreme point solution of QUBO(C) then* $(\mathbf{x}^*, \mathbf{x}^*)$ *is an optimal solution to HBLP.*

Proof Let $(\mathbf{x}^*, \mathbf{y}^*)$ be an optimal extreme point solution of HBLP and $\mathbf{x}^0$ be an optimal extreme point solution of QUBO(C). Then,

$$f(\mathbf{x}^o) \geq f(\mathbf{x}^*) \text{ and } f(\mathbf{x}^0) \geq f(\mathbf{y}^*). \tag{1.1}$$

Further,

$$\eta(\mathbf{x}^*, \mathbf{y}^*) = \max\{\eta(\mathbf{x}, \mathbf{y}) : \mathbf{x}, \mathbf{y} \in [0, 1]^n\}$$
$$\geq \max\{\eta(\mathbf{x}, \mathbf{x}) : \mathbf{x} \in [0, 1]^n\} = f(\mathbf{x}^0) \tag{1.2}$$

Since $(\mathbf{x}^*, \mathbf{y}^*)$ is an optimal solution to HBLP,

$$\eta(\mathbf{x}^*, \mathbf{y}^*) - \eta(\mathbf{x}^*, \mathbf{x}^*) \geq 0 \text{ and } \eta(\mathbf{x}^*, \mathbf{y}^*) - \eta(\mathbf{y}^*, \mathbf{y}^*) \geq 0 \tag{1.3}$$

But

$$\eta(\mathbf{x}^*, \mathbf{y}^*) - \eta(\mathbf{x}^*, \mathbf{x}^*) = (\mathbf{x}^*)^T\mathbf{Q}(\mathbf{y}^* - \mathbf{x}^*) + \frac{1}{2}\mathbf{c}^T(\mathbf{y}^* - \mathbf{x}^*) \text{ and} \tag{1.4}$$

$$\eta(\mathbf{x}^*, \mathbf{y}^*) - \eta(\mathbf{y}^*, \mathbf{y}^*) = (\mathbf{y}^*)^T\mathbf{Q}(\mathbf{x}^* - \mathbf{y}^*) + \frac{1}{2}\mathbf{c}^T(\mathbf{x}^* - \mathbf{y}^*) \tag{1.5}$$

From (1.3), (1.4) and (1.5) we have,

$$(\mathbf{x}^*)^T\mathbf{Q}(\mathbf{y}^* - \mathbf{x}^*) + \frac{1}{2}\mathbf{c}^T(\mathbf{y}^* - \mathbf{x}^*) \geq 0 \text{ and} \tag{1.6}$$

$$(\mathbf{y}^*)^T\mathbf{Q}(\mathbf{x}^* - \mathbf{y}^*) + \frac{1}{2}\mathbf{c}^T(\mathbf{x}^* - \mathbf{y}^*) \geq 0. \tag{1.7}$$

Adding (1.6) and (1.7), we get $(\mathbf{x}^* - \mathbf{y}^*)^T\mathbf{Q}(\mathbf{x}^* - \mathbf{y}^*) \leq 0$. Since $\mathbf{Q}$ is symmetric and positive semidefinite, from Lemma 1.3, $\mathbf{Q}(\mathbf{x}^* - \mathbf{y}^*) = 0$. Substituting this in (1.6) and (1.7) we get $\mathbf{c}^T(\mathbf{x}^* - \mathbf{y}^*) = 0$. Thus from (1.4) and (1.5) we have

$$\eta(\mathbf{x}^*, \mathbf{y}^*) = \eta(\mathbf{x}^*, \mathbf{x}^*) = \eta(\mathbf{y}^*, \mathbf{y}^*).$$

But $\eta(\mathbf{x}^*, \mathbf{x}^*) = f(\mathbf{x}^*)$ and $\eta(\mathbf{y}^*, \mathbf{y}^*) = f(\mathbf{y}^*)$. Thus,

$$\eta(\mathbf{x}^*, \mathbf{y}^*) = f(\mathbf{x}^*) = f(\mathbf{y}^*) \tag{1.8}$$

From (1.1) and (1.8), $\eta(\mathbf{x}^*, \mathbf{y}^*) \leq f(\mathbf{x}^0)$ and the result follows in view of (1.2) and (1.8). □

1.4.2 The Maximum Cut Problem and QUBO

Let $G = (V, E)$ be a graph with $V = \{1, 2, \ldots, n\}$ and for each edge $(i, j) \in E$ a weight w_{ij} is prescribed. For any $S \subseteq V$, the pair $(S, V \setminus S)$ is called a *cut* in G.

The weight of the cut $(S, V \setminus S)$ is

$$w(S, V \setminus S) = \sum_{i \in S, j \in V \setminus S, (i,j) \in E} w_{ij}.$$

In general, our definition of a cut allows the possibility that $S = \emptyset$ or $S = V$ and in either case $w(S, V \setminus S) = 0$. The *minimum cut* problem seeks a cut $(S, V \setminus S)$ with minimum $w(S, V \setminus S)$ value. When $w_{ij} \geq 0$ for all $(i, j) \in E$, we normally do not permit $S = \emptyset$ or V for the minimum cut problem since otherwise the trivial solution $S = \emptyset$ is optimal. Likewise, the *maximum cut problem* seeks a cut $(S, V \setminus S)$ with maximum $w(S, V \setminus S)$ value. When $w_{ij} \geq 0$ for all $(i, j) \in E$, the minimum cut problem is solvable in polynomial time [44] but the maximum cut problem is NP-hard [23]. When w_{ij} is allowed to take positive and negative values, both minimum cut and maximum cut problems are NP-hard. We now observe that QUBO and the maximum cut problem are essentially the same, presented under different frameworks [7, 33, 38].

Let $\mathbf{x} \in \{0, 1\}^n$ and $\bar{\mathbf{x}} = \mathbf{e} - \mathbf{x}$ where $\mathbf{e}$ is the all-one vector in $\mathbb{R}^n$. An $\mathbf{x} \in \{0, 1\}^n$ is an *incidence vector* of $S \subseteq V$ if and only if $x_i = 1$ for all $i \in S$. Let $\mathbf{Q}$ be the *weighted incidence matrix* of G. Define $q_{ij} = q_{ji} = w_{ij}$ if $(i, j) \in E$ (i.e. i and j are adjacent in G) and zero, otherwise. Note that $\mathbf{Q}$ is symmetric. Then, for any cut $(S, \bar{S})$ (with $\bar{S} = V \setminus S$) and the incidence vector $\mathbf{x}$ of S, we have $w(S, V \setminus S) = \mathbf{x}^{\mathrm{T}}\mathbf{Q}\bar{\mathbf{x}}$. For example, consider the graph below with the associated edge weights and the corresponding matrix $\mathbf{Q}$ (Fig. 1.1).

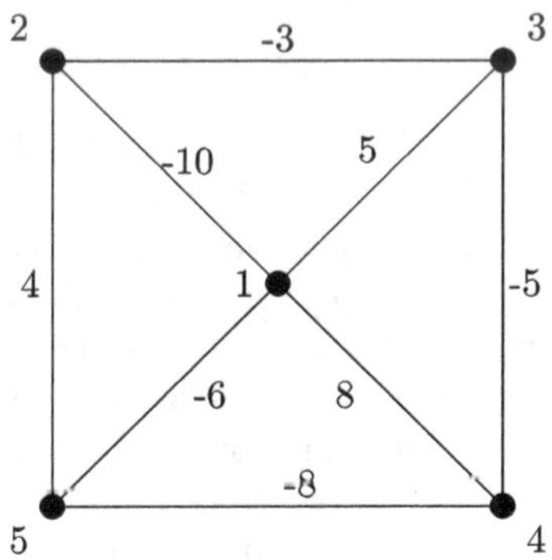

$$\mathbf{Q} = \begin{pmatrix} 0 & -10 & 5 & 8 & -6 \\ -10 & 0 & -3 & 0 & 4 \\ 5 & -3 & 0 & -5 & 0 \\ 8 & 0 & -5 & 0 & -8 \\ -6 & 4 & 0 & -8 & 0 \end{pmatrix}$$

Fig. 1.1 An instance of WMCP

Now, choose the cut $(S, \bar{S})$ where $S = \{1, 4\}$ and the incidence vector $\mathbf{x} = (1, 0, 0, 1, 0)^T$ associated with S. Then, $w(S, \bar{S}) = -24 = \mathbf{x^T Q \bar{x}}$.

Thus, the maximum cut problem can be written as

$$\max_{S \subseteq V} w(S, V \setminus S) = \max_{\mathbf{x} \in \{0,1\}^n} \mathbf{x^T Q \bar{x}} \tag{1.9}$$

For calculating $\mathbf{x^T Q \bar{x}}$, the diagonal elements of $\mathbf{Q}$ are irrelevant. For any $n \times n$ matrix $\mathbf{Q}$,

$$\mathbf{x^T Q \bar{x}} = \sum_{i=1}^{n}\sum_{j=1}^{n} q_{ij} x_i \bar{x}_j = \sum_{i=1}^{n}\sum_{j=1}^{n} q_{ij} x_i (1 - x_j) = \sum_{i=1}^{n} r_i x_i - \sum_{i=1}^{n}\sum_{j=1}^{n} q_{ij} x_i x_j \tag{1.10}$$

where $r_i = \sum_{j=1}^{n} q_{ij}$. Define the $n \times n$ matrix $\hat{\mathbf{Q}} = (\hat{q}_{ij})$ where

$$\hat{q}_{ij} = \begin{cases} -q_{ij} & \text{if } i \neq j \\ r_i - q_{ii} & \text{if } i = j. \end{cases}$$

Then $\mathbf{x^T Q \bar{x}} = \mathbf{x^T \hat{Q} x}$, establishing that the maximum cut problem can be formulated as a QUBO. We now observe that any instance, say $(\mathbf{Q}, \mathbf{c})$, of QUBO can be formulated as a maximum cut problem. To see this, from Eq. (1.10),

$$\mathbf{x}^T\mathbf{Q}\mathbf{x} + \mathbf{c}^T\mathbf{x} = \mathbf{r}^T\mathbf{x} - \mathbf{x}^T\mathbf{Q}\bar{\mathbf{x}} + \mathbf{c}^T\mathbf{x} = (\mathbf{r} + \mathbf{c})^T\mathbf{x} - \mathbf{x}^T\mathbf{Q}\bar{\mathbf{x}} \tag{1.11}$$

where $\mathbf{r} = (r_1, r_2, \ldots, r_n)$.

Now, consider the graph $G' = (V', E')$ where $V' = \{0, 1, 2, \ldots, n\}$. The edge set $E' = E \cup E^0$ where $E = \{(i, j) : q_{ij} \neq 0; i, j = 1, 2, \ldots, n, i \neq j\}$ and $E^0 = \{(0, i) : r_i + c_i \neq 0\}$. Now, define the weight w_{ij} of the edge (i, j) as $-q_{ij}$ if $(i, j) \in E$ and $-(r_j + c_j)$ for $(0, j) \in E^0$. Let $\alpha = \sum_{i=1}^{n}(r_i + c_i)$. For any cut $(S, \bar{S})$ in G', without loss of generality assume that the node $0 \in S$ and consider the binary variables $x_1, x_2, \ldots, x_n$ with $x_i = 1$ if and only if $i \in S$. Let $w(S, \bar{S})$ be the value of the cut $(S, \bar{S})$ in G' with the edge weight function w. Then, it can be verified that $w(S, \bar{S}) + \alpha = \mathbf{x}^T\mathbf{Q}\mathbf{x} + \mathbf{c}^T\mathbf{x}$. Thus, any instance $(\mathbf{Q}, \mathbf{c})$ of the QUBO with n variables can be formulated as a maximum weight cut problem on a graph with $n + 1$ nodes.

We now illustrate this construction using the following example. Let

$$\mathbf{Q} = \begin{pmatrix} 0 & 10 & -5 & -8 & 6 \\ 10 & 0 & 3 & 2 & -4 \\ -5 & 3 & 0 & 5 & 0 \\ -8 & 2 & 5 & 0 & 8 \\ 6 & -4 & 0 & 8 & 0 \end{pmatrix} \text{ and } \mathbf{c} = \begin{pmatrix} -3 \\ -5 \\ 11 \\ -3 \\ -2 \end{pmatrix}$$

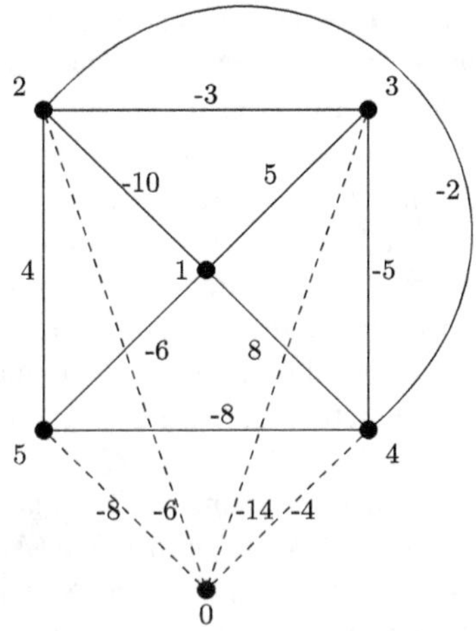

Fig. 1.2 The instance of WMCP equivalent to $(\mathbf{Q}, \mathbf{c})$

Then, $(\mathbf{r} + \mathbf{c})^T = (0, 6, 14, 4, 8)$ and $\alpha = 32$. The graph G' constructed following the procedure discussed above, is given in Fig. 1.2. The numbers on the edges are the corresponding weights. Edges with weight zero are removed from the graph.

The QUBO formulation of the maximum cut problem is also evident from the sum of squares formulation of the maximum cut problem given by

$$\max_{S \subseteq V} w(S, V \setminus S) = \max_{\mathbf{x} \in \{0,1\}^n} \sum_{(i,j) \in E} w_{ij}(x_i - x_j)^2.$$

Another popular formulation of the maximum cut problem is obtained by using variables that take values -1 or 1. Define $y_i = 1$ if $i \in S$ and $y_i = -1$ if $i \in V \setminus S$. Assume that the edges of the graph G are labelled such that $(i, j) \in E$ implies $i < j$ and $w_{ij} = 0$ if $(i, j) \notin E$. Consider the symmetric matrix $\mathbf{W}$. Then,

$$w(S, V \setminus S) = \sum_{\substack{i \in S, j \in V \setminus S \\ (i,j) \in E}} w_{ij} = \frac{1}{2} \sum_{i=1}^{n} \sum_{j=i+1}^{n} w_{ij}(1 - y_i y_j)$$

$$= \frac{1}{4} \sum_{i=1}^{n} \sum_{j=1}^{n} w_{ij}(1 - y_i y_j) = \frac{1}{4} \sum_{i=1}^{n} \sum_{j=1}^{n} w_{ij} \left(\frac{y_i^2 + y_j^2}{2} - y_i y_j \right)$$

$$= \frac{1}{4} \left(- \sum_{i=1}^{n} \sum_{j=1}^{n} w_{ij} y_i y_j + \frac{1}{2} \sum_{i=1}^{n} \Big(\sum_{j=1}^{n} w_{ij} \Big) y_i^2 + \frac{1}{2} \sum_{j=1}^{n} \Big(\sum_{i=1}^{n} w_{ij} \Big) y_j^2 \right)$$

$$= \frac{1}{4} \left(- \sum_{i=1}^{n} \sum_{j=1}^{n} w_{ij} y_i y_j + \sum_{i=1}^{n} \Big(\sum_{j=1}^{n} w_{ij} \Big) y_i^2 \right) = \frac{1}{4} \mathbf{y}^T \mathbf{L} \mathbf{y}$$

where $\mathbf{L} = \mathbf{D} - \mathbf{W}$ and $\mathbf{D}$ is a diagonal matrix such that $d_{ii} = \sum_{j=1}^{n} w_{ij}$ assuming $w_{ij} = 0$ for $(i, j) \notin E$. Thus, the maximum cut problem can be formulated as the Ising QUBO $(\frac{1}{4}\mathbf{L}, \mathbf{0})$.

Let $G = (V, E)$ be a weighted graph with edge weight w_{ij} for each $(i, j) \in E$ and $V = \{1, 2, \ldots, n\}$. Choose $w_{ij} = 0$ if $(i, j) \notin E$ and define the weight matrix $\mathbf{W}$ as the $n \times n$ matrix with its (i, j)th element w_{ij}. Let $\mathbf{D}$ be the weighted degree matrix which is the diagonal matrix with $d_{ii} = \sum_{j=1}^{n} w_{ij}$. Then the weighted Laplacian matrix of G is $L = D - W$. The matrix L is positive semidefinite and the row and columns sums of the matrix is zero. Note that the matrix L constructed in our reduction from the maximum weight cut problem to the Ising QUBO is precisely the Laplacian matrix of G.

In fact, any Ising QUBO can be written as a maximum cut problem as well. To see this, consider an instance $(\mathbf{A}, \mathbf{b})$ of the Ising QUBO in n variables. Without loss of generality, assume that $\mathbf{A}$ is symmetric and $\mathbf{b}$ is the zero vector. Now, consider the graph $G = (V, E)$ where $V = \{1, 2, \ldots, n\}$ and $E = \{(i, j) : a_{ij} \neq 0\}$. Choose the weight of the edge $(i, j) \in E$ as $-a_{ij}$ and let $\mathbf{L}$ be the corresponding weighted Laplacian matrix. Then, $\mathbf{A} = \mathbf{L} + \mathbf{D}$ where $\mathbf{D}$ is a diagonal matrix with $d_{ii} = -l_{ii} + a_{ii}$. For any $\mathbf{y} \in \{-1, 1\}^n$,

$$\mathbf{y}^{\mathrm{T}}\mathbf{A}\mathbf{y} = \mathbf{y}^{\mathrm{T}}\mathbf{L}\mathbf{y} + \mathbf{y}^{\mathrm{T}}\mathbf{D}\mathbf{y} = \mathbf{y}^{\mathrm{T}}\mathbf{L}\mathbf{y} + \mathrm{tr}(\mathbf{D})$$

where $\mathrm{tr}(\mathbf{D})$ is the trace of $\mathbf{D}$ which is a constant. Thus an optimal solution to the Ising QUBO $(\mathbf{A}, \mathbf{0})$ is obtained by solving the Ising QUBO $(\mathbf{L}, \mathbf{0})$ which is precisely the maximum cut problem on G with weights $-a_{ij}$ for $(i, j) \in E$.

1.4.3 Equivalence with the Stable Set Problem

We have seen that the maximum weight stable set problem can be formulated as a QUBO. We now show that any instance of QUBO can be formulated as a maximum weight stable set problem. Our discussion here follows the paper by Hammer, Hansen and Simeone [35]. First, let us rewrite the objective function of a QUBO where all the quadratic terms have non-negative coefficients. After rewriting, the quadratic terms could involve both the original binary variables and their complements. Recall that

$$f(\mathbf{x}) = \mathbf{x}^T\mathbf{Q}\mathbf{x} + \mathbf{c}^T\mathbf{x} = \sum_{(i,j)\in P} q_{ij}x_ix_j + \sum_{(i,j)\in N} q_{ij}x_ix_j + \sum_{i=1}^{n}(q_{ii} + c_i)x_i,$$

where $P = \{(i, j) : q_{ij} > 0, i \neq j\}$ and $N = \{(i, j) : q_{ij} < 0, i \neq j\}$. Also, let $N_j = \{i : (i, j) \in N\}$ and $\rho_j = \sum_{i\in N_j} q_{ij}$. Let $\bar{x}_i = 1 - x_i$ be the complement of

x_i. i.e. $x_i \in \{0, 1\}$ if and only if $\bar{x}_i \in \{0, 1\}$. Now,

$$\begin{aligned} f(\mathbf{x}) &= \sum_{(i,j)\in P} q_{ij}x_ix_j + \sum_{(i,j)\in N} q_{ij}(1-\bar{x}_i)x_j + \sum_{i=1}^{n}(q_{ii}+c_i)x_i \\ &= \sum_{(i,j)\in P} q_{ij}x_ix_j + \sum_{(i,j)\in N} (-q_{ij})\bar{x}_ix_j + \sum_{(i,j)\in N} q_{ij}x_j + \sum_{i=1}^{n}(q_{ii}+c_i)x_i, \\ &= \sum_{(i,j)\in P} q_{ij}x_ix_j + \sum_{(i,j)\in N} (-q_{ij})\bar{x}_ix_j + \sum_{i=1}^{n}(q_{ii}+c_i+\rho_i)x_i \qquad (1.12) \\ &= h(\mathbf{x}), \text{ say.} \end{aligned}$$

Note that the coefficients of x_ix_j and $\bar{x}_ix_j$ in $h(\mathbf{x})$ are positive. This representation $h(\mathbf{x})$ of $f(\mathbf{x})$ is sometimes referred to as Rhys form [35] indicating an early work on a special case of QUBO [72] and is one of the posiform representations of $f(\mathbf{x})$ [7, 8, 35] when viewed as a pseudo-Boolean function. For interesting properties of posiforms of the QUBO objective function, we refer to Chap. 5. Note that maximizing $f(\mathbf{x})$ over $x_i \in \{0, 1\}$ is equivalent to maximizing $h(\mathbf{x})$ over $x_i, \bar{x}_i \in \{0, 1\}$. We now show that the problem of maximizing $h(\mathbf{x})$ over $x_i, \bar{x}_i \in \{0, 1\}$ is equivalent to solving a maximum weight stable set problem. The proof is based on a construction given in [35].

For each $(i, j) \in P \cup N$ introduce a node v_{ij} with weight $|q_{ij}|$. Also, for $i = 1, 2, \ldots, n$, introduce two nodes i and i' with respective weights $w_i = q_{ii} + c_i + \rho_i + M$ and $w_{i'} = M$, where M is a large positive number. Now introduce the edges (v_{ij}, i'), (v_{ij}, j') for each $(i, j) \in P$, the edges (v_{ij}, i), (v_{ij}, j') for each $(i, j) \in N$. Finally connect each node i with node i' by an edge (i, i'), for $i = 1, 2, \ldots n$. Let $G' = (V', E')$ be the resulting graph.

Define the product node variables y_{ij} where

$$y_{ij} = \begin{cases} 1 & \text{if vertex } v_{ij} \text{ is selected} \\ 0 & \text{if vertex } v_{ij} \text{ is not selected.} \end{cases}$$

Similarly, define the selection variables for nodes i and i' as

$$x_i = \begin{cases} 1 & \text{if vertex } i \text{ is selected} \\ 0 & \text{if vertex } i \text{ is not selected} \end{cases} \quad \text{and} \quad x_i' = \begin{cases} 1 & \text{if vertex } i' \text{ is selected} \\ 0 & \text{if vertex } i' \text{ is not selected} \end{cases}$$

Then, the standard integer programming formulation of the maximum weight stable set problem (MWSSP) on G' is

$$\text{SSIP: Maximize} \sum_{(i,j)\in P} q_{ij}y_{ij} + \sum_{(i,j)\in N} -q_{ij}y_{ij} + \sum_{i=1}^{n}(q_{ii}+c_i+r_i+M)\,x_i + \sum_{i=1}^{n} Mx_i'$$

Subject to:

$$\left.\begin{aligned} y_{ij} + x'_i &\le 1 \\ y_{ij} + x'_j &\le 1 \end{aligned}\right\} \text{ for } (i, j) \in P \tag{1.13}$$

$$\left.\begin{aligned} y_{ij} + x_i &\le 1 \\ y_{ij} + x'_j &\le 1 \end{aligned}\right\} \text{ for } (i, j) \in N \tag{1.14}$$

$$x_i + x'_i \le 1, i = 1, 2, \ldots, n, \tag{1.15}$$

$$y_{ij}, x_i, x'_i \in \{0, 1\}. \tag{1.16}$$

Theorem 1.2 *If* $(\mathbf{y}, \mathbf{x}, \mathbf{x}')$ *is an optimal solution to the SSIP defined above, then* $\mathbf{x}$ *is an optimal solution to the QUBO with objective function* $f(\mathbf{x})$.

Proof Recall that optimizing $f(\mathbf{x})$ over $\mathbf{x} \in \{0, 1\}^n$ is equivalent to optimizing $h(\mathbf{x})$ over $\mathbf{x}, \bar{\mathbf{x}} \in \{0, 1\}^n$. Let $\phi(\mathbf{y}, \mathbf{x}, \mathbf{x}')$ denote the objective function of SSIP. Since M is large enough, $x'_i = 1 - x_i$ in every optimal solution to SSIP on G' for otherwise, a better solution can be obtained. The inequalities (1.13) imply that $y_{ij} \le \min\{x_i, x_j\}$ in every optimal solution and hence $y_{ij} = 0$ if at least one of x_i or x_j is zero. If $x_i = x_j = 1$ then, $y_{ij} = 1$ in an optimal solution since $q_{ij} > 0$ for $(i, j) \in P$. Thus $y_{ij} = x_i x_j$ in an optimal solution of SSIP on G'. Similarly, we can show that $y_{ij} = x'_i x_j$ in an optimal solution. Noting that $x'_i = 1 - x_i = \bar{x}_i$, we have

$$\phi(\mathbf{y}, \mathbf{x}, \mathbf{x}') = h(\mathbf{x}) + nM \tag{1.17}$$

for every optimal solution $(\mathbf{y}, \mathbf{x}, \mathbf{x}')$ of SSIP on G'.
Conversely, for any $\mathbf{x} \in \{0, 1\}^n$ define $y_{ij} = x_i x_j$ for all $(i, j) \in P$, $y_{ij} = (1 - x_i)x_j$ for all $(i, j) \in N$, and $x'_i = \bar{x}_i = 1 - x_i$. The solution $(\mathbf{y}, \mathbf{x}, \mathbf{x}')$ constructed above is indeed a feasible solution to SSIP on G' satisfying

$$\phi(\mathbf{y}, \mathbf{x}, \mathbf{x}') = h(\mathbf{x}) + nM \tag{1.18}$$

and the result follows. □

Example 1.1 Consider the QUBO with objective function

$$f(\mathbf{x}) = 7x_1x_2 - 3x_1x_3 - 12x_1x_4 + 4x_2x_3 + 8x_2x_4 + 3x_1 - 10x_2 + 5x_4.$$

Then, $P = \{(1, 2), (2, 3), (2, 4)\}$ and $N = \{(1, 3), (1, 4)\}$. The cost matrix $\mathbf{Q}$, the sets N_i, $i = 1, 2, 3, 4$, and the values ρ_i and $q_{ii} + c_i + \rho_i$, for $i = 1, 2, 3, 4$ are given

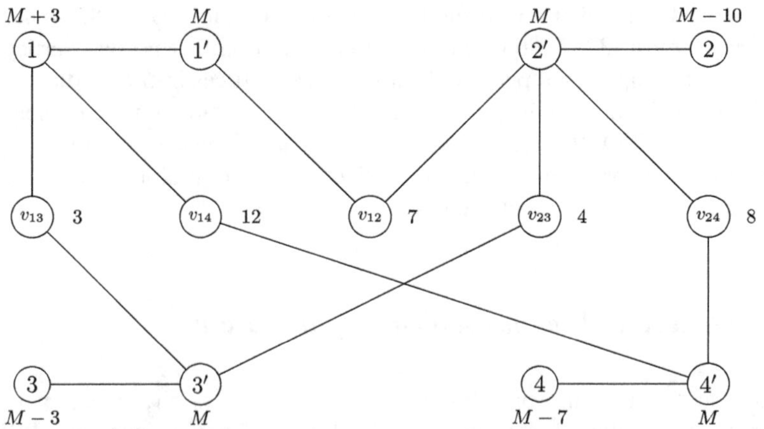

Fig. 1.3 The graph for the maximum weight stable set problem constructed from the QUBO in Example 1.1.

below

$$\mathbf{Q} = \begin{pmatrix} 0 & 7 & -3 & -12 \\ 0 & 0 & 5 & 0 \\ 0 & 0 & 0 & 0 \\ 0 & 0 & 0 & 0 \end{pmatrix}$$

i	N_i	ρ_i	$q_{ii} + \rho_i + c_i$
1	$\emptyset$	0	3
2	$\emptyset$	0	−10
3	{1}	−3	−3
4	{1}	−12	−7

Now construct graph $G' = (V, E)$ for the stable set problem as discussed above. The resulting graph is given in Fig. 1.3. In the figure, the numbers shown outside the nodes represent the weight of the node.

It can be verified that $x_1 = 0, x_2 = x_3 = x_4 = 1$ is an optimal solution to the QUBO with optimal objective function value 7. An optimal solution to the constructed maximum weight stable set problem is $S = \{v_{13}, 1', v_{14}, 4, v_{23}, v_{24}, 2, 3\}$ with value equal to $7 + 4M$. Also, the QUBO solution recovered from S is $x_1 = 0, x_2 = x_3 = x_4 = 1$ and this is optimal with optimal objective function value 7.

Considering the equivalence between the maximum weight stable set problem and the maximum weight clique problem, we can see that QUBO is equivalent to the maximum weight clique problem as well.

1.4.4 QUBO and the Generalized Stable Set Problem

Let $G = (V, E)$ be a graph and c_i be the weight associated with vertex $i \in V$. Also, for each edge $(i, j) \in E$ a cost q_{ij} is given. Recall that a stable set in G is a subset S of V such that no two nodes in S are adjacent in G. In the *generalized stable set problem* [37, 39, 65], we relax the restriction that no two vertices in S are adjacent by imposing a penalty. That is, if two vertices in S are adjacent in G, a penalty q_{ij} is incurred. The generalized stable set problem on a complete graph is precisely a QUBO. If G is not complete, we can define $q_{ij} = 0$ for $(i, j) \notin E$ to yield a QUBO formulation. The generalized stable set problem and some of its variations are studied by many authors [4, 37, 39, 65].

1.4.5 Quadratic Pseudo-Boolean Optimization

A quadratic function in variables $X = \{x_1, x_2, \ldots, x_n, \bar{x}_1, \bar{x}_2, \ldots, \bar{x}_n\}$ where $x_i, \bar{x}_i \in \{0, 1\}$ is called a *quadratic pseudo-Boolean function* (quadratic PBF) [8]. The objective function $\mathbf{x}^T\mathbf{Q}\bar{\mathbf{x}}$ of the maximum-cut problem is a quadratic PBF and so is the objective function $f(\mathbf{x}) = \mathbf{x}^T\mathbf{Q}\mathbf{x} + \mathbf{c}^T\mathbf{x}$ of QUBO. A quadratic PBF is in *posiform* if all associated coefficients (except possibly the constant term) are positive. A posiform is *homogeneous* if the associated constant is zero. A homogeneous posiform ζ can be written as

$$\zeta(\mathbf{x}, \bar{\mathbf{x}}) = \sum_{i<j} \left(a_{ij}x_i x_j + b_{ij}\bar{x}_i x_j + c_{ij}x_i\bar{x}_j + d_{ij}\bar{x}_i\bar{x}_j\right) + \sum_{i=1}^{n} (\alpha_i x_i + \beta_i\bar{x}_i)$$

The function $h(\mathbf{x})$ given in Eq. (1.12) has positive coefficients for quadratic terms but the associated linear terms still have positive and negative coefficients. Note that $-x_i = \bar{x}_i - 1$ and $-\bar{x}_i = x_i - 1$. Using these transformations, we can construct a posiform $h^1(\mathbf{x}, \bar{\mathbf{x}})$ such that $f(\mathbf{x}) = \alpha_1 + h^1(\mathbf{x}, \bar{\mathbf{x}})$ for all $\mathbf{x} \in \{0, 1\}$ and the coefficients of h^1 are positive. Likewise, we can construct a posiform $h^2(\mathbf{x}, \bar{\mathbf{x}})$ such that $f(\mathbf{x}) = \alpha_2 - h^2(\mathbf{x}, \bar{\mathbf{x}})$ for all $\mathbf{x} \in \{0, 1\}^n$.

Lemma 1.4 ([35]) $f(\mathbf{x}) = \alpha_1 - \zeta^1(\mathbf{x}, \bar{\mathbf{x}}) = \alpha_2 + \zeta^2(\mathbf{x}, \bar{\mathbf{x}})$, *where* ζ^1 *and* ζ^2 *are in homogeneous posiform and* α_1 *and* α_2 *are constants.*

Proof Without loss of generality, we assume that the matrix $\mathbf{Q}$ is lower triangular and the diagonal entries are zero (see Sect. 1.3). Thus

$$f(\mathbf{x}) = \sum_{i<j} q_{ij}x_i x_j + \sum_{i=1}^{n} c_i x_i.$$

Note that when x_i and x_j are binary variables,

$$x_i x_j = \frac{1}{2}\left(x_i x_j + \bar{x}_i\bar{x}_j + x_i + x_j - 1\right)$$

and

$$-x_i x_j = \frac{1}{2}\left(x_i \bar{x}_j + \bar{x}_i x_j - x_i - x_j\right)$$

Substitute these values in $f(\mathbf{x})$ and simplify. Then, in the linear terms, if coefficient of x_i or $\bar{x}_i$ is negative, make a substitution using the equalities $-x_i = \bar{x}_i - 1$ and $-\bar{x}_i = x_i - 1$ and we get the first equality in the lemma. An analogous proof can be given for the second part also. □

The homogeneous posiforms ζ^1 and ζ^2 are not unique. The function $h(x)$ constructed in Eq. (1.12) have positive coefficients for all quadratic terms and can be converted into posiform by applying the transformation for linear terms used in the proof above. There are many other ways to obtain the required posiforms in the lemma above.

Since $\zeta^1(\mathbf{x}, \bar{\mathbf{x}})$ and $\zeta^2(\mathbf{x}, \bar{\mathbf{x}})$ are positive for all $\mathbf{x} \in \{0, 1\}^n$, we have

$$\alpha_2 \leq f(\mathbf{x}) \leq \alpha_1 \text{ for all } \mathbf{x} \in \{0, 1\}^n.$$

Let α be the smallest value real number for which there exist a homogeneous posiform $\zeta(\mathbf{x}, \bar{\mathbf{x}})$ such that $f(\mathbf{x}) = \alpha - \zeta(\mathbf{x}, \bar{\mathbf{x}})$. Then, $f(\mathbf{x}) \leq \alpha$ for all $\mathbf{x} \in \{0, 1\}^n$ and $\zeta(\mathbf{x}, \bar{\mathbf{x}})$ is called the complement of $f(\mathbf{x})$. Thus, by computing the complement of $f(\mathbf{x})$ we get an immediate upper bound on $f(\mathbf{x})$. For interesting discussions on computing the best upper bound of this type and other related results, we refer to [35].

1.5 Roof Duality

Consider the objective function of QUBO in Rhys form $h(\mathbf{x})$. i.e.,

$$\sum_{(i,j)\in P} q_{ij} x_i x_j + \sum_{(i,j)\in N} (-q_{ij}) \bar{x}_i x_j + \sum_{i=1}^{n} l_i x_i.$$

Without loss of generality assume that $q_{ij} = 0$ if $i > j$. Recall that $P = \{(i, j) : q_{ij} x_i x_j \in h(\mathbf{x}), i \neq j, q_{ij} \neq 0\}$ and $N = \{(i, j) : q_{ij} \bar{x}_i x_j \in h(\mathbf{x}), i \neq j, q_{ij} \neq 0\}$. The set inclusion notation $q_{ij} x_i x_j \in h(\mathbf{x})$ used here simply indicates that $q_{ij} x_i x_j$ is a term in $h(\mathbf{x})$.

For $q_{ij} \geq 0$ and $x_i, x_j \in \{0, 1\}$, $q_{ij} x_i x_j \leq \lambda_{ij} x_i + \mu_{ij} x_j$ and $q_{ij} \bar{x}_i x_j \leq \lambda_{ij}(1 - x_i) + \mu_{ij} x_j$ for all $\lambda_{ij} + \mu_{ij} = q_{ij}, \lambda_{ij} \geq 0, \mu_{ij} \geq 0$. A *roof* of $h(\mathbf{x})$ is a linear 0-1 function of the form [35]

$$r(\mathbf{x}, \boldsymbol{\lambda}, \boldsymbol{\mu}) = \sum_{i=1}^{n} l_i x_i + \sum_{(i,j)\in P} \left(\lambda_{ij} x_i + \mu_{ij} x_j\right) + \sum_{(i,j)\in N} \left(\lambda_{ij}(1 - x_i) + \mu_{ij} x_j\right) \tag{1.19}$$

where $\lambda_{ij} + \mu_{ij} = q_{ij}$ and $\lambda_{ij} \geq 0$, $\mu_{ij} \geq 0$ for all $(i, j) \in P \cup N$. By construction

$$h(\mathbf{x}) \leq r(\mathbf{x}, \boldsymbol{\lambda}, \boldsymbol{\mu}) \text{ for all } \mathbf{x} \in \{0, 1\}^n.$$

Let $G = (V, E)$ be a graph with $V = \{1, 2, \ldots, n\}$ and $E = P \cup N$. For each $i \in V$, define the set of outgoing and incoming arcs at node i by

$$O(i) = \{j \in V : (i, j) \in E\} \text{ and } I(i) = \{k \in V : (k, i) \in E\}.$$

Then the coefficient, $p_i(\boldsymbol{\lambda}, \boldsymbol{\mu})$, of x_i in $r(\mathbf{x}, \boldsymbol{\lambda}, \boldsymbol{\mu})$ is given by

$$p_i(\boldsymbol{\lambda}, \boldsymbol{\mu}) = l_i + \sum_{k \in I(i)} \mu_{ki} + \sum_{j \in O(i)} \delta_{ij} \lambda_{ij},$$

where

$$\delta_{ij} = \begin{cases} 1 & \text{if } (i, j) \in P \\ -1 & \text{if } (i, j) \in N \end{cases}$$

Thus, $r(\mathbf{x}, \boldsymbol{\lambda}, \boldsymbol{\mu})$ can be written as

$$r(\mathbf{x}, \boldsymbol{\lambda}, \boldsymbol{\mu}) = \sum_{i=1}^{n} p_i(\boldsymbol{\lambda}, \boldsymbol{\mu}) x_i + \sum_{(i,j) \in N} \lambda_{ij}$$

where $\lambda_{ij} + \mu_{ij} = q_{ij}$, $\lambda_{ij} \geq 0$, $\mu_{ij} \geq 0$. Then the *roof dual* of QUBO [7, 35], when given in terms of $h(\mathbf{x})$, is

$$\begin{aligned} &\text{Minimize } \max\{r(\mathbf{x}, \boldsymbol{\lambda}, \boldsymbol{\mu}) : \mathbf{x} \in \{0, 1\}^n\} \\ &\text{Subject to: } \lambda_{ij} + \mu_{ij} = q_{ij} \text{ for all } (i, j) \in P \cup N \\ &\qquad \lambda_{ij} \geq 0, \mu_{ij} \geq 0 \text{ for all } (i, j) \in P \cup N. \end{aligned}$$

Since $r(\mathbf{x}, \boldsymbol{\lambda}, \boldsymbol{\mu})$ is linear in $\mathbf{x}$, its maximum is attained when $x_i = 1$ for all $p_i(\boldsymbol{\lambda}, \boldsymbol{\mu}) \geq 0$ and $x_i = 0$ otherwise. i.e.,

$$\max\{r(\mathbf{x}, \boldsymbol{\lambda}, \boldsymbol{\mu}) : \mathbf{x} \in \{0, 1\}^n\} = \sum_{i=1}^{n} \max\{0, p_i(\boldsymbol{\lambda}, \boldsymbol{\mu})\} + \sum_{(i,j) \in N} \lambda_{ij}.$$

Thus the *roof dual* [6, 35] can be written as

$$\begin{aligned} \text{RD: } &\text{Minimize} \sum_{i=1}^{n} u_i + \sum_{(i,j) \in N} \lambda_{ij} \\ &\text{Subject to: } u_i \geq p_i(\boldsymbol{\lambda}, \boldsymbol{\mu}) \end{aligned}$$

$$\lambda_{ij} + \mu_{ij} = q_{ij} \text{ for all } (i, j) \in P \cup N$$
$$\lambda_{ij} \geq 0, \mu_{ij} \geq 0 \text{ for all } (i, j) \in P \cup N$$
$$u_i \geq 0 \text{ for } i = 1, 2, \ldots, n.$$

Note that RD is a linear program and it is closely linked to network flows on the support graph G. The optimal objective function value of RD provides an upper bound for the optimal objective function value of QUBO. The dual of RD is precisely the continuous relaxation of the linearization of the Rhys form of QUBO. Thus, the upper bound obtained by the roof dual is precisely the upper bound obtained by solving the continuous relaxation of the linearization of Rhys form of QUBO. Let us take an example to illustrate the concept of roof dual.

Example 1.2 Let $f(\mathbf{x}) = 10x_1x_2 - 5x_1x_3 + 20x_1x_4 - 12x_2x_4 - 2x_3x_4 - 6x_1 - 5x_2 + 8x_3 + 5x_4$. Then $P = \{(1, 2), (1, 4)\}$ and $N = \{(1, 3), (2, 4), (3, 4)\}$. Using the transformation $x_i = 1 - \bar{x}_i$ for $(i, j) \in N$ we obtain the Rhys form $h(x)$ given by

$$h(\mathbf{x}) = 10x_1x_2 + 5\bar{x}_1x_3 + 20x_1x_4 + 12\bar{x}_2x_4 + 2\bar{x}_3x_4 - 6x_1 - 5x_2 + 3x_3 - 9x_4.$$

The support graph associated with $h(\mathbf{x})$ is given below (Fig. 1.4).
Note that, $O(1) = \{2, 3, 4\}, O(2) = \{4\}, O(3) = \{4\}, O(4) = \emptyset$ and $I(1) = \emptyset, I(2) = \{1\}, I(3) = \{1\}, I(4) = \{1, 2, 3\}$. Therefore,

$$p_1(\boldsymbol{\lambda}, \boldsymbol{\mu}) = -6 + \lambda_{12} - \lambda_{13} + \lambda_{14}$$
$$P_2(\boldsymbol{\lambda}, \boldsymbol{\mu}) = -5 - \lambda_{24} + \mu_{12}$$
$$P_3(\boldsymbol{\lambda}, \boldsymbol{\mu}) = 3 - \lambda_{34} + \mu_{13}$$
$$P_4(\boldsymbol{\lambda}, \boldsymbol{\mu}) = -9 + \mu_{14} + \mu_{24} + \mu_{34}.$$

Then, the roof dual RD is given by,

$$\text{RD:} \quad \text{Minimize} \sum_{i=1}^{n} u_i + \lambda_{13} + \lambda_{24} + \lambda_{34}$$
$$\text{Subject to:} \quad u_1 - \lambda_{12} + \lambda_{13} - \lambda_{14} \geq -6$$
$$u_2 + \lambda_{24} - \mu_{12} \geq -5$$

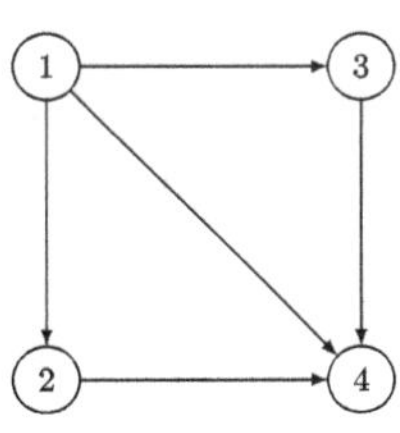

Fig. 1.4 The support graph of $h(\mathbf{x})$

$$
\begin{aligned}
& u_3 + \lambda_{34} - \mu_{13} \geq 3 \\
& u_4 - \mu_{14} - \mu_{24} - \mu_{34} \geq -9 \\
& \lambda_{ij} + \mu_{ij} = q_{ij} \text{ for all } (i, j) \in P \cup N \\
& \lambda_{ij} \geq 0, \mu_{ij} \geq 0 \text{ for all } (i, j) \in P \cup N \\
& u_i \geq 0 \text{ for } i = 1, 2, \ldots, 4.
\end{aligned}
$$

The dual of this linear program is

$$
\begin{aligned}
\text{Maximize} \; & -6x_1 - 5x_2 + 3x_3 - 9x_4 + 10y_{12} + 5y_{13} + 20y_{14} + 12y_{24} + 2y_{34} \\
\text{Subject to:} \; & y_{12} \leq x_1, y_{12} \leq x_2 \\
& y_{13} \leq 1 - x_1, y_{13} \leq x_3 \\
& y_{14} \leq x_1, y_{14} \leq x_4 \\
& y_{24} \leq 1 - x_2, y_{24} \leq x_4 \\
& y_{34} \leq 1 - x_3, y_{34} \leq x_3 \\
& 0 \leq x_i \leq 1, i = 1, 2, 3, 4 \\
& y_{ij} \text{ unrestricted for all } (i, j) \in P \cup N.
\end{aligned}
$$

and this is the continuous relaxation of the linearization of $h(\mathbf{x})$. (The notion of linearization is discussed in Sect. 1.7 and Chap. 4 and 6.)

1.6 Model Building Using QUBO

Let us now look at some motivating applications of the QUBO model. For a detailed discussion on the applications of QUBO and its power of providing a unifying framework to represent a large class of combinatorial optimization problems we refer to Chap. 2.

1.6.1 Person Detection and Tracking in a Crowded Environment

Identifying and tracking people is an important problem within a variety of application scenarios. These include, examination and exploration of group behaviour, video surveillance, pedestrian detection systems, disaster management, among others. The problem however is very complex, particularly due to the high level of occlusions, and researchers in computer vision developed various techniques to solve this problem making use of statistical machine learning, mathematical modelling, and optimization [16, 73, 74]. Let us now discuss one such application modelled as a QUBO by Rodriguez et al. [73].

Let $N = \{1, 2, \ldots, n\}$ be a set of points identified in an image as possible person detection locations and let c_i be an associated confidence score for location i, for $i \in N$. The point i and the score c_i can be estimated in different ways, as by appropriately trained preprocessing algorithms. We are also given person density information d_i (i.e. the number of people per pixel) estimated in a window of size σ at location i, for $i \in N$. We want to find locations of people in the image such that the sum of detector confidence score is maximized while making use of the density information to minimize selection of locations with significantly overlapping neighborhoods, which in turn minimizes potential counting errors accumulated due to multiple detection of the same person. Figure 1.5a gives a sample shot of a crowd and (b) represents the density contours. Part (c) of the figure gives potential locations and (d) gives the filtered solution produced by QUBO. Note

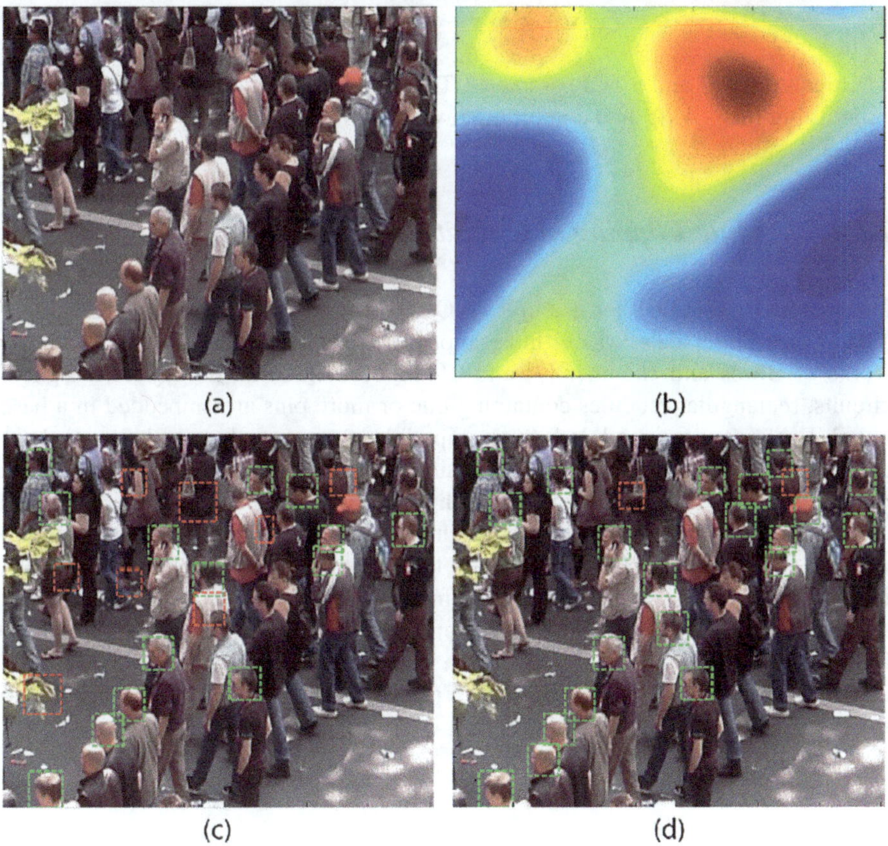

Fig. 1.5 Person detection sample and crowd density contours

that some of the red rectangles, that are potential locations, are either removed as irrelevant or confirmed in the QUBO solution.[1]

Let $\mathbf{x} = (x_1, x_2, \ldots, x_n) \in \{0, 1\}^n$, where $x_i = 1$ implies a detection at location i is confirmed and 0, otherwise. Then $\mathbf{cx}$ measures the total confidence score of 'confirmed' locations. To make sure that only valid configurations of non-overlapping locations are selected, we construct a penalty matrix $\mathbf{W}$, where $w_{ij} = -\infty$ if detections at locations i and j have significant area overlap ratio, and 0 otherwise. Then, maximizing $\mathbf{c}^T\mathbf{x} + \mathbf{x}^{\mathbf{T}}\mathbf{W}\mathbf{x}$ provides a meaningful model to represent the person detection problem. This, in fact, is a variation of a model proposed in [16].

To improve the accuracy of the model, Rodriguez et al. [73] introduced another quadratic term penalizing the difference between the density values estimated in two ways: (i) the vector $\mathbf{d}$ obtained using a regression-based density estimator and (ii) the vector $\mathbf{Ax}$ counting the density of active detections for an appropriately defined $\mathbf{A}$ matrix. This leads to the penalty term $||\mathbf{d} - \mathbf{Ax}||_1^2$. Then, the objective function to be maximized is

$$\mathbf{c}^T\mathbf{x} + \mathbf{x}^{\mathbf{T}}\mathbf{W}\mathbf{x} - \alpha||\mathbf{d} - \mathbf{Ax}||_1^2,$$

where $\mathbf{x} \in \{0, 1\}^n$ and α is a parameter. This problem can be written as QUBO by simplifying $||\mathbf{d} - \mathbf{Ax}||_1^2$ and combining it with $\mathbf{x}^T\mathbf{W}\mathbf{x}$.

1.6.2 Module Flipping in Circuit Layout Design

Let us now look at another example of QUBO that arises in the layout design of VLSI (very large scale system integration). Our discussion here is based on the works of Boros and Hammer [7] and Cheng et al. [12]. In the layout design of circuits, rectangular modules containing one or more pins are embedded in a base board. Each pin on a module has a compatible pin on another module and these compatible pairs need to be connected using horizontal and/or vertical wiring. The reliability has an inverse relationship with wire length and hence we are interested in minimizing the total wire length. Each module can be placed in four different ways on its designated space on the base board by flipping horizontally, vertically, or both. In Fig. 1.6a, we give a layout of modules and in Fig. 1.6b we give a layout after vertically flipping module 1, horizontally flipping module 2, and a vertical flip followed by a horizontal flip on module 5. The compatible pairs of pins are labelled using the same alphabets. For example, a pin with label a needs to be connected to another with the same label a and so on.

[1] Figure 1.5 was provided by Mikel Rodriguez and reproduced here with his permission.

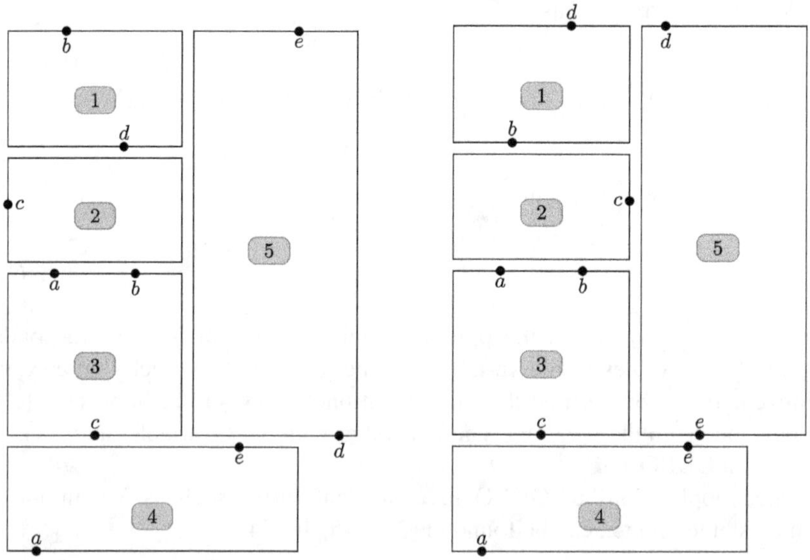

(a) (b)

Fig. 1.6 Module flipping example. Applied a vertical flip on module 1, horizontal flip on module 2, vertical flip followed by horizontal flip on module 5. (**a**) Original position. (**b**) After flip operations are applied on modules 1, 2, and 5

The module flipping problem is to find the orientations of the placement of each module on the base board, identified by one or more flipping operations, so that the total wire length is minimized. This problem was originally proposed by Cheng et al. [12] and they presented a maximum cut formulation, which as we know, is equivalent to QUBO. Boros and Hammer [7] presented a direct QUBO formulation of the problem and we discuss this model here.

Let $N = \{1, 2, \ldots, n\}$ be the collection of all modules placed on the base board with an initial layout. Consider the binary decision variable x_i which takes value 1 if module i is flipped horizontally, for $i = 1, 2, \ldots, n$. Likewise, consider the binary decision variable y_i which takes value 1 if module i is flipped vertically, for $i = 1, 2, \ldots, n$. For each pair (i, j) of modules and $(r, s) \in \{0, 1\} \times \{0, 1\}$, let $H_{ij}^{(r,s)}$ denotes the total length of horizontal wire segments which connects the pins of modules i and j under the flipping sequence (r, s). For example, if $(r, s) = (0, 0)$ no horizontal flipping occurs for modules i and j, if $(r, s) = (0, 1)$, only module j is flipped horizontally, and so on. Similarly, let $V_{ij}^{(r,s)}$ denotes the total length of vertical wire segments which connects the pins of modules i and j under the flipping sequence (r, s). Now, the total wire-length, as a function of flipping operation of the modules, can be expressed as

$$\phi(\mathbf{x}, \mathbf{y}) = \sum_{i=1}^{n-1} \sum_{j=i+1}^{n} \left(H_{ij}^{(0,0)} \bar{x}_i \bar{x}_j + H_{ij}^{(0,1)} \bar{x}_i x_j + H_{ij}^{(1,0)} \mathbf{x}_i \bar{x}_j + H_{ij}^{(1,1)} x_i x_j \right)$$
$$+ \sum_{i=1}^{n-1} \sum_{j=i+1}^{n} \left(H_{ij}^{(0,0)} \bar{y}_i \bar{y}_j + H_{ij}^{(0,1)} \bar{y}_i y_j + H_{ij}^{(1,0)} y_i \bar{y}_j + H_{ij}^{(1,1)} y_i y_j \right) \tag{1.20}$$

where $\bar{x}_i = 1 - x_i$. Then, the optimal flipping of the modules corresponds to assigning $0 - 1$ values to the variables x_i and y_i for $i \in N$ such that $\phi(\mathbf{x}, \mathbf{y})$ is minimized. It can be verified that minimization of $\phi(\mathbf{x}, \mathbf{y})$ decomposes into two problems, one with the **x**-variables and the other with the **y**-variables and each such problem is a QUBO.

Related applications of QUBO in equivalent forms such as Maximum Cut, weighted stable set etc., can be found in [3, 9, 15, 46, 54].

1.6.3 Side-Chain Positioning in Protein Design

The *side chain positioning problem* arises as a subproblem in protein structure prediction which has a natural QUBO formulation. Our discussion here is based on the works [11, 30, 31, 47]. While the formulation itself is straightforward, to make the discussion clearer, let us very briefly review some related concepts, terminology, and background information and for this purpose, we follow the articles [11, 31].

A *protein molecule* is composed of a chain of amino acids and each amino acid consists of a centralized single carbon atom along with a hydrogen atom, an amino group NH_2, a carboxyl group COOH, and a *side chain* which characterizes the amino acid. Carbon, hydrogen, the amino group, and the carboxyl group are called the *main atoms* and the *protein backbone* is formed by a repeating sequence of the main atoms (main chain) and a *side chain* is attached to the backbone for each element of this sequence (Fig. 1.7).

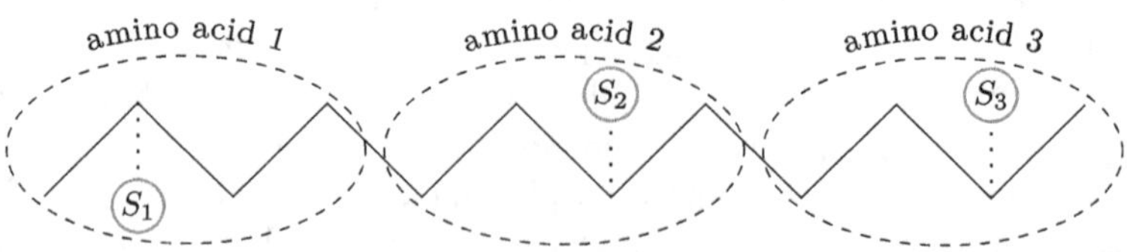

Fig. 1.7 A chain of three amino acids. The labels S_1, S_2 and S_3 indicate the corresponding side-chains

The chemical composition of a protein molecule is specified by the sequence of the associated amino acids. Every amino acid main chain has the freedom to *rotate* at specified points. A side chain of amino acids (with the exception of glycine) can also rotate at different points and the three dimensional structure of a protein is identified by the location of its backbone (main chain) atoms and the combined rotations of the main chain and the side chains. For an appropriate notion of *energy*, a protein structure folds into a minimal energy state to reach chemical stability. Quoting from [31], "Chemical stability of a protein comes from four sources: the internal stability of the three-dimensional structure of its backbone; the internal stability of each rotamer; the interaction of each rotamer with the main chain of the amino acid it is attached to; and the chemical interactions of the rotamers that are positioned close to each other, either on the protein sequence, or in three-dimensional space." In protein synthesis, we want to maximize the chemical stability using appropriate quantitative stability measures.

The backbone structure is assumed to be given and our aim is to select one rotamer (side-chain molecule) to be associated with each amino acid sequence in the backbone. Let e_0 be the stability measure for the backbone structure, which is constant. Suppose that the backbone consists of m amino acid sequences and for amino acid k, we select a rotamer from the candidate list $V_k, k = 1, 2, \ldots, m$. For rotamer $i \in V_k$ let e_i^k be the contribution of i to the stability measure which consists of the sum of the interaction stability measure of candidate rotamer i with its associated amino acid k and the internal stability measure of rotamer i. For each $i \in V_k$ and $j \in V_t$, let e_{ir}^{kt} be the stability measure due to the interaction between rotamer $i \in V_k$ and $j \in V_t$. This value depends on the nature of the rotamers involved, the nature of the amino acids they attach to, and the proximity level of i and j. Since the backbone is already known, the proximity level is also known. The values e_0, e_i^k, and e_{ir}^{kt} are estimated either by experiments or by theoretical considerations. For our modeling purpose, it is not highly relevant the way these values are obtained or their magnitudes. The side-chain positioning problem is to select an i_k from V_k for each $k = 1, 2, \ldots m$ such that

$$e_0 + \sum_{k=1}^{m} e_{i_k}^k + \frac{1}{2} \sum_{k=1}^{m} \sum_{t=1}^{m} e_{i_k i_t}^{kt}$$

is maximized. Let us now formulate this problem as QUBO.

Let $V = \cup_{k=1}^{m} V_k$. Without loss of generality, assume $V = \{1, 2, \ldots, n\}$. Construct the complete m-partite undirected graph $G = (V, E)$ where $(i, j) \in E$ if $i \in V_k, j \in V_t, t \neq k$. Note that we need to select precisely one element for each V_k for $k = 1, 2 \ldots, m$. For each $i \in V$ consider the binary decision variable x_i where $x_i = 1$ if rotamer i is selected and, 0 otherwise. Define the $n \times n$ cost matrix Q with its (i, j)th element q_{ij} as

$$q_{ij} = \begin{cases} 0 & \text{if } i = j \\ -M & \text{if } i \neq j \text{ and } i, j \in V_k \text{ for some } k, \\ \frac{1}{2}e^{kt}_{rs} & \text{if } (i, j) \in E, \end{cases}$$

where M is a large positive number and the n vector $\mathbf{c}^T = (c_1, c_2, \ldots, c_n)$ where $c_i = e^k_j$ if $i \in V^k$ and $i = j$. Now an optimal solution to the QUBO instance $(\mathbf{Q}, \mathbf{c}, \max)$ gives an optimal solution to the side chain positioning problem. Note that, no two elements from the same set V_k will be selected in an optimal solution for any k because of the penalty value and since the objective function is of maximization type, $c_i \geq 0$ for all i and $e^{kt}_{rs} \geq 0$ for all k, t, r, s at least one elements from each V_k will be selected, establishing the validity of the model.

1.6.4 QUBO and Machine Scheduling

We now discuss a general framework of a special case of QUBO, which can be used to model several important machine scheduling problems. A matrix $\mathbf{Q}$ is called a *half product matrix* if it is upper triangular with diagonal elements zero and there exist vectors $\mathbf{a} = (a_1, a_2, \ldots, a_n)$ and $\mathbf{b} = (b_1, b_2, \ldots, b_n)$ such that $q_{ij} = a_i b_j$ for $i = 1, 2, \ldots, n$ and $j > i$. A *half-product* QUBO is a special case of QUBO where $\mathbf{Q}$ is a half-product matrix. Thus, a QUBO with a half-product matrix can be written as

$$\begin{aligned} \text{Maximize} \quad & \sum_{1 \leq i < j \leq n} a_i b_j x_i x_j + \sum_{j=1}^{n} c_j x_j \\ \text{Subject to} \quad & \\ & \mathbf{x} \in \{0, 1\}^n \end{aligned}$$

The half-product QUBO was introduced by Badics and Boros [2] and independently by Kubiak [52]. Despite the apparently simple special structure, the half-product QUBO is still NP-hard since the subset sum problem is a special case of it [2].

Various single machine scheduling problems can be formulated as a half-product QUBO. This includes, the single-machine variance minimization problem [2, 52], Single machine scheduling to minimize total weighted earliness and tardiness[43] and scheduling with controllable processing times [41]. The Ising version of the half-product QUBO was considered by Kubiak [52], Amit [1], and Mattis [58]. Scheduling jobs on two machines to minimize makespan [43] as well as scheduling on two machines to minimize total weighted completion time [43] can also be formulated as a half-product QUBO. Let us discuss one such formulations using the example of *single machine scheduling with controllable processing times* (SCPT) [41].

There are n jobs to be processed on a single machine where the processing time of jobs are variables. For job j, let the processing time $p_j \in [0, \alpha_j]$, $j = 1, 2, \ldots, n$. To schedule the jobs, one needs to identify the optimal values of the processing times p_j and then order the jobs so that the sum of the total weighted completion time $\sum_{j=1}^{n} w_j C_j$ and the total weighted processing time compression $\sum_{j=1}^{n} v_j(u_j - p_j)$ is minimized. Here, C_j denote the completion time of job j. This scheduling problem, denoted by SCPT, was first studied by Vickson [78, 79] and the half-product QUBO formulation discussed here is from [41]. Vickson [78] showed that there exists an optimal processing time vector $\mathbf{p} = (p_1, p_2, \ldots, p_n)$ such that $p_j = 0$ or α_j. Combining this observation with the well-known weighted shortest processing time rule (SWPT) [76], we can see that there exists an optimal solution to SCPT such that jobs processing times $p_j = \alpha_j$ are sequenced in the non-decreasing order of the ratios $\frac{\alpha_j}{w_j}$ [41]. With this observation, define the decision variables

$$x_j = \begin{cases} 0 & \text{if } p_j = 0 \\ 1 & \text{if } p_j = \alpha_j \end{cases}$$

and the problem SCPT is well defined when the vectors $\mathbf{w} = (w_1, w_2, \ldots, w_n)$, $\boldsymbol{\alpha} = (\alpha_1, \alpha_2, \ldots, \alpha_n$, and $\mathbf{v} = (v_1, v_2, \ldots, v_n)$ are given. Then, the objective function of SCPT is to minimize $f(\mathbf{x})$, where

$$\begin{aligned} f(\mathbf{x}) &= \sum_{j=1}^{n} w_j x_j \sum_{i=1}^{j} \alpha_i x_i + \sum_{j=1}^{n} v_j \alpha_j (1 - x_j) \\ &= \sum_{1 \le i < j \le n} \alpha_i w_j x_i x_j - \sum_{j=1}^{n} \alpha_j (v_j - w_j) x_j + \sum_{j=1}^{n} v_j \alpha_j \end{aligned}$$

Note that $\sum_{j=1}^{n} v_j \alpha_j$ is a constant and can be discarded. Choosing, $a_i = \alpha_i, b_i = -w_i$ and $c_i = \alpha_i(v_i - w_i)$, we can see that minimization of $f(\mathbf{x})$ can be accomplished by solving the half-product QUBO.

1.7 Mixed Integer Programming Formulations

Various mixed integer linear programming (MILP) formulations of QUBO are discussed in detail in other chapters of this book. Most of the basic formulations are based on writing the product of two variables (not necessarily both are of binary type) as a new variable and force the definition of the underlying product by imposing additional constraints, that may or may not exploit the sign of the q_{ij} values. Early research work on these types of reductions goes back to 1960s [13, 20, 29, 80, 81] where product of two binary variables are 'linearized'. Later, Glover [25, 26] and Glover and Woolsey [27] considered linearization of the product of a binary variable and a continuous variable. Most of these linearization techniques can be presented under a general approximation framework introduced by McCormick [59, 60].

Consider the bounded variables

$$l_1 \leq x_1 \leq u_1 \text{ and } l_2 \leq x_2 \leq u_2.$$

Then,

$$\begin{aligned}(x_1 - l_1)(u_2 - x_2) \geq 0, \quad & (u_1 - x_1)(x_2 - l_2) \geq 0 \\ (x_1 - l_1)(x_2 - l_2) \geq 0, \quad & \text{and } (u_1 - x_1)(u_2 - x_2) \geq 0.\end{aligned}$$

Expanding and rearranging the terms, we have

$$\begin{aligned}x_1x_2 &\leq l_1x_2 + u_2x_1 - l_1u_2 \\ x_1x_2 &\leq l_2x_1 + u_1x_2 - l_2u_1 \\ x_1x_2 &\geq l_1x_2 + l_2x_1 - l_1l_2 \\ x_1x_2 &\geq u_2x_1 + u_2x_2 - u_1u_2\end{aligned}$$

Now, let us replace the term x_1x_2 by a new variable w in the above inequalities and we get

$$w \leq l_1x_2 + u_2x_1 - l_1u_2 \tag{1.21}$$

$$w \leq l_2x_1 + u_1x_2 - l_2u_1 \tag{1.22}$$

$$w \geq l_1x_2 + l_2x_1 - l_1l_2 \tag{1.23}$$

$$w \geq u_2x_1 + u_2x_2 - u_1u_2 \tag{1.24}$$

The inequalities (1.21) and (1.22) are called *lower McCormick envelops* and (1.23), and (1.24) are called *upper McCormick envelops* [59, 60]. The variable w along with the McCormick envelop inequalities provide an approximation to the product x_1x_2. This approximation becomes exact when one of the variables x_1 or x_2 is binary and it becomes the linearizations proposed by Glover [25, 26] and Glover and Woolsey [27]. To see the validity of the linearization, without loss of generality, assume x_2 is binary. Then $l_2 = 0$ and $u_2 = 1$. When $x_2 = 0$, the inequalities (1.22) and (1.23) guarantees that $w = 0$. Likewise, when $x_2 = 1$, inequalities (1.21) and (1.24) guarantees that $w = x_1$. The special case of McCormick envelop inequalities involving binary variables leads to a natural MILP formulation of QUBO which is the well-known Glover's linearization Glover [25, 26]. When both variables involved in a product are binary, McCormick envelop inequalities reduces to Glover-Woolsey linearization [27]. We also want to highlight that the linearization of product of two binary variables are reported in literature either explicitly or implicitly, even prior to [27]. For example, Goldman [29] pointed out the works of Fortet [20, 21] and Dantzig [13] in this context.

Replace the product $x_i x_j$ with a new variable w_{ij} in the objective function of QUBO and force the equality $x_i x_j = w_{ij}$ by introducing the McCormick envelop inequalities. The resulting MILP formulation, known as Glover-Woolsey linearization or the *standard linearization*, is given below.

$$\text{GW:} \quad \text{Maximize} \quad \sum_{i=1}^{n}\sum_{j=1}^{n} q_{ij}w_{ij} + \sum_{i=1}^{n} c_i x_i$$

$$\text{Subject to:} \quad w_{ij} - x_i \leq 0 \text{ for } i, j \in \{1, 2, \ldots, n\} \tag{1.25}$$

$$w_{ij} - x_j \leq 0 \text{ for } i, j \in \{1, 2, \ldots, n\} \tag{1.26}$$

$$x_i + x_j - w_{ij} \leq 1 \text{ for } i, j \in \{1, 2, \ldots, n\} \tag{1.27}$$

$$w_{ij} \geq 0 \text{ for } i, j \in \{1, 2, \ldots, n\} \tag{1.28}$$

$$x_i \in \{0, 1\} \text{ for } i = 1, 2 \ldots, n. \tag{1.29}$$

Recall the definition of the sets P and N introduced in Sect. 1.4.3. For $(i, j) \in P$ constraints (1.27) and (1.28) can be removed and for $(i, j) \in N$ constraints (1.25) and (1.26) can be removed. Thus GW becomes the *reduced Glover-Woolsey linearization*

$$\text{RGW:} \quad \text{Maximize} \quad \sum_{i=1}^{n}\sum_{j=1}^{n} q_{ij}w_{ij} + \sum_{i=1}^{n} c_i x_i$$

$$\text{Subject to:} \quad w_{ij} - x_i \leq 0 \text{ for } (i, j) \in P \tag{1.30}$$

$$w_{ij} - x_j \leq 0 \text{ for } (i, j) \in P \tag{1.31}$$

$$x_i + x_j - w_{ij} \leq 1 \text{ for } (i, j) \in N \tag{1.32}$$

$$w_{ij} \geq 0 \text{ for } (i, j) \in N \tag{1.33}$$

$$x_i \in \{0, 1\} \text{ for } i = 1, 2, \ldots, n. \tag{1.34}$$

The convex hull of solutions of GW is called the *Boolean quadric polytope*[64] which is discussed in detail in Chap. 4. Various other MILP formulations of QUBO are known in literature and most of them uses Glover-Woolsey linearization or Glover's linearization in one form or another. For a through discussion of this, we refer to Chap. 6 and the recent papers [67, 68].

The continuous relaxation of GW (and RGW) is denoted by CGW (CRGW) and it is obtained by replacing constraint (1.29) (constraint (1.34)) by $x_i \in [0, 1]$ for $i = 1, 2 \ldots, n$. CGW and CRGW are linear programs and hence can be solved using general purpose linear programming solvers. However, their special structure also brings us additional structural properties and efficient algorithms. Some of these connections and linkages with roof duality is discussed in Chap. 5.

The polytope defined by the constraints of CGW is called the *fractional Boolean quadric polytope*. Let us first present a useful necessary condition for a solution $(\mathbf{x}, \mathbf{w})$ to be an extreme point of the fractional Boolean quadric polytope.

Lemma 1.5 *If* $(\mathbf{x}, \mathbf{w})$ *is an extreme point of the fractional Boolean quadric polytope then* $w_{ij} \in \{0, \min\{x_i, x_j\}, x_i + x_j - 1\}$ *for all* i, j.

Proof Suppose $(\mathbf{x}, \mathbf{w})$ is an extreme point of the fractional Boolean quadric polytope. Then, choose a cost matrix $\mathbf{Q}$ and a cost vector $\mathbf{c}$ such that $(\mathbf{x}, \mathbf{w})$ is an optimal solution to CGW. Without loss of generality assume that $q_{ij} \neq 0$ for $i \neq j$ and $c_i \neq 0$ for all i. Clearly, by feasibility $0 \leq w_{ij} \leq \min\{x_i, x_j\}$ for every $i, j, i \neq j$. If possible let $0 < w_{hk} < \min\{x_h, x_k\}$ for some $h, k, h \neq k$ and $x_h + x_k - 1 \neq w_{hk}$. Define the solution $(\mathbf{x}, \hat{\mathbf{w}})$ where

$$\hat{w}_{ij} = \begin{cases} w_{ij} + \epsilon & \text{if } (i, j) = (h, k) \\ w_{ij} & \text{Otherwise,} \end{cases}$$

where $\epsilon \neq 0$. If $q_{hk} > 0$ then $(\mathbf{x}, \hat{\mathbf{w}})$ is a feasible solution to CGW for sufficiently small $\epsilon > 0$ which is better that $(\mathbf{x}, \mathbf{w})$, contradicting the optimality of $(\mathbf{x}, \mathbf{w})$. Similarly, if $q_{hk} < 0$ then $(\mathbf{x}, \hat{\mathbf{w}})$ is a feasible solution to CGW for sufficiently large $\epsilon < 0$ which is better than $(\mathbf{x}, \mathbf{w})$, contradicting the optimality of $(\mathbf{x}, \mathbf{w})$. This completes the proof. □

We now prove the *half integrality property* of extreme points of the fractional Boolean quadric polytope [64].

Theorem 1.3 *If* $(\mathbf{x}, \mathbf{w})$ *is an extreme point of the fractional Boolean quadric polytope the* $x_i \in \{0, 1, \frac{1}{2}\}$ *for all* i *and* $w_{ij} \in \{0, 1, \frac{1}{2}\}$ *for all* $(i, j) \in P \cup N$.

Proof Let $\mathfrak{F}$ denote the fractional Boolean quadric polytope and $(\mathbf{x}, \mathbf{w})$ be an extreme point. Let

$$E^1 = \{i : 0 < x_i < \frac{1}{2}\} \text{ and } E^2 = \{i : \frac{1}{2} < x_i < 1\}.$$

If $E^1 \cup E^2 = \emptyset$ then from Lemma 1.5, $(\mathbf{x}, \mathbf{w})$ is of the required type. So assume that $E^1 \cup E^2 \neq \emptyset$. Now construct the solutions $(\mathbf{x^1}, \mathbf{w^1})$ and $(\mathbf{x^2}, \mathbf{w^2})$ as follows: For $\epsilon > 0$ let

$$x_i^1 = \begin{cases} x_i + \epsilon & \text{if } i \in E^1 \\ x_i - \epsilon & \text{if } i \in E^2 \\ x_i & \text{Otherwise} \end{cases} \quad \text{and} \quad x_i^2 = \begin{cases} x_i - \epsilon & \text{if } i \in E^1 \\ x_i + \epsilon & \text{if } i \in E^2 \\ x_i & \text{Otherwise.} \end{cases}$$

From Lemma 1.5, $w_{ij} \in \{0, \min\{x_i, x_j\}, x_i + x_j - 1\}$. Define

$$w_{ij}^1 = \begin{cases} \min\{x_i^1, x_j^1\} & \text{if } w_{ij} = \min\{x_i, x_j\} \\ x_i^1 + x_j^1 - 1 & \text{if } w_{ij} = x_i + x_j - 1 \\ 0 & \text{Otherwise} \end{cases}$$

$$w_{ij}^2 = \begin{cases} \min\{x_i^2, x_j^2\} & \text{if } w_{ij} = \min\{x_i, x_j\} \\ x_i^2 + x_j^2 - 1 & \text{if } w_{ij} = x_i + x_j - 1 \\ 0 & \text{Otherwise} \end{cases}$$

For sufficiently small $\epsilon > 0$, $(\mathbf{x^1}, \mathbf{w^1})$ and $(\mathbf{x^2}, \mathbf{w^2})$ are feasible solutions. It can be verified that $(\mathbf{x}, \mathbf{w}) = \frac{1}{2}\{(\mathbf{x^1}, \mathbf{w^1}) + (\mathbf{x^1}, \mathbf{w^1})\}$, contradicting the fact that $(\mathbf{x}, \mathbf{w})$ is an extreme point and the result follows. □

In Sect. 1.4 we observed that the maximum cut problem and QUBO are equivalent. In this sense, an integer programming formulation of the maximum cut problem provides an indirect integer programming representation of QUBO. Consider a complete graph $K_n = (V, E)$ and let q_{ij} be the weight of edge $(i, j) \in E$. Then the maximum cut problem in K_n can be written as an MILP

$$\begin{aligned} \text{MCUT:} \quad & \text{Minimize} \quad \sum_{i \leq i < j \leq n} q_{ij} x_{ij} \\ & \text{Subject to:} \quad x_{ij} + x_{ik} + x_{jk} \leq 2 \text{ for } 1 \leq i < j \leq n \\ & \qquad x_{ij} - x_{ik} - x_{jk} \leq 0 \text{ for } 1 \leq i < j \leq n, k \neq i, j \\ & \qquad x_{ij} \in \{0, 1\} \text{ for } 1 \leq i < j \leq n. \end{aligned}$$

The convex hull of feasible solutions of MCUT is called the *cut polytope*. The relationship between the Boolean quadric polytope and the cut polytope is discussed in Chap. 4.

1.8 Conclusion

The primary goal of this chapter was to give a brief introduction to QUBO along with basic definitions and motivating examples. Each of the chapters that follows is more focussed and is on a specific aspect of QUBO providing a comprehensive and in-depth analysis of the topic.

Acknowledgments I am thankful to Fred Glover and Navpreet Kaur for their feedback on an earlier version of this chapter which improved the presentation. This work was partially supported by an NSERC discovery grant.

References

1. D.J. Amit, *Modeling Brain Function: The World of Attractor Neural Networks* (Cambridge University Press, Cambridge, 1989)
2. T. Badics, E. Boros, Minimization of half-products. Math. Oper. Res. **23**, 649–660 (1998)
3. F. Barahona, M. Grotschel, M. Junger, G. Reinelt, An application of combinatorial optimization to statistical physics and circuit layout design. Oper. Res. **36**, 493–513 (1988)
4. R. Bar-Yehuda, S. Even, A local-ratio theorem for approximating the weighted vertex cover problem. Ann. Discrete Math. **25**, 27–45 (1985)
5. Z. Bian, F. Chudak, W.G. Macready, G. Rose, The Ising model: teaching an old problem new tricks. D-Wave systems technical report, Aug 2010
6. E. Boros, P.L. Hammer, A max-flow approach to improved roof-duality in quadratic $0-1$ minimization. RUTCOR Research Report RRR 15-1989, RUTCOR (1989)
7. E. Boros, P.L. Hammer, The Max-Cut problem and quadratic 0-1 optimization; polyhedral aspects, relaxations and bounds. Ann. Oper. Res. **33**, 151–180 (1991)
8. E. Boros, P.L. Hammer, Pseudo-Boolean optimization. Discrete Appl. Math. **123**, 155–225 (2002)
9. E. Boros, P.L. Hammer, M. Minoux, D.J. Rader Jr., Optimal cell flipping to minimize channel density in VLSI design and pseudo-Boolean optimization. Discrete Appl. Math. **90**, 69–88 (1999)
10. S.G. Brush, History of the Lenz-Ising model. Rev. Mod. Phys. **39**, 883–893 (1967)
11. B. Chazelle, C. Kingsford, M. Singh, A semidefinite programming approach to side chain positioning with new rounding strategies. INFORMS J. Comput. **16**, 380–392 (2004)
12. C.K. Cheng, S.Z. Yao, T.C. Hu, The orientation of modules based on graph decomposition. IEEE Trans. Comput. **C-40**, 774–780 (1991)
13. G.B. Dantzig, On the significance of solving linear programming problems with some integer variables. Econometrica **28**, 30–44 (1960)
14. G.B. Dantzig, Linear programming. Oper. Res. **50**, 42–47 (2002)
15. B. Das, A.K. Mahato, A.K. Khan, Via minimization for multi-layer channel routing in VLSI design, in *Fourth International Conference on Communication Systems and Network Technologies* (2014)
16. C. Desai, D. Ramanan, C.C. Fowlkes, Discriminative models for multi-class object layout. Int. J. Comput. Vis. **95**, 1–12 (2011)
17. P. Erdos, G. Szekeres, A combinatorial problem in geometry. Compos. Math. **2**, 463–470 (1935)
18. J.-A. Ferrez, K. Fukuda, Th.M. Liebling, Solving the fixed rank convex quadratic maximization in binary variables by a parallel zonotope construction algorithm. Eur. J. Oper. Res. **166**, 35–50 (2005)
19. L. Festinger, The analysis of sociograms using matrix algebra. Hum. Relat. **2**, 153–158 (1949)
20. R. Fortet, Applications de l'algèbre de boole en recherche opérationelle. Rev. Fr. Rech. Opér. **4**, 5–36 (1959)
21. R. Fortet, L'algèbre de boole et ses applications en recherche opérationnelle. Cahiers Centre d'Etudes Rech. Oper. **4**, 17–26 (1960)
22. E. Forsyth, L. Katz, A matrix approach to the analysis of sociometric data: preliminary report. Sociometry **9**, 340–347 (1946)
23. M.R. Garey, D.S. Johnson, *Computers and Intractability: A Guide to the Theory of NP-Completeness* (W. H. Freeman, San Francisco, 1979)
24. S. Galam, *Sociophysics: A Physicist's Modeling of Psycho-Political Phenomena* (Springer, New York, 2012)
25. F. Glover, Improved linear integer programming formulations of nonlinear integer problems. Manag. Sci. **22**, 455–460 (1975)
26. F. Glover, An improved MIP formulation for products of discrete and continuous variables. J. Inf. Optim. Sci. **5**, 469–471 (1984)

27. F. Glover, E. Woolsey, Further reduction of zero-one polynomial programming problems to zero-one linear programming problems. Oper. Res. **21**, 141–161 (1973)
28. F. Glover, G. Kochenberger, Y. Du, A tutorial on formulating and using QUBO models. University of Colorado, Denver, 2019
29. A.J. Goldman, Linearization in 0-1 variables: a clarification. Oper. Res. **31**, 946–947 (1983)
30. R.F. Goldstein, Efficient rotamer elimination applied to protein side-chains and related spin glasses. Biophys. J. **66**, 1335–1340 (1994)
31. D. Gusfield, *Integer Linear Programming in Computational and Systems Biology: An Entry-Level Text and Course* (Cambridge University Press, Cambridge, 2019)
32. P.L. Hammer, A.A. Rubin, Some remarks on quadratic programming with 0-1 variables. RAIRO-Oper. Rese. Rech. Opér. **3**, 67–79 (1970)
33. P.L. Hammer, S. Rudeanu, *Boolean Methods in Operations Research and Related Areas* (Springer, Berlin, 1968)
34. P.L. Hammer, I. Rosenberg, S. Rudeanu, On the determination of the minima of pseudo-Boolean functions. Stud. Cerc. Mat. **14**, 359–364 (1963)
35. P.L. Hammer, P. Hansen, B. Simone, Roof duality, complementations, and persistency in quadratic 0-1 optimization. Math. Program. **28**, 121–155 (1984)
36. F. Harary, I.C. Ross, A procedure for clique detection using the group matrix. Sociometry **20**, 205–215 (1957)
37. R. Hassin, A. Levin, The minimum generalized vertex cover problem, in *European Symposium on Algorithms* (2003), pp. 289–300
38. C. Helmberg, F. Rendl, Solving quadratic (0,1)-problems by semidefinite programs and cutting planes. Math. Program. **82**, 291–315 (1998)
39. D.S. Hochbaum, A. Pathria, Forest harvesting and minimum cuts: a new approach to handling spatial constraints. For. Sci. **43**, 544–554 (1997)
40. T. Ising, R. Folk, R. Kenna, B. Berche, Y. Holovatch, The fate of Ernst Ising and the fate of his model. J. Phys. Stud. **21**(3), 3002, 19 pp. (2017)
41. A. Janiak, M.Y. Kovalyov, W. Kubiak, F. Werner, Positive half-products and scheduling with controllable processing times. Eur. J. Oper. Res. **165**, 416–422 (2005)
42. M. Jünger, T. Liebling, D. Naddef, G. Nemhauser, W. Pulleyblank, G. Reinelt, G. Rinaldi, L. Wolsey, *50 Years of Integer Programming 1958–2008: From the Early Years to the State-of-the-Art* (Springer, Berlin, 2010)
43. B. Jurisch, W. Kubiak, J. Józefowska, Algorithms for minclique scheduling problems. Discret. Appl. Math. **72**, 115–139 (1997)
44. D.R. Karger, C. Stein, A new approach to the minimum cut problem. J. ACM **43**, 601–640 (1996)
45. G. Kochenberger, F. Glover, A unified framework for modeling and solving combinatorial optimization problems: a tutorial, in *Multiscale Optimization Methods and Applications*, ed. by W. Hager, S.-J. Huang, P. Pardalos, O. Prokopyev (Springer, Berlin, 2006), pp. 101–124
46. H. Kim, An application algorithm for the via minimization problem in channel routing, in *Proceedings of the 1990 Symposium on Applied Computing* (1990)
47. C.L. Kingsford, B. Chazelle, M. Singh, Solving and analyzing side-chain positioning problems using linear and integer programming. Bioinformatics **21**, 1028–1036 (2005)
48. G. Kochenberger, J.-K. Hao, F. Glover, M. Lewis, Z. Lu, H. Wang, Y. Wang, The unconstrained binary quadratic programming problem: a survey. J. Comb. Optim. **28**, 58–81 (2014)
49. G. Kohring, Ising models of social impact: the role of cumulative advantage. J. Phys. I **6**, 301–308 (1996)
50. H. Konno, Maximization of a convex quadratic function under linear constraints. Math. Program. **11**, 117–127 (1976)
51. H. Konno, Maximizing a convex quadratic function over a hypercube. J. Oper. Res. Soc. Jpn. **23**, 171–188 (1980)
52. W. Kubiak, New results on the completion time variance minimization. Discrete Appl. Math. **58**, 157–168 (1995)
53. R. Lazmy, Mixed integer quadratic programming. Math. Program. Study **22**, 332–349 (1982)

54. F. Liers, G. Pardella, Partitioning planar graphs: a fast combinatorial approach for max-cut. Comput. Optim. Appl. **51**, 323–344 (2012)
55. R.D. Luce, A.D. Perry, A method of matrix analysis of group structure. Psychometrika **14**, 95–116 (1949)
56. H.M. Markowitz, Portfolio selection. J. Financ. **7**, 77–91 (1952)
57. H.M. Markowitz, The optimization of a quadratic function subject to linear constraints. Nav. Res. Logist. Q. **3**, 111–133 (156)
58. D.C. Mattis, Solvable spin systems with random interaction. Phys. Lett. **6A**, 412 (1976)
59. G.P. McCormick, Converting general nonlinear programming problems to separable nonlinear programming problems. Report T—267, The George Washington University, 1972
60. G.P. McCormick, Computability of global solutions to factorable nonconvex programs: part I - Convex underestimating problems. Math. Program. **10**, 147–175 (1976)
61. M. Niss, History of the Lenz-Ising model 1920–1950: from ferromagnetic to cooperative phenomena. Arch. Hist. Exact Sci. **59**, 267–318 (2005)
62. M. Niss, History of the Lenz-Ising model 1950–1965: from irrelevance to relevance. Arch. Hist. Exact Sci. **63**, 243–287 (2009)
63. M. Niss, History of the Lenz-Ising model 1965–1971: the role of a simple model in understanding critical phenomena. Arch. Hist. Exact Sci. **65**, 625–658 (2011)
64. M. Padberg, The Boolean quadric polytope: some characteristics, facets and relatives. Math. Program. **45**, 134–172 (1989)
65. P. Pandey, A.P. Punnen, The generalized vertex cover problem Discret. Optim. **30**, 121–143 (2018)
66. S.S. Petrova, A.D. Solov́e, The origin of the method of steepest descent. Hist. Math. **24**, 361–375 (1997)
67. A.P. Punnen, N. Kaur, Revisiting some classical explicit linearizations for the quadratic binary optimization problem. Research Report, Department of Mathematics, Simon Fraser University, 2021
68. A.P. Punnen, N. Kaur, On compact linearizations of the quadratic binary optimization problem. Research Report, Department of Mathematics, Simon Fraser University, 2021
69. A.P. Punnen, P. Pandey, M. Friesen, Representations of quadratic combinatorial optimization problems: a case study using the quadratic set covering problem. Comput. Oper. Res. **112**, 104769 (2019)
70. M. Raghavachari, On connections between zero-one integer programming and concave programming under linear constraints. Oper. Res. **17**, 680–684 (1969)
71. M. Raghavachari, Supplement. Oper. Res. **18**, 564–565 (1970)
72. J.M.W. Rhys, A selection problem of shared fixed costs and network flows. Manag. Sci. **17**, 200–207 (1970)
73. M. Rodriguez, I. Laptev, J. Sivic, J.-Y. Audibert, Density-aware person detection and tracking in crowds, in *2011 International Conference on Computer Vision, Barcelona* (2011), pp. 2423–2430
74. S. Rujikietgumjorn, R.T. Collins, Optimized pedestrian detection for multiple and occluded people, in *CVPR '13: Proceedings of the 2013 IEEE Conference on Computer Vision and Pattern Recognition*, June 2013, pp. 3690–3697
75. B. Simon, *The Statistical Mechanics of Lattice Gases*, vol. I (Princeton University Press, Princeton, 2014)
76. W.E. Smith, Various optimizers for single-stage production. Nav. Res. Logist. Q. **3**, 59–66 (1956)
77. D. Sornette, Physics and financial economics (1776–2014): puzzles, Ising and agent-based models. Rep. Prog. Phys. **77**(6), 062001, 28 pp. (2014)
78. R.G. Vickson, Two single machine sequencing problems involving controllable job processing times. AIIE Trans. **12**, 258–262 (1980)
79. R.G. Vickson, Choosing the job sequence and processing times to minimize total processing plus flow cost on single machines. Oper. Res. **28**, 1155–1167 (1980)

80. L.J. Watters, Reduction of integer polynomial programming problems to zero-one linear programming problems. Oper. Res. **15**, 1171–1174 (1967)
81. W.I. Zangwill, Media selection by decision programming. J. Advert. Res. **5**, 30–36 (1965)

Chapter 2
Applications and Computational Advances for Solving the QUBO Model

Fred Glover, Gary Kochenberger, and Yu Du

Abstract QUBO models have proven to be remarkable for their ability to function as an alternative modeling framework for a wide variety of combinatorial optimization problems. Many studies have underscored the usefulness of the QUBO model to serve as an effective approach for modeling and solving important combinatorial problems. The significance of this unifying nature of the QUBO model is enhanced by the fact that the model can be shown to be equivalent to the Ising model that plays a prominent role in physics and is a major focus of the quantum computing community. Consequently, the broad range of optimization problems approached as QUBO models from the traditional Operations Research community are joined by an important domain of problems with connection to the physics community. Across the board, the QUBO model is used today as an alternative modeling and solution approach for a growing number of important problems found in industry and government. We describe important new applications of this model and sketch fundamental ways to create effective QUBO formulations. We also report computational experience showing the power of recent algorithmic advances. (The introduction section draws on material from Glover et al., 4OR 17:335–371, 2019.)

2.1 Introduction

The field of Combinatorial Optimization (CO) is one of the most important areas in the field of optimization, with practical applications found in every industry, including both the private and public sectors. Generally, these problems are concerned with

F. Glover (✉)
Meta-Analytics, Boulder, CO, USA

G. Kochenberger
Entanglement, Inc., Westminster, CO, USA
e-mail: gary.kochenberger@ucdenver.edu

Y. Du
Business School, University of Colorado at Denver, Denver, CO, USA
e-mail: yu.du@ucdenver.edu

A. P. Punnen (ed.), *The Quadratic Unconstrained Binary Optimization Problem*,
https://doi.org/10.1007/978-3-031-04520-2_2

making wise choices in settings where a large number of yes/no decisions must be made and each set of decisions must satisfy certain constraints while optimizing a corresponding objective function value—like a cost or profit value. Finding good solutions in these settings is extremely difficult as these problems are typically NP-hard. The traditional approach is for the analyst to develop a solution algorithm that is tailored to the mathematical structure of the problem at hand. While this approach has produced good results in certain problem settings, it has the disadvantage that the diversity of applications arising in practice requires the creation of a diversity of solution techniques, each with limited application outside their original intended use.

Early articles such as those by Hammer and Rudeanu [33], Rosenberg [68], Hansen et al. [34], and Boros and Hammer [13] suggested the possible use of what has become known today as the QUBO model. However, only in recent years, through extensive computational work, has the research community demonstrated that the QUBO model can in fact be successfully employed to model and solve an exceptional variety of important CO problems found in industry, science and government, as documented in studies such as Glover et al. [30], Kochenberger et al. [39, 45], Lucas [54] and Anthony et al. [5]. Through special reformulation techniques that are easy to apply, the power of QUBO solvers can be used to efficiently solve many important problems once they are put into the QUBO framework. This approach has proven not only to be competitive with traditional methods, but often superior in performance in terms of both solution time and quality.

It is interesting to note that in recent years the QUBO model has emerged as an underpinning of the quantum computing area known as quantum annealing and Fujitsu's digital annealing, and has become a subject of study in neuromorphic computing. Through these connections, QUBO models lie at the heart of experimentation carried out with quantum computers developed by D-Wave Systems and neuromorphic computers developed by IBM. The consequences of these new discoveries linking QUBO models to quantum computing are being explored in initiatives by organizations such as IBM, Google, Amazon, Microsoft, D-Wave and Lockheed Martin in the commercial realm and Los Alamos National Laboratory, Oak Ridge National Laboratory, Lawrence Livermore National Laboratory and NASA's Ames Research Center in the public sector. Computational experience is being amassed by both the classical and the quantum computing communities that highlights not only the potential of the QUBO model but also its effectiveness as an alternative to traditional modeling and solution methodologies.

In Sect. 2.2 of this chapter, we highlight a diversity of applications that have been reported using the QUBO model. Different from the discussion of QUBO in other chapters, we consider the general form of the problem described in Sect. 2.3 to have a minimization objective, which is equivalent to the maximization objective treated in other parts of this volume. Then, in Sect. 2.3, we illustrate in the transformation process enabling a classically formulated model to be recast into the form of a QUBO model and computational experience showing the power of recent algorithmic advances. Section 2.4 follows with a summary and some conclusions.

2.2 Applications of the QUBO Model

Chapter 1 of this book mentioned several applications including the stable set problem, the Ising Spin Glass problem, the circuit layout design problem, and an application regarding detecting and tracking people in a crowded environment. These applications serve to illustrate the diversity of problems that fall under the QUBO umbrella.

In recent years, many other applications of the QUBO model have been reported in the operations research literature. In addition, quantum and quantum inspired computer companies with their QUBO solvers have encouraged exploration of applications, leading to many more accounts of important uses of the QUBO model. All told, the literature in general reports many interesting applications in wide variety of application settings.

In this section we summarize four applications that showcase the usefulness of the QUBO model for modeling and solving important combinatorial optimization problems. Section 2.2 that follows mentions many additional applications recently reported in the literature. References to these applications are provided to enable readers to follow up according to their interests.

Application # 1: QUBO and the RNA Folding Problem

Lewis et al. [52], report advances derived from the QUBO model applied to the RNA folding problem, by extension of the work of Forester and Greenberg [26] on quadratic binary models in computational biology.

RNA molecules, which play informational, structural and metabolic roles in all cells, are chains of nucleotides that interact through bases A, C, G, U to determine cell functionality and structure. The associated optimization problem seeks to minimize the thermodynamic free energy of a structure by selecting which bases will be paired provided certain constraints are satisfied. Binary variables are associated with potential base pairings and constraints are imposed to limit a variable to one base pair. Additional constraints are included to prohibit selecting base pairs that cross as well as promoting the formation of long RNA stems.

Lewis et al. [52], approach this problem with a series of QUBO models that have proved extremely successful in predicting how base pairings determine RNA secondary structures, and thus facilitating biological functions related to information flow and metabolism. Extensive testing with standard benchmarks underscores the effectiveness of the QUBO approach whose results not only compare favorably with traditional RNA folding programs, but offer an alternative methodology with the potential to lead to improved predictions of RNA folding.

Application # 2: QUBO and Vehicle Routing Problem

The Vehicle Routing Problem (VRP) is an important combinatorial optimization problem in which the goal is to find an optimal set of routes for a fleet of vehicles that deliver goods from an origin (depot) to a set of destinations (customers). The VRP is a generalization of the Travelling Salesman Problem (TSP) in which a single vehicle visits all destinations in one continuous path, starting and ending at the depot. Due to their computational challenge, VRPs and TSPs are typically solved by heuristic methods in practice.

Recently the quantum affiliated research community has investigated solving vehicle routing problems as well as TSPs using quantum and quantum inspired technology. For example, Feld et al. [25] report results obtained by formulating the vehicle routing problem as a QUBO model to be solved by the D-WAVE quantum annealer. Computational experiments show results comparable to conventional solvers in reasonable amounts of time.

Building on the work of Feld and co-workers, Borowski et al. [14] report results obtained from using D-Wave's Leap framework on well-established benchmark test cases, including problem instances based on realistic road networks. They also compared new quantum and hybrid methods with well-known classical algorithms for solving VRP. The experiments indicate that the hybrid quantum annealing methods give promising outcomes and are able to find solutions of similar or even better quality than the classical algorithms.

The application reported by Borowski et al. [14] and the solution procedure they employ is limited to small instances compared to practical application. This shortcoming may be overcome by the advances being explored by Wang et al. [86] in their work on TSP and related problems using the AlphaQUBO [4] solver which can accommodate QUBO models with up to 1,000,000 variables.

Application # 3: QUBO and Machine Learning

Optimizing Gaussian Process Sampling has great promise for enhancing the performance of machine learning methods by improving training set selection. Bottarelli and Farinelli [15] reported a successful use of the QUBO model for variance reduction in such settings. Building on this work, Sargent et al. [72] at the University of Toronto are developing machine learning models where training set selection is made by optimizing a QUBO model formulated to represent the Gaussian Process posterior function. The overall objective of the research project is to develop enhanced machine learning models to accelerate material discovery with a particular focus on predicting the stability of acidic Oxygen Evolution Reaction electrocatalysts. Early testing based on QUBO solutions obtained from both the Fujitsu Digital Annealer and the AlphaQUBO [4] solver from Meta-Analytics, Inc. show very encouraging outcomes. Preliminary results indicate that for a database containing 4000 metal oxides structures, the energy mean absolute error (MAE) obtained from the QUBO model was 0.1 eV/atom less than that given by the conventional method. Further work will address the challenge of developing other QUBO-enhanced regression models like support vector regression.

Application # 4: QUBO and Large-scale Set Partitioning Problems

The set partitioning problem (SPP) seeks to partition a set of items into subsets such that each item appears in exactly one subset and the cost of the subsets chosen is minimized. This problem appears in many application settings including the airline and other industries. The traditional formulation for SPP is given by

$$\text{Minimize} \sum_{j=1}^{n} c_j x_j$$

$$\text{Subject to} \sum_{j=1}^{n} a_{ij} x_j = 1 \text{ for } i = 1, \dots, m$$

$$x_j \in \{0, 1\}, j = 1, 2, \dots, n.$$

where x_j denotes whether or not subset j is chosen, c_j is the cost of subset j, and the a_{ij} coefficients are 0 or 1 denoting whether or not variable x_j explicitly appears in constraint i. Such models are easily re-cast into an equivalent QUBO model, without adding any new variables, using the standard reformulation procedure given in Sect. 2.3 of this chapter.

In a recent paper, Du et al. [21] report on a study involving large-scale instances of SPP ranging in size from 10,000 to 100,000 variables and 2000 to 20,000 constraints. Extensive computational experiments were conducted comparing the traditional model solved by CPLEX and the equivalent QUBO model solved by a modern metaheuristic QUBO solver AlphaQUBO [4] developed by Meta-Analytics. These are some of the largest QUBO models reported on in the literature. (Detailed computational results from this study are given at the end of Sect. 2.3 in this chapter.)

Both approaches were successful in solving the smaller instances considered. For larger problems, however, CPLEX was unable to find optimal solutions in an allotted time frame of 6 hours and typically reported large gaps when terminating due to the time limit. The QUBO approach on these larger problems, in contrast, found better solutions than the best results produced by CPLEX, often in a few minutes and always within one hour. This study confirms earlier successes of the QUBO model on smaller test problems for SPP reported by Lewis et al. [50].

2.2.1 Additional Applications by Category

The vast amount of work going on today extending the use of the QUBO model is creating an ever growing list of QUBO applications. Below we list some to the key applications that have been reported in various categories of problem settings.

Classical Combinatorial Optimization

- Graph Coloring
  - Kochenberger et al. [41].
  - Wang et al. [83].

- Capital Budgeting Problems
  - Laughhunn [48].
- Asset Exchange Problems
  - Glover et al. [32].
- Task Allocation Problems (distributed computer systems)
  - Lewis et al. [49].
  - Tomasiewicz et al. [77].
- Warehouse Location and Product Distribution Problems
  - Ding et al. [20].
- Multiple Knapsack Problems
  - Glover et al. [29].
  - Forrester and Hunt-Isaak [27].
- Maximum Independent Set Problems
  - Pardalos and Xue [64].
  - Kochenberger et al. [43].
  - Yarkoni et al. [88].
- Maximum Cut Problems
  - Boros and Hammer [12].
  - Kochenberger et al. [44].
  - Wang et al. [85].
  - Dunning et al. [22].
- Maximum Clique Problems
  - Pelofske et al. [65].
  - Pardalos and Xue [64].
  - Chapuis et al. [17]
- Constraint Satisfaction Problems (CSPs)
  - Vyskocil and Djidjev [80].
- Number Partitioning Problems
  - Alidaee et al. [2].
- Set Packing Problems
  - Alidaee et al. [3].
- Linear Ordering Problems
  - Lewis et al. [51].

- Quadratic Assignment Problems
  - Wang et al. [84].
- Clique Partitioning Problems
  - Wang et al. [81].
  - Kochenberger et al. [42].
  - Shaydulin et al. [73].
  - Kochenberger et al. [46].
- Satisfiability (SAT) and Max Sat Problems
  - Kochenberger et al. [40].
  - Bonet et al. [11].
  - Santra et al. [71].
  - Bian et al. [10].
- Clustering Problems
  - Kochenberger et al. [42].
  - Mniszewski et al. [58].
  - Ushijima-Mwesigwa et al. [78].
  - Mniszewski et al. [57]
  - Kumar et al. [47]
  - Mniszewski et al. [58].
  - Bauckhage et al. [8].
  - Negre et al. [59].

Financial Services

- Portfolio optimization
  - Elsokkary et al. [24].
  - Kalra et al. [37].
  - Kochenberger et al. [38].
  - Cohen et al. [19].
- Arbitrage/currency exchange
  - Rosenberg [69].
- Credit risk assessment & scoring
  - Milne [56].
  - Egger et al. [23].

Transportation

- Route & traffic optimization
  - Neukart et al. [60].
  - Feld et al. [25].
  - Clark et al. [18].

  - Ohzeki et al. [61].
  - Inoue et al. [35].
  - Borowski et al. [14].

- Satellite coverage and surveillance
  - Bass et al. [7].

Manufacturing

- Product assembly optimization
  - Yarkoni et al. [89].
- Autonomous/robotic paths opt.
  - Mehta [55].
- Job scheduling
  - Alidaee et al. [1].
  - Venturelli et al. [79]
- Group technology
  - Wang et al. [82].

Pharmaceuticals and Related

- Molecular similarity/composition
  - Sahner [70].
- New drugs and materials discovery
  - Snelling et al. [76].
- Computational biology
  - Lewis et al. [52].

Network and Energy

- Cybersecurity problems
  - Berwald et al. [9].
  - Reinhardt [66].
- Power system design
  - Jones et al. [36].

Machine Learning

- Classical machine learning
  - Glover and Kochenberger [28].
  - Li et al. [53].
  - Willsch et al. [87].

- Deep learning
  - Sleeman et al. [74].

Miscellaneous

- Smelyanskiy et al. [75]
- Pakin [63].
- O'Malley et al. [62].
- Aramon et al. [6].
- Bottarelli and Farinelli [15].
- Chang et al. [16].
- Rogers and Singleton [67].

Taken together, the applications highlighted above indicate the widespread diversity of the QUBO model and give a glimpse of what is being reported today and what is to come in the near term The focus on "quantum readiness" as well as the growing appreciation for the usefulness of the QUBO model in general will lead to a continued growth in notable applications in the coming years.

In the next section we provide an overview of the methodology used to convert a traditionally modeled problem into the unified QUBO framework.

2.3 Creating QUBO Models

While some problems, like the number partitioning problem and the famous Ising spin glass problem, appear naturally in the form of a QUBO model, by far the largest number of problems of interest include additional constraints that must be satisfied as the optimizer searches for good solutions. Such problems can be effectively re-formulated as a QUBO model by introducing *quadratic penalties* in the objective function as an alternative to explicitly imposing constraints in the classical sense. The penalties are chosen so that the influence of the original constraints on the solution process can alternatively be achieved by the natural functioning of the optimizer as it looks for solutions that avoid incurring the penalties. That is, the penalties are formulated so that they equal zero for feasible solutions and equal some positive penalty amount for infeasible solutions. For a minimization problem, these penalties are added to create an augmented objective function to be minimized. If the penalty terms can be driven to zero, the augmented objective function becomes the original function to be minimized.

For certain types of constraints, quadratic penalties useful for creating QUBO models are known in advance and readily available to be used in transforming a given constrained problem into a QUBO model. To illustrate the main idea, consider a traditionally constrained problem of the form:

$$\text{Minimize } x_0 = f(\mathbf{x})$$

$$\text{Subject to } x_1 + x_2 \leq 1$$

where x_1 and x_2 are binary variables. Note that this constraint allows either or neither **x** variable to be chosen. However, it explicitly precludes both from being chosen (i.e., both cannot be set to 1).

A quadratic penalty that corresponds to our constraint is

$$Px_1x_2$$

where P is a positive scalar. As can be seen, this penalty function is equal to P when both x_1 and x_2 are equal to 1. Otherwise, it is equal to 0. For P chosen sufficiently large, the unconstrained problem

$$\text{Minimize } x_0 = f(\mathbf{x}) + Px_1x_2$$

has the same optimal solution as the original constrained problem. If $f(\mathbf{x})$ is linear or quadratic, then this unconstrained model will be in the form of a QUBO model, i.e., Minimize $x_0 = \mathbf{x}^T\mathbf{Q}\mathbf{x}$.

In this example, any optimizer trying to minimize x_0 will tend to avoid solutions having both x_1 and x_2 equal to 1, else a large positive amount will be added to the objective function. This simple constraint $(x_1 + x_2 \leq 1)$ arises in many important QUBO applications where the penalty (Px_1x_2) is used as an alternative. See, for instance, the work on the maximum clique and related problems by Pardalos and Xue [64].

While the example above illustrates the approach of adding penalties to the objective function to create a QUBO model, simple penalties won't always be known in advance and will have to be discovered. The procedure for this is straightforward as shown below.

2.3.1 *Creating QUBO Models: A General Purpose Approach*

Consider the general traditionally modeled combinatorial problem

$$\text{Minimize } x_0 = \mathbf{x}^T\mathbf{C}\mathbf{x}$$

$$\text{Subject to } \mathbf{A}\mathbf{x} = \mathbf{b}, \ \mathbf{x} \in \{0, 1\}^n$$

This model accommodates both quadratic and linear objective functions since the linear case results when C is a diagonal matrix (observing that $x_j^2 = x_j$ when x_j is a 0-1 variable). Under the assumption that **A** and **b** have integer components, problems with inequality constraints can always be put in this form by including slack variables and then representing the slack variables by a binary expansion. (For example, this would introduce a slack variable s to convert the inequality $4x_1 + 5x_2 - x_3 \leq 6$ into $4x_1 + 5x_2 - x_3 + s = 6$, and since clearly $s \leq 7$ (for the case where $x_1 = x_2 = 0$ and $x_3 = 1$), s could be represented by the binary expansion $s_1 + 2s_2 + 4s_3$ where s_1, s_2 and s_3 are additional binary variables. If it is additionally known that not both x_1 and x_2 can be 0, then s can be at most 3 and can be represented by the expansion $s_1 + 2s_2$. These constrained quadratic optimization models are converted into equivalent unconstrained QUBO models by converting the constraints $\mathbf{Ax} = \mathbf{b}$ (representing slack variables as x variables) into quadratic penalties to be added to the objective function, following the same re-casting as we illustrated in the discussion that precedes Sect. 3.1.

Specifically, for a positive scalar P, we add a quadratic penalty $P(\mathbf{Ax}-\mathbf{b})^T(\mathbf{Ax}-\mathbf{b})$ to the objective function to get

$$\begin{aligned} y &= \mathbf{x}^T\mathbf{Cx} + P(\mathbf{Ax} - \mathbf{b})^T(\mathbf{Ax} - \mathbf{b}) \\ &= \mathbf{x}^T\mathbf{Cx} + \mathbf{x}^T\mathbf{Dx} + c \\ &= \mathbf{x}^T\mathbf{Qx} + c \end{aligned}$$

where the matrix $\mathbf{D}$ and the additive constant c result directly from the matrix multiplication indicated. Dropping the additive constant, the equivalent unconstrained version of the constrained problem becomes

$$\text{Minimize} \quad \mathbf{x}^T\mathbf{Qx}, \ \text{Subject to: } \mathbf{x} \in \{0, 1\}^n$$

This general transformation procedure can in principle be applied to any 0/1 model with a linear or quadratic objective function subject to linear constraints.

Remarks

1. A suitable choice of the penalty scalar P, as we commented earlier, can always be chosen so that the optimal solution to QUBO is the optimal solution to the original constrained problem. Solutions obtained can always be checked for feasibility to confirm whether or not appropriate penalty choices have been made. Boros and Hammer [13] give a discussion of this approach which is the basis for establishing the generality of QUBO.
2. For realistic applications, a program, perhaps in Python, will need to be written implementing the transformation and producing the Q matrix needed for the QUBO model. For small problems we can usually proceed manually as we'll do in example to follow.
3. Note that the additive constant, c, does not impact the optimization and can be ignored during the optimization process. Once the QUBO model has been solved, the constant c can be used to recover the original objective function value. Alternatively, the original objective function value can always be determined by using the optimal x_j found when QUBO is solved.

This general transformation procedure is illustrated by the following example. Consider the set partitioning problem

$$\text{Min } x_0 = 3x_1 + 2x_2 + x_3 + x_4 + 3x_5 + 2x_6$$

$$\text{s.t. } x_1 + x_3 + x_6 = 1$$

$$x_2 + x_3 + x_5 + x_6 = 1$$

$$x_3 + x_4 + x_5 = 1$$

$$x_1 + x_2 + x_4 + x_6 = 1$$

and $x = (x_1, x_2, .., x_6)$ binary. Normally, the general transformation would be embodied in a supporting computer routine and employed to re-cast this problem into an equivalent instance of a QUBO model. For this small example, however, we can proceed manually as follows: The conversion to an equivalent QUBO model involves forming quadratic penalties and adding them to the original objective function. In general, the quadratic penalties to be added (for a minimization problem) are given by $P \sum_i \left(\sum_{j=1}^{n} a_{ij} x_{ij} - b_i \right)^2$ where the outer summation is taken over all constraints in the system $\mathbf{Ax} = \mathbf{b}$.

For our example we have

$$\text{Minimize } x_0 = 3x_1 + 2x_2 + x_3 + x_4 + 3x_5 + 2x_6 + P(x_1 + x_3 + x_6 - 1)^2 +$$
$$P(x_2 + x_3 + x_5 + x_6 - 1)^2 + P(x_3 + x_4 + x_5 - 1)^2 + P(x_1 + x_2 + x_4 + x_6 - 1)^2$$

Arbitrarily taking P to be 10, and recalling that $x_j^2 = x_j$ since our variables are binary, this becomes

$$\text{Minimize } y = -17x_1^2 - 18x_2^2 - 29x_3^2 - 19x_4^2 - 17x_5^2 - 28x_6^2 + 20x_1x_2 + 20x_1x_3 +$$
$$20x_1x_4 + 40x_1x_6 + 20x_2x_3 + 20x_2x_4 + 20x_2x_5 + 40x_2x_6 + 20x_3x_4 + 40x_3x_5 +$$
$$40x_3x_6 + 20x_4x_5 + 20x_4x_6 + 20x_5x_6 + 40$$

Dropping the additive constant 40, we then have our QUBO model

$$\text{Minimize } \mathbf{x}^T\mathbf{Q}\mathbf{x}, \text{ Subject to: } \mathbf{x} \in \{0, 1\}^n$$

where the $\mathbf{Q}$ matrix is

$$\mathbf{Q} = \begin{bmatrix} -17 & 10 & 10 & 10 & 0 & 20 \\ 10 & -18 & 10 & 10 & 10 & 20 \\ 10 & 10 & -29 & 10 & 20 & 20 \\ 10 & 10 & 10 & -19 & 10 & 10 \\ 0 & 10 & 20 & 10 & -17 & 10 \\ 20 & 20 & 20 & 10 & 10 & -28 \end{bmatrix}$$

Solving this QUBO formulation gives an optimal solution $x_1 = x_5 = 1$ (with all other variables equal to 0) to yield $x_0 = 6$.

As noted in Sect. 2.2 of this chapter, using the QUBO model for solving large-scale set partitioning problems has proven to be very successful. For detailed examples of re-casting other traditional models into the form of a QUBO model, the reader is referred to Glover et al. [19, 31].

2.3.2 *Illustrative Computational Experience*

As mentioned earlier in this chapter, the QUBO modeling and solution approach has proven to be successful in representing and solving a wide variety of difficult combinatorial optimization problems. We commented in section 3.1 and earlier in section 2, for instance, on the recent study by Du et al. [21] illustrating such success on large set partitioning problems. The following tables present the main results of this study undertaken on medium, large and very large test problems. Note that the results shown for CPLEX were obtained from the standard linear model for set partitioning while the AlphaQUBO results were obtained from the equivalent QUBO representation.

The bold values in Tables 2.1, 2.2, and 2.3 are the objective function values for optimal solutions or best known solutions. As shown in Table 2.1, both CPLEX and AlphaQUBO found optimal solutions for the first 7 of these modest sized problems. AlphaQUBO obtained a better solution than CPLEX on the last problem and outperformed CPLEX on "time to best" by a wide margin on all 8 problems.

Table 2.2 show that AlphaQUBO quickly found best known solutions for all 24 problems. CPLEX was able to find best known solutions for only 11 of the 24 problems. AlphaQUBO had a "time to best" advantage over CPLEX that typically ranged from 1 to 3 orders of magnitude.

For the very large problems of Table 2.3, CPLEX was unable to find the best known solution to any of these problems in the time limit of 6 hours. AlphaQUBO quickly provided best known solutions for all problems, outperforming CPLEX in terms of solution quality and time. Note that these QUBO models are the largest reported on in the literature to date.

Table 2.1 Medium sized problems: comparing AlphaQUBO and CPLEX

				CPLEX		AlphaQUBO	
ID	Vars	Constraints	Density %	OFV	Time(s)	OFV	Time(s)
SPP01a	6000	1500	25	**10,872**	7851	**10,872**	53
SPP01b	6000	1500	50	**6975**	3180	**6975**	6
SPP01c	6000	3000	25	**22,860**	2598	**22,860**	275
SPP01d	6000	3000	50	**14,793**	14,115	**14,793**	12
SPP02a	8000	2000	25	**14,959**	15,348	**14,959**	34
SPP02b	8000	2000	50	**9621**	18,071	**9621**	74
SPP02c	8000	4000	25	**30,425**	16,423	**30,425**	95
SPP02d	8000	4000	50	19,882	10,191	**19,816**	19

Table 2.2 Large sized problems

ID	Vars	Constraints	Density %	Cplex OFV	Cplex Time(s)	AlphaQUBO OFV	AlphaQUBO Time(s)
SPP1	10,000	1000	25%	**7292**	947	**7292**	378.2
SPP2	10,000	1000	50	**4543**	1543	**4543**	11.6
SPP3	10,000	5000	25	**37,968**	284	**37,968**	5.4
SPP4	10,000	5000	50	**24,297**	8683	**24,297**	1337
SPP5	15,000	1500	25	**10,930**	66	**10,930**	17.5
SPP6	15,000	1500	50	7174	2176	**7047**	77
SPP7	15,000	7500	25	57,834	993	**57,419**	706.7
SPP8	15,000	7500	50	37,962	2373	**37,671**	556.8
SPP9	20,000	2000	25	**14,900**	9833	**14,900**	1535.1
SPP10	20,000	2000	50	**9412**	369	**9412**	3.6
SPP11	20,000	10, 000	25	77,448	2119	**77,198**	1729.1
SPP12	20,000	10, 000	50	**50,188**	4786	**50,188**	5.7
SPP13	25,000	2500	25	18,517	589	**18,498**	10
SPP14	25,000	2550	50	12,008	1847	**11,923**	813.8
SPP15	25,000	12, 500	25	96,690	548	**96,445**	1474.3
SPP16	25,000	12, 500	50	63,173	18,903	**63,156**	98
SPP17	30,000	3000	25	**22,405**	859	**22,405**	14.1
SPP18	30,000	3000	50	**14,457**	2507	**14,457**	1551.7
SPP19	30,000	15, 000	25	115,950	16,031	**115,687**	410.3
SPP20	30,000	15, 000	50	76,276	17,393	**75,684**	11.5
SPP21	40,000	4000	25	30,592	2532	**30,445**	894.1
SPP22	40,000	4000	50	19,815	7184	**19,558**	11.5
SPP23	40,000	20, 000	25	155,162	19,152	**155,069**	393.4
SPP24	40,000	20, 000	50	**101,835**	10,286	101,835	12.9

Table 2.3 Very large instances

ID	Vars	Constraints	Density %	Cplex OFV	Cplex Time(s)	AlphaQUBO OFV	AlphaQUBO Time(s)
SPP50k	50, 000	10,000	25	76,903	18,144	**76,402**	2315
SPP60k	60, 000	12,000	25	92,293	16,060	**91,912**	2876
SPP70k	70, 000	14,000	25	109,168	19,084	**108,112**	272
SPP80k	80, 000	16,000	25	125,139	18,600	**123,890**	175
SPP90k	90, 000	18,000	25	140,223	15,278	**139,269**	1804
SPP100k	100, 000	20,000	25	154,694	19,509	**154,351**	622

2.4 Summary and Conclusion

The remarkable diversity of QUBO applications and the documented successes in solving them effectively highlight the practical importance of these models. The relevance of this area to quantum computing and the discovery of new applications in machine learning, biotechnology, supply chains, portfolio analysis and a host of other realms promises to stimulate further advances in the months ahead. Some of these will mimic and build upon past QUBO successes while others will be completely new as the research community continues to devise creative ways to recast important problems into the QUBO framework.

The applications reported here underscore the success of the QUBO model as a useful alternative to traditional approaches for solving combinatorial optimization problems. As the performance of QUBO solution methods continues to advance, both on the conventional and the quantum side, the practice of employing the QUBO model may be expected to expand as well.

References

1. B. Alidaee, G. Kochenberger, A. Ahmadian, 0–1 Quadratic programming approach for optimum solutions of two scheduling problems. Int. J. Syst. Sci. **25**, 401–408 (1994)
2. B. Alidaee, F. Glover, G. Kochenberger, C. Rego, A new modeling and solution approach for the number partitioning problem. J. Appl. Math. Decis. Sci. **9**, 135–145 (2005)
3. B. Alidaee, G. Kochenberger, K. Lewis, M. Lewis, H. Wang, A new approach for modeling and solving set packing problems. Eur. J. Oper. Res. **186**, 504–512 (2008)
4. AlphaQUBO (2020). https://ma-website.azurewebsites.net/
5. M. Anthony, E. Boros, Y. Crama, A. Gruber, Quadratic reformulations of nonlinear binary optimization problems. Math. Program. **162**, 115–144 (2017)
6. M. Aramon, G. Rosenberg, E. Valiante, T. Miyazawa, H. Tamura, H.G. Katzgraber, Physics-inspired optimization for quadratic unconstrained problems using a digital annealer. Front. Phys. **7**, 48 (2019)
7. G. Bass, C. Tomlin, V. Kumar, P. Rihaczek, J. Dulny, Heterogeneous quantum computing for satellite constellation optimization: solving the weighted k-clique problem (2017). https://arxiv.org/abs/1709.05381
8. C. Bauckhage, N. Piatkowski, R. Sifa, D. Hecker, S. Wrobel, A QUBO Formulation of the k-medoids problem, in *LWDA 2019 Proceedings* (2019)
9. J.J. Berwald, J.M. Gottlieb, E. Munch, Computing Wasserstein distance for persistence diagrams on a quantum compute (2018). arXiv:1809.06433
10. Z. Bian, F. Chudak, W. Macready, A. Roy, R. Sebastiani, S. Varotti, Solving SAT and MaxSAT with a quantum annealer: foundations and a preliminary report, in *Frontiers of Combining Systems, FroCoS 2017*. Lecture Notes in Computer Science, vol. 10483 (Springer, Cham, 2017)
11. M.L. Bonet, J. Levy, F. Manyà, Resolution for Max-SAT. Artif. Intell. **171**, 606–618 (2007)
12. E. Boros, P.L. Hammer, The Max-Cut problem and quadratic 0-1 optimization; polyhedral aspects, relaxations and bounds. Ann. Oper. Res. **33**, 151–180 (1991)
13. E. Boros, P.L. Hammer, Pseudo-Boolean optimization. Discrete Appl. Math. **123**, 155–225 (2002)
14. M. Borowski, P. Gora, K. Karnas, M. Blajda, K. Krol, A. Matyjasek, D. Burczyk, M. Szewczyk, M. Kutwin, New hybrid quantum annealing algorithms for solving vehicle routing problem, in *International Conference on Computer Science, ICCS2020* (2020), pp. 546–561
15. L. Bottarelli, A. Farinelli, A Qubo model for Gaussian process variance reduction (2019). Preprint, arXiv:1901.10982
16. C.C. Chang, A. Gambhir, T.S. Humble, S. Sota, Quantum annealing for systems of polynomial equations. Sci. Rep. **9**, 10258 (2019)
17. G. Chapuis, H. Djidjev, G. Hahn, G. Rizk, Finding maximum cliques on the D-wave quantum annealer. J. Signal Process. Syst. **91**, 363–377 (2019)

18. J. Clark, T. West, J. Zammit, X. Guo, L. Mason, D. Russell, Towards real time multi-robot routing using quantum computing technologies, in *HPC Asia 2019, Proceedings of the International Conference on High Performance Computing in Asia-Pacific Region* (2019), pp. 111–119
19. J. Cohen, A. Khan, C. Alexander, Portfolio optimization of 60 stocks using classical and quantum algorithms (2020). Preprint, arXiv:2008.08669
20. Y. Ding, X. Chen, L. Lamata, E. Solano, M. Sanz, Implementation of a hybrid classical-quantum annealing algorithm for logistic network design. SN Comput. Sci. **2**, 68 (2021)
21. Y. Du, G. Kochenberger, F. Glover, H. Wang, R. Hennig, Optimal solutions to the set partitioning problem: a comparison of alternative models. Working paper, University of Colorado Denver, 2020
22. I. Dunning, S. Gupta, J. Silberholz, What works best when? A systematic evaluation of heuristics for max-cut and qubo. INFORMS J. Comput. **30**, 608–624 (2018)
23. D.J. Egger, R.G. Gutierrez, J.C. Mestre, S. Woerner, Credit risk analysis using quantum computers. IEEE Trans. Comput. **70**, 2136–2145 (2021)
24. N. Elsokkary, F.S. Khan, T.S. Humble, D.L. Torre, J. Gottlieb, Financial portfolio management using D-Wave's quantum optimizer: the case of Abu Dhabi securities exchange, in *IEEE High-performance Extreme Computing Conference* (HPEC) (2017)
25. S. Feld, C. Roch, T. Gabor, C. Seidel, F. Neukart, I. Galter, W. Mauerer, C. Linnhoff-Popien, A hybrid solution method for the capacitated vehicle routing problem using a quantum annealer. Front. ICT **6**, 13 (2019)
26. R.J. Forrester, H.J. Greenberg, Quadratic binary programming models in computational biology. Algorithmic Oper. Res. **3**, 110–129 (2008)
27. R.J. Forrester, N. Hunt-Isaak, Computational comparison of exact solution methods for 0-1 quadratic programs: recommendations for practitioners. J. Appl. Math. **2020**, Article ID 5974820, 21 pages (2020)
28. F. Glover, G. Kochenberger, New optimization models for data mining. Int. J. Inf. Technol. Decis. Mak. **5**, 605–609 (2006)
29. F. Glover, G. Kochenberger, B. Alidaee, M. Amini, Solving quadratic knapsack problems by reformulation and tabu search, in *Combinatorial and Global Optimization*, ed. by P.M. Pardalos, A. Megados, R. Burkard (World Scientific Publishing, Singapore, 2002), pp. 272–287
30. F. Glover, G. Kochenberger, Y. Du, Quantum bridge analytics I: a tutorial on formulating and using QUBO models. 4OR **17**, 335–371 (2019)
31. F. Glover, G. Kochenberger, Y. Du, A tutorial on formulating and using QUBO models (2019). arXiv:1811.11538
32. F. Glover, G. Kochenberger, M. Ma, Y. Du, Quantum bridge analytics II: QUBO-Plus, network optimization and combinatorial chaining for asset exchange. 4OR **18**, 387–417 (2020)
33. P.L. Hammer, S. Rudeanu, *Boolean Methods in Operations Research and Related Areas* (Springer, Berlin, 1968)
34. P. Hansen, B. Jaumard, V. Mathon, State-of-the-art survey—constrained nonlinear 0–1 programming. ORSA J. Comput. **5**, 97–119 (1993)
35. D. Inoue, A. Okada, T. Matsumori, K. Aihara, H. Yoshida, Traffic signal optimization on a square lattice with quantum annealing. Sci. Rep. **11**, 3303 (2021)
36. E.B. Jones, E. Kapit, C. Chang, D. Biagioni, D. Vaidhynathan, P. Graf, W. Jones, On the computational viability of quantum optimization for PMU placement. IEEE Power & Energy Society General Meeting (PESGM), 2020
37. A. Kalra, F. Qureshi, M. Tisi, Portfolio asset identification using graph algorithms on a quantum annealer (2018). http://www.henryyuen.net/fall2018/projects/qfinance.pdf
38. G. Kochenberger, M. Ma, Quantum computing applications of QUBO models to portfolio optimization. White paper, University of Colorado, Denver, Sept 2019
39. G. Kochenberger, F. Glover, B. Alidaee, C. Rego, A unified modeling and solution framework for combinatorial optimization problems. OR Spectr. **26**, 237–250 (2004)

40. G. Kochenberger, F. Glover, B. Alidaee, K. Lewis, Using the unconstrained quadratic program to model and solve Max 2-Sat problems. Int. J. OR **1**, 89–100 (2005)
41. G. Kochenberger, F. Glover, B. Alidaee, C. Rego, An unconstrained quadratic binary programming approach to the vertex coloring problem. Ann. Oper. Res. **139**, 229–241 (2005)
42. G. Kochenberger, F. Glover, B. Alidaee, H. Wang, Clustering of microarray data via clique partitioning. J. Comb. Optim. **10**, 77–92 (2005)
43. G. Kochenberger, B. Alidaee, F. Glover, H. Wang, An effective modeling and solution approach for the generalized independent set problem. Optim. Lett. **1**, 111–117 (2007)
44. G. Kochenberger, J.-K. Hao, S. Lu, H. Wang, F. Glover, Solving large scale max cut problems via Tabu search. J. Heuristics **19**, 565–571 (2013)
45. G. Kochenberger, J.-K. Hao, F. Glover, M. Lewis, Z. Lu, H. Wang, Y. Wang, The unconstrained binary quadratic programming problem: a survey. J. Comb. Optim. **28**, 58–81 (2014)
46. G. Kochenberger, Y. Du, F. Glover, H. Wang, M. Lewis, T. Tsuyuguchi, Solving clique partitioning problems: a comparison of models and commercial solvers. Working paper (2020)
47. V. Kumar, G. Bass, C. Tomlin, Quantum annealing for combinatorial clustering. Quantum Inf. Process. **17**, Article 39 (2018)
48. D.J. Laughhunn, Quadratic binary programming with applications to capital budgeting problems. Oper. Res. **18**, 454–461 (1970)
49. M. Lewis, B. Alidaee, G. Kochenberger, Using xQx to model and solve the uncapacitated task allocation problem. Oper. Res. Lett. **33**, 176–182 (2005)
50. M. Lewis, G. Kochenberger, B. Alidaee, A new modeling and solution approach for the set partitioning problem. Comput. Oper. Res. **35**, 807–813 (2008)
51. M. Lewis, B. Alidaee, F. Glover, G. Kochenberger, A note on xQx as a modeling and solution framework for the linear ordering problem. Int. J. OR **5**, 152–162 (2009)
52. M. Lewis, A. Verma, T. Eckdahl, Qfold: a new modeling paradigm for the RNA folding problem. Working paper, 2020
53. R.Y. Li, R. Di Felice, R. Rohs, D.A. Lidar, Quantum annealing versus classical machine learning applied to a simplified computational biology problem. NPJ Quantum Inf. **4**, Article 14 (2018)
54. A. Lucas, Ising formulations of many NP problems. Front. Phys. **2**, 5 (2014)
55. A. Mehta, Quantum annealing based optimization of robotic movement in manufacturing. White paper, 2019
56. A. Milne, Optimal feature selection for credit scoring and classification. White paper, 1Qbit, 2017
57. S. Mniszewski, C.F.A. Negre, H. Ushijima-Mwesigwa, Graph partitioning using the D-Wave for electronic structure problems. Los Alamos National Lab (LANL), Los Alamos, NM (United States), LA-UR-16-27873 (2016), pp. 1–21
58. S.M. Mniszewski, C.F.A. Negre, H. Ushijima-Mwesigwa, Graph clustering approaches using nearterm quantum computing. Argonne Quantum Computing Workshop (2018)
59. C.F.A. Negre, H. Ushijima-Mwesigwa, S.M. Mniszewsk, Detecting multiple communities using quantum annealing on the D-Wave system. PLoS ONE **15**, 1–14 (2020)
60. F. Neukart, G. Compostella, C. Seidel, D. Dollen, S. Yarkoni, B. Parney, Traffic flow optimization using a quantum annealer. Front. ICT **4**, 29 (2017)
61. M. Ohzeki, A. Miki, M.J. Miyama, M. Terabe, Control of automated guided vehicles without collision by quantum annealer and digital devices. Front. Comput. Sci. **1**, 9 (2019)
62. D. O'Malley, V.V. Vesselinov, B.S. Alexandrov, L.B. Alexandrov, Nonnegative/binary matrix factorization with a D-Wave quantum annealer. PLoS ONE **13**, 12 (2018)
63. S. Pakin, Navigating a maze using a quantum annealer, in *Proceedings of the Second International Workshop on Post Moores Era Supercomputing* (2017), pp. 30–36
64. P.M. Pardalos, J. Xue, The maximum clique problem. J. Global Optim. **4**, 301–328 (1994)
65. E. Pelofske, G. Hahn, H. Djidjev, Solving large maximum clique problems on a quantum annealer. First International Workshop, QTOP 2019, Munich, Germany, 18 March 2019
66. S. Reinhardt, Detecting lateral movement with a compute-intense graph kernel (2018). http://www.clsac.org/uploads/5/0/6/3/50633811/reinhardt-clsac-2018.pdf

67. M.L. Rogers, R.L. Singleton, Floating-point calculations on a quantum annealer: division and matrix inversion. Front. Phys. **8**, 265 (2020)
68. I. Rosenberg, Reduction of bivalent maximization to the quadratic case. Cahiers Centre d'Etudes Rech. Oper. **17**, 71–74 (1975)
69. G. Rosenberg, Finding optimal arbitrage opportunities using a quantum annealer. White paper, 2016, 1Qbit
70. D. Sahner, A potential role for quantum annealing in the enhancement of patient outcomes? (2018). https://www.dwavesys.com/sites/default/files/Sahner.2018.pdf
71. S. Santra, G. Quiroz, G.V. Steeg, D.A. Lidar, Max 2-SAT with up to 108 qubits. New J. Phys. **16**(4), 045006 (2014)
72. E. Sargent, Y. Chang, H. Choubisa, Personal communication with authors of Chapter 2 (2020)
73. R. Shaydulin, H. Ushijima-Mwesigwa, I. Safro, S. Mniszewski, Y. Alexeev, Community detection across emerging quantum architectures (2018). Preprint, arXiv:1810.07765
74. J. Sleeman, J. Dorband, M. Halem, A hybrid quantum enabled RBM advantage: convolutional autoencoders for quantum image compressing and generative learning, in *Proceedings Volume 11391*, Quantum Information Science, Sensing, and Computation XII; 113910B (2020)
75. V.N. Smelyanskiy, E.G. Rieffel, S.I. Knysh, A near-term quantum computing approach for hard computational problems in space exploration (2012). arXiv:1204.2821 [quant-ph]
76. D. Snelling, G. Shahane, W.J. Shipman, A. Balaff, M. Pearce, S. Keinan, A quantum-inspired approach to de-novo drug design. Whitepaper, Fujitsu, 2020
77. D. Tomasiewicz, M. Pawlik, M. Malawski, K. Rycerz, Foundations for workflow application scheduling on D-wave system, in *Computational Science – ICCS 2020: 20th International Conference, Amsterdam, The Netherlands, Proceedings, Part VI, 12142* 3–5 June 2020, pp. 516–530
78. H. Ushijima-Mwesigwa, C.F.A. Negre, S.M. Mniszewsk, Graph partitioning using quantum annealing on the D-Wave system, in *Proceedings of the Second International Workshop on Post Moores Era Supercomputing* (2017), pp. 22–29
79. D. Venturelli, D.J.J. Marchand, G. Rojo, Quantum annealing implementation of job-shop scheduling (2016). arXiv:1506.08479 [quant-ph]
80. T. Vyskocil, H.N. Djidjev, Constraint embedding for solving optimization problems on quantum annealers, in *2019 IEEE International Parallel and Distributed Processing Symposium Workshops* (2019), pp. 635–644
81. H. Wang, B. Alidaee, G. Kochenberger, Evaluating a clique partitioning problem model for clustering high-dimensional data mining, in *AMCIS 2004 Proceedings*, paper 234 (2004)
82. H. Wang, B. Alidaee, F. Glover, G. Kochenberger, Solving group technology problems via clique partitioning. Int. J. Flex. Manuf. Syst. **18**, 77–87 (2006)
83. Y. Wang, J.-K. Hao, F. Glover, Z. Lu, Solving the minimum sum coloring problem via binary quadratic programming (2013). arXiv:1304.5876 [cs.DS]
84. H. Wang, Y. Wang, M. Resende, G. Kochenberger, A QUBO approach to solving QAP problems. Unpublished manuscript, 2016
85. Z. Wang, S. Hadfield, Z. Jiang, E. G. Rieffel, The quantum approximation optimization algorithm for MaxCut: a fermionic view. Phys. Rev. A **97**, 022304 (2018)
86. H. Wang, Y. Du, R. Hennig, G. Kochenberger, F. Glover, Solutions to the traveling salesman problem: a comparison of quantum and heuristic solvers. Working Paper, Texas A & M International University, 2020
87. D. Willsch, M. Willsch, H.D. Raedt, K. Michielsen, Support vector machines on the D-Wave quantum annealer. Comput. Phys. Commun. **248**, 107006 (2020)
88. S. Yarkoni, A. Plaat, T. Back, First results solving arbitrarily structured maximum independent set problems using quantum annealing, in *2018 IEEE Congress on Evolutionary Computation (CEC), Rio de Janeiro* (2018), pp. 1–6
89. S. Yarkoni, M. Leib, A. Skolik, M. Streif, F. Neukart, D. von Dollen, Volkswagen and quantum computing: an industrial perspective. Digitale Welt **3**, 34–37 (2019)

Chapter 3
Complexity and Polynomially Solvable Special Cases of QUBO

Eranda Çela and Abraham P. Punnen

Abstract The quadratic unconstrained binary optimization problem (QUBO) is equivalent to a number of prominent combinatorial and discrete optimization problems and generalizes many others. In this chapter we will discuss the computational complexity aspects of the problem along with tractable special cases. Many of those results are derived from the related properties of the optimization problems which are equivalent to or generalized by QUBO.

3.1 Introduction and Notations

In the previous chapters, we have seen that QUBO can be used to model various applied and theoretical optimization problems. In this chapter, we primarily focus on the computational complexity of the QUBO model and identify various special cases of the model that can be solved in polynomial time. Recall that QUBO can be stated as the mathematical programming problem

$$\begin{aligned} \text{Maximize} \quad & \mathbf{x}^T\mathbf{Q}\mathbf{x} + \mathbf{c}^T\mathbf{x} \\ \text{Subject to:} \quad & \mathbf{x} \in \{0, 1\}^n, \end{aligned}$$

where $\mathbf{Q} = (q_{ij})$ is an $n \times n$ matrix and $\mathbf{c}^\mathrm{T} = (c_1, c_2, \ldots, c_n)$ is an n-vector. When $\mathbf{Q}$ and $\mathbf{c}$ are given, an instance of QUBO is well defined and hence we sometimes represent an instance of QUBO by the ordered pair $(\mathbf{Q}, \mathbf{c})$. The size n of the QUBO model is implicit in the dimensions of $\mathbf{Q}$ and $\mathbf{c}$. The QUBO model is also represented in an equivalent form, called the *Ising QUBO*, where the variables take values -1

E. Çela (✉)
Department of Discrete Mathematics, TU Graz, Graz, Austria
e-mail: cela@opt.math.tu-graz.ac.at

A. P. Punnen
Department of Mathematics, Simon Fraser University, Surrey, BC, Canada
e-mail: apunnen@sfu.ca

A. P. Punnen (ed.), *The Quadratic Unconstrained Binary Optimization Problem*,
https://doi.org/10.1007/978-3-031-04520-2_3

or 1. The Ising QUBO leads to a natural formulation of the maximum weight cut problem and hence it is sometimes called as the cut version of the QUBO. The Ising QUBO can be stated as

$$\begin{aligned} \text{Maximize} \quad & \mathbf{y}^T\mathbf{A}\mathbf{y} + \mathbf{b}^T\mathbf{y} \\ \text{Subject to:} \quad & \mathbf{y} \in \{-1, 1\}^n, \end{aligned}$$

where $\mathbf{A} = (a_{ij})$ is an $n \times n$ matrix and $\mathbf{b}^\mathrm{T} = (b_1, b_2, \ldots, b_n)$ is an n-vector. When $\mathbf{A}$ and $\mathbf{b}$ are given, an instance of the Ising QUBO is well defined and hence such an instance is represented by the ordered pair $(\mathbf{A}, \mathbf{b})$.

In the definition of QUBO, without loss of generality, one may assume that $\mathbf{Q}$ is symmetric or upper triangular and/or the diagonal entries are zeros. Similar assumptions can be made on $\mathbf{A}$ in the definition of the Ising QUBO. See Chap. 1 for more details on the representations of QUBO. In this chapter, we however make no assumptions on $\mathbf{Q}$ or $\mathbf{A}$ and whenever such assumptions are required, we will state them explicitly. For a review of some solvable cases of QUBO, we refer to the survey paper [61] and the book chapter [66]. More citations on polynomially solvable special cases will be included when we discuss the corresponding results, later in this chapter.

It is well known that the problems QUBO and the Ising QUBO are equivalent [6] and one can be obtained from the other using a linear transformation. Here, the equivalence is understood as follows: There is a bijection ϕ between feasible solutions $\mathbf{x}$ of the QUBO with input $(\mathbf{Q}, \mathbf{c})$ and the feasible solutions $\mathbf{y}$ of the Ising QUBO with Input $(\mathbf{A}^{(q)}, \mathbf{b}^{(q)})$, given by

$$\mathbf{y} = \phi(\mathbf{x}) = 2\mathbf{x} - \mathbf{e} \tag{3.1}$$

where $\mathbf{e}$ is the all-one vector. Moreover, $\mathbf{A}^\mathbf{q} = \frac{1}{4}\mathbf{Q}$, $\mathbf{b}^q = \frac{1}{4}(\mathbf{Q}\mathbf{e} + \mathbf{Q}^T\mathbf{e} + 2\mathbf{c})$ and the objective function value of $\mathbf{x}$ and $\mathbf{y}$ differ just on a constant $K^{(q)} = \frac{1}{4}\mathbf{e}^T\mathbf{Q}\mathbf{e} + \frac{1}{2}\mathbf{c}^T\mathbf{e}$. Conversely, the bijection $\phi^{(-1)}$ given by

$$\mathbf{x} = \phi^{(-1)}(\mathbf{y}) = \frac{1}{2}(\mathbf{y} + \mathbf{e}) \tag{3.2}$$

maps the feasible solutions $\mathbf{y}$ of the Ising QUBO instance $(\mathbf{A}, \mathbf{b})$ to the feasible solutions $\mathbf{x}$ of the QUBO instance $(\mathbf{Q}^{(a)}, \mathbf{c}^{(a)})$, where $\mathbf{Q}^{(a)} = 4\mathbf{A}$ and $\mathbf{c}^{(a)} = 2(\mathbf{b} - \mathbf{A}\mathbf{e} - \mathbf{A}^T\mathbf{e})$. Analogously the objective function values of $\mathbf{y}$ and $\mathbf{x}$ differ just on a constant $K^{(a)} = \mathbf{e}^T\mathbf{A}\mathbf{e} - \mathbf{b}^T\mathbf{e}$. See Chaps. 1 and 8 for further discussion on these transformations and their implications.

Given any symmetric $n \times n$ matrix $\mathbf{Q}$, we can associate a graph $G = (V, E)$ where $V = \{1, 2, \ldots, n\}$ and $E = \{(i, j) : q_{ij} \neq 0\}$. The graph G is called *the support graph* of $\mathbf{Q}$. Whenever we talk about the support graph of $\mathbf{Q}$ or $\mathbf{A}$, either implicitly or explicitly, these matrices are assumed to be symmetric, without loss of generality. Note that the transformations (3.1) and (3.2) preserve the structure of the support graphs of $\mathbf{Q}$ and $\mathbf{A}$. This property is useful in relating the polynomial time solvability of some special cases of QUBO to that of the Ising QUBO and vice versa.

Let us now briefly introduce some notational conventions used in this chapter, keeping consistency with those discussed in Chap. 1. Matrices and vectors are represented using bold capital letters and their elements by the corresponding small letters, along with accents, if any, and the location coordinates. The zero vector in any dimension is represented by $\mathbf{0}$. The objective function of the instance $(\mathbf{Q}, \mathbf{c})$ of QUBO is denoted by $f_{\mathbf{Q},\mathbf{c}}$. That is, $f_{\mathbf{Q},\mathbf{c}}(\mathbf{x}) = \mathbf{x}^T\mathbf{Q}\mathbf{x} + \mathbf{c}^T\mathbf{x}$. If the inputs $\mathbf{Q}$ and $\mathbf{c}$ are unambiguously clear from the context we will drop the subscripts of $f_{\mathbf{Q},\mathbf{c}}$ and simply denote the objective function by f. Similarly, the objective function of the Ising QUBO $(\mathbf{A}, \mathbf{b})$ is denoted by $g_{\mathbf{A},\mathbf{b}}$. That is $g_{\mathbf{A},\mathbf{b}}(\mathbf{y}) = \mathbf{y}^T\mathbf{A}\mathbf{y} + \mathbf{b}^T\mathbf{y}$. If the inputs $\mathbf{A}, \mathbf{b}$ are unambiguously clear from the context we will drop the subscripts of $g_{\mathbf{A},\mathbf{b}}$ and simply denote the objective function by g. For any undirected graph, and edge joining vertices i and j is represented by $\{i, j\}$.

The chapter is organized as follows. In Sect. 3.2, we discuss the computational complexity of QUBO and show that various special cases of the problem are NP-hard. Section 3.3 deals with polynomially solvable special cases of the QUBO and the Ising QUBO identified by exploiting the properties of the cost matrices $\mathbf{Q}$ and $\mathbf{A}$. These include strip matrices and low rank matrices with additional properties, as well as some special matrices admitting high rank but closely related to low rank matrices. In Sect. 3.4, we present polynomial time algorithms for QUBO and the Ising QUBO using the structure of the underlying support graphs of $\mathbf{Q}$ and $\mathbf{A}$. These algorithms exploit connections with polynomially solvable special cases of the maximum weight cut problem, maximum weight stable set problem, and the maximum weight clique problem, along with other optimization problems on graphs. Section 3.5 discusses special cases of QUBO and the Ising QUBO that can be solved using pseudo-polynomial algorithms, followed by concluding remarks in Sect. 3.6.

3.2 Computational Complexity

Let us now explore the computational complexity of QUBO and the Ising QUBO. Before discussing the results on these problems, we briefly discus some of the closely related graph theoretic optimization problems and their computational complexity.

The *maximum weight cut problem* (MWCP) is one of the fundamental combinatorial optimization problems studied extensively in literature. The MWCP is equivalent to QUBO (see Hammer [53]) and the Ising QUBO (see Barahona [6], Barahona et al. [7]). The input for the MWCP is the pair (G, w), where $G = (V, E)$ is an undirected simple graph with vertex set V, edge set $E \subseteq \{\{i, j\}: i, j \in V, i \neq j\}$, and the weight function $w: E \to \mathbb{R}$. In particular, the weight of the edge $\{i, j\}$ is denoted by w_{ij}. The MWCP searches for a subset S of V such that the weight

$w(\delta_G(S)) = \sum_{\{i,j\}\in\delta(S)} w_{ij}$ of the cut $\delta_G(S) = \{\{i, j\} \in E : i \in S, j \notin S\}$ is maximized. If G is unambiguously clear we will write $\delta(S)$ instead of $\delta_G(S)$. An instance of the MWCP on G with weight function w is denoted by the ordered pair (G, w). If G' is the graph obtained from G after removing the edges with weight zero and w' is the mapping w restricted to the edge set E' of G', then the instances (G, w) and (G', w') are equivalent (i.e. $w(\delta_G(S)) = w'(\delta_{G'}(S))$ holds for all $S \subseteq V$). The cut $\delta(S)$ is also denoted by $(S, \bar{S})$, where $\bar{S}$ is the complement of S with respect to V.

The special case of the MWCP where all edge weights fulfill $w_{ij} = 1$ is called the *maximum cardinality cut problem* (MCCP). The maximum weight (cardinality) cut problem is well-known to be strongly NP-hard [42, 43, 79]. In the following theorem, we summarize various special cases of the MWCP and the MCCP that remain NP-hard in spite of the specific properties of the input.

Theorem 3.1 *The maximum weight cut problem (G, w) is NP-hard under the following conditions.*

1. *G is a cubic graph and $w_{ij} = 1$ for all $\{i, j\} \in E$ [91].*
2. *G is a two-level grid and $w_{ij} \in \{-1, 0, 1\}$ for all $\{i, j\} \in E$ [6].*
3. *G is a unit disk graph and $w_{ij} = 1$ for all $\{i, j\} \in E$ [34]*
4. *G is a bipartite graph [71]*
5. *G is a chordal graph, an undirected path graph, a double interval graph, a tripartite graph, or a co-bipartite graph with $w_{ij} = 1$ for all $(i, j) \in E$ [16].*

The Euclidean version of the maximum weight cut problem (EMWCP) can be defined as follows. Given a set $P = \{\mathbf{x}^1, \mathbf{x}^2, \ldots, \mathbf{x}^n\}$ of n points in $\mathbb{R}^m$, find a partition $S, \bar{S}$ of $\{1, 2, \ldots, n\}$ such that $\sum_{i\in S, j\in\bar{S}} ||\mathbf{x}^i - \mathbf{x}^j||$ is maximized. If we take the square $||\mathbf{x}^i - \mathbf{x}^j||^2$ of the Euclidean distance instead of $||\mathbf{x}^i - \mathbf{x}^j||$, we get an instance of the *squared Euclidean maximum cut problem* (SEMWCP).

Theorem 3.2 ([1]) *EMWCP and SEMWCP are strongly NP-hard.*

Let us now consider the maximum weight stable set problem (MWSSP). The MWSSP takes as input a pair (G, d), where $G = (V, E)$ is an undirected graph with vertex set V, edge set $E \subseteq \{\{i, j\} : i, j \in V\}$, and $d : V \to \mathbb{R}$ is a weight function on the vertices of G. Thus, the weight of the vertex $i \in V$ is d_i. Then, the MWSSP searches for a subset of S of the vertex set V such that the weight $d(S) = \sum_{i\in S} d_i$ is maximized and the subgraph of G induced by S is a collection of isolated vertices. An instance of the MWSSP on the graph G with vertex weight function d is denoted by by the ordered pair (G, d). Without loss of generality we assume that the weights d_i of the vertices are non-negative since vertices of negative weight would never belong to an optimal solution and hence can be removed from G together with the edges incident on them. The MWSSP is well-known to be strongly NP-hard [42]. The theorem below summarizes some special cases of the MWSSP that remain NP-hard in spite of the specific properties of the input.

Theorem 3.3 *The maximum weight stable set problem (G, w) is NP-hard under the following conditions.*

1. *G is a planar graph with maximum vertex degree 3 and $d_j = 1$ for all $j \in V$ [43].*
2. *G is a triangle free graph and $d_j = 1$ for all $j \in V$ [86].*
3. *G is a 3-regular Hamiltonian graph and $d_j = 1$ for all $j \in V$ [39]*
4. *G is a planar graph with large girth and $d_j = 1$ for all $j \in V$ [73].*
5. *G is an H-free graph [2]*

Note that item (4) of Theorem 3.3 is a strengthened version of item (a) of the theorem.

Another problem closely related to the MWSSP is the *maximum weight clique problem* (MWClP). Note that S is a stable set in a graph G if and only if it is a clique in the complement $\bar{G}$ of G. As in the case of the MWSSP, the *maximum weight clique problem* (MWClP) also takes as input the pair (G, d) and searches for a subset S of V such that the subgraph of G induced by S is a complete graph and $\sum_{i \in S} d_i$ is maximized. For a thorough discussion on the maximum weight clique problem, we refer to [18]. MWClP is well-known to be strongly NP-hard [42]. The maximum weight clique problem is NP-hard on the complements of the graphs indicated in Theorem 3.3. In the theorem below we summarizes some additional special cases of the MWClP that remain NP-hard in spite of the specific properties of the input.

Theorem 3.4 *The maximum weight clique problem is NP-hard on the following graphs, even if $d_i = 1$ for all $i \in V$.*

1. *2-interval and unit 3-track graphs [40]*
2. *Ray intersection graphs, segment intersection graph (intersection graphs of segments in plane) [24]*
3. *String graphs [63, 72]*
4. *Ball graphs and unit 4-dimensional graphs [19]*

It may be noted that the MWSSP is NP-hard on the complements of the graphs discussed in Theorem 3.4.

Let us now explore the complexity of QUBO. The equivalence between QUBO and MWCP immediately establishes that QUBO and the Ising QUBO are strongly NP-hard and also APX hard. Let us examine now the complexity of QUBO and the Ising QUBO for various special cases.

Theorem 3.5 ([6]) *Let $\mathbf{A}$ be an $n \times n$ symmetric matrix and $G = (V, E)$ be the support graph of $\mathbf{A}$ with $V = \{1, 2, \ldots, n\}$. Then, G with the weight function $w_{ij} = a_{ij}$ has a cut $(S, \bar{S})$ of value K if and only if there exists $y_j \in \{-1, 1\}, j = 1, 2, \ldots, n$ such that $\sum_{i=1}^{n} \sum_{j=1}^{n} a_{ij} y_i y_j = \sum_{i=1}^{n} \sum_{j=1}^{n} a_{ij} - 4K$.*

Proof Suppose that G has a cut $(S, \bar{S})$ such that $\sum_{i \in S, j \in \bar{S}} a_{ij} = K$. Now, define $y_i = 1$ if $i \in S$ and $y_i = -1$ is $i \in \bar{S}$. Then,

$$\sum_{i=1}^{n}\sum_{j=1}^{n} a_{ij} y_i y_j = \sum_{i \in S}\sum_{j \in S} a_{ij} y_i y_j + \sum_{i \in \bar{S}}\sum_{j \in S} a_{ij} y_i y_j + 2\sum_{i \in S}\sum_{j \in \bar{S}} a_{ij} y_i y_j$$

$$= \sum_{i \in S}\sum_{j \in S} a_{ij} + \sum_{i \in \bar{S}}\sum_{j \in \bar{S}} a_{ij} - 2\sum_{i \in S}\sum_{j \in \bar{S}} a_{ij} \tag{3.3}$$

$$= \sum_{i=1}^{n}\sum_{j=1}^{n} a_{ij} - 4K \tag{3.4}$$

The equality (3.4) follows by adding and subtracting $2\sum_{i \in S}\sum_{j \in \bar{S}} a_{ij}$ to RHS of Eq. (3.3). Conversely, suppose there exists $y_j \in \{-1, 1\}$, $j = 1, 2, \ldots, n$ such that $\sum_{i=1}^{n}\sum_{j=1}^{n} a_{ij} y_i y_j = \sum_{i=1}^{n}\sum_{j=1}^{n} a_{ij} - 4K$. Now define $S = \{j : y_j = 1\}$. Then we have

$$\sum_{i=1}^{n}\sum_{j=1}^{n} a_{ij} y_i y_j = \sum_{i=1}^{n}\sum_{j=1}^{n} a_{ij} - 4\sum_{i \in S}\sum_{j \in \bar{S}} a_{ij}$$

and hence $\sum_{i \in S}\sum_{j \in \bar{S}} a_{ij} = K$. □

From Theorem 3.5 it follows that the Ising QUBO $(\mathbf{A}, \mathbf{0})$ is NP-hard if the maximum weight cut problem is NP-hard on the support graph $G = (V, E)$ of $\mathbf{A}$ with weight $-a_{ij}$ for $\{i, j\} \in E$. In particular, we have

Theorem 3.6 *The Ising QUBO* $(\mathbf{A}, \mathbf{0})$ *is NP-hard if* $a_{ij} \in \{0, -1\}$ *and the support graph of* $\mathbf{A}$ *is a chordal graph, an undirected path graph, a double interval graph, tripartite graphs, a co-bipartite graph, a tripartite graph, and a cubic graph.*

The proof of Theorem 3.6 follows from Theorem 3.1. Further, it also follows that

Theorem 3.7 *If* $a_{ij} \in \{0, 1, -1\}$ *then the Ising QUBO* $(\mathbf{A}, \mathbf{0})$ *is NP-hard if the support graph of* $\mathbf{A}$ *is a two-level grid.*

Corollary 3.1 *The QUBO* $(\mathbf{Q}, \mathbf{c})$ *is NP-hard if* $q_{ij} \in \{0, -1\}$ *and the support graph of* $\mathbf{Q}$ *is a chordal graph, an undirected path graph, a double interval graph, tripartite graphs, a co-bipartite graph, a tripartite graph, and a cubic graph. Further, the QUBO* $(\mathbf{Q}, \mathbf{c})$ *is NP-hard if* $q_{ij} \in \{0, 1, -1\}$ *and the support graph of* $\mathbf{Q}$ *is a two-level grid.*

Proof From Eq. (3.1), it follows that the Ising QUBO $(\mathbf{A}, \mathbf{0})$ is equivalent to the QUBO $(\mathbf{Q}^a, \mathbf{c}^a)$ where $\mathbf{e}$ is the all-one vector, $\mathbf{Q}^a = 4\mathbf{A}$ and $\mathbf{c}^a = 2(-\mathbf{e}^T\mathbf{A} - \mathbf{e}^T\mathbf{A}^T)$. But QUBO $(\mathbf{Q}^a, \mathbf{c}^a)$ is equivalent to QUBO $(\frac{1}{4}\mathbf{Q}^a, \frac{1}{4}\mathbf{c}^a)$. Further, if $a_{ij} \in \{0, -1\}$ then elements of $\frac{1}{4}\mathbf{Q}^a$ also belongs to $\{0, -1\}$ and the support graph of $\frac{1}{4}\mathbf{Q}^a$ is the same as the support graph of $\mathbf{A}$. Now, the first part of the corollary follows from

Theorem 3.6. Using similar arguments, the second part of the corollary follows from Theorem 3.7. □

Analogous to the Euclidean maximum cut problem, we have the Euclidian version of the Ising QUBO which is defined as follows. Given a set $P = \{\mathbf{x}^1, \mathbf{x}^2, \ldots, \mathbf{x}^n\}$ of n points in $\mathbb{R}^m$, find a $y_j \in \{-1, 1\}, j = 1, 2, \ldots, n$ such that $\sum_{i=1}^{n}\sum_{j=1}^{n} ||\mathbf{x}^i - \mathbf{x}^j||y_i y_j$ is minimized. Instead of the Euclidean distance $||\mathbf{x}^i - \mathbf{x}^j||$, if we take the square $||\mathbf{x}^i - \mathbf{x}^j||^2$ of the Euclidean distance, we get an instance of the *squared Euclidean Ising QUBO*. Note that the problem is defined in the minimization form. The maximization version of the problem is trivial since $y_j = 1$ for all j is an optimal solution. From Theorems 3.5 and 3.2 we have

Theorem 3.8 *The Euclidean Ising QUBO and the squared Euclidean Ising QUBO are strongly NP-hard.*

It may be noted that the QUBO with a 0-1 matrix $\mathbf{Q}$ (or any non-negative matrix $\mathbf{Q}$) and non-negative vector $\mathbf{c}$ is trivial since $x_i = 1$ for all i is an optimal solution. When $\mathbf{Q}$ is a 0-1 matrix (or any non-negative matrix $\mathbf{Q}$) and $\mathbf{c}$ has negative entries, the associated QUBO can be solved polynomial time as a maximum flow problem. (See Sect. 3.3 for a more general result of this type.) Interestingly, when $q_{ij} \in \{-1, 0\}$ and c_i is non-negative for all i, QUBO is no longer easy.

Theorem 3.9 *The QUBO* $(\mathbf{Q}, \mathbf{c})$ *is strongly NP-hard if* $q_{ij} \in \{0, -1\}$, $c_i = 1$ *for all* $i = 1, 2, \ldots, n$ *and the maximum cardinality stable set problem is NP-hard on the support graph of* $\mathbf{Q}$.

Proof Let $G = (V, E)$ be a graph on which the maximum cardinality stable set problem is NP-hard. Now construct the matrix $\mathbf{Q}$ such that $q_{ij} = -1$ if $\{i, j\} \in E$ and $q_{ij} = 0$ if $\{i, j\} \notin E$. Also, choose $\mathbf{c}$ as the all-one vector in $\mathbb{R}^{|V|}$. The QUBO instance $(\mathbf{Q}, \mathbf{c})$ constructed here satisfies the conditions of the theorem. Let $\mathbf{x}$ be an optimal solution to the QUBO $(\mathbf{Q}, \mathbf{c})$. Note that the optimal objective function value of the QUBO $(\mathbf{Q}, \mathbf{c})$ is at least one and at most $|V|$. We claim that the optimal objective function value of $(\mathbf{Q}, \mathbf{c})$ is precisely the size of a maximum cardinality stable set in G. First observe that if the edge $\{i, j\} \in E$ then, in an optimal solution $\mathbf{x}^*$ of the QUBO, at least one of x_i^* or x_j^* is zero. On contrary, assume that $x_i^* = x_j^* = 1$ in an optimal $\mathbf{x}^*$. Since $q_{ij} = q_{ji} = -1, c_i = c_j = 1$, and other entries of Q are no more than zero, setting x_i^* or x_j^* equal to zero results in an improved solution, violating the optimality of $\mathbf{x}^*$. Thus the set $S^* = \{j : x_j^* = 1\}$ is a stable set in G. But any stable set S in G generates a feasible solution to the QUBO by defining $x_j = 1$ if $j \in S$ and $x_j = 0$, otherwise with QUBO objective function value is precisely $|S|$. Therefore S^* must be an optimal solution to the maximum cardinality stable set problem on G. Since the maximum cardinality stable set problem on G is assumed to be NP-hard, the result follows. □

The above proof is based on the QUBO formulation of the maximum cardinality stable set problem and maximum clique problem given in [81]. In particular, Theorem 3.9 shows that QUBO $(\mathbf{Q}, \mathbf{c})$ is NP-hard even if $q_{ij} \in \{0, -1\}$ and the support graph of $\mathbf{Q}$ is any of the graphs listed in Theorem 3.3 or complements of the graphs listed in Theorem 3.4.

Corollary 3.2 *The Ising QUBO* $(\mathbf{A}, \mathbf{b})$ *is strongly NP-hard if* $a_{ij} \in \{0, -1\}$ *and the maximum cardinality stable set problem is NP-hard on the support graph of* $\mathbf{A}$.

The proof of this corollary follows from Theorem 3.9 and is similar to that of Corollary 3.1 except that we will be using Eq. (3.2) to develop the arguments.

Let us now look at the special case of QUBO when $\mathbf{Q}$ is symmetric and of rank one. If the symmetric matrix $\mathbf{Q}$ is of rank one, then, there exist vectors $\mathbf{a}^T = (a_1, a_2, \ldots, a_n)$ and $\mathbf{b}^T = (b_1, b_2, \ldots, b_n)$ such that $q_{ij} = a_i b_j$ for $i, j = 1, 2, \ldots, n$ and $b_i = a_i$ for all i or $b_i = -a_i$ for all i. Such a representation of $\mathbf{Q}$ can be obtained using the reduced echelon form of $\mathbf{Q}$ or by using singular value decomposition. An instance of QUBO where $\mathbf{Q}$ is symmetric and is of rank one is called *symmetric rank-one QUBO*. The objective function of a symmetric rank one QUBO can be written as $f(\mathbf{x}) = (\mathbf{a}^{\mathrm{T}}\mathbf{x})(\mathbf{b}^T\mathbf{x}) + \mathbf{c}^T\mathbf{x}$ where $\mathbf{a}$ and $\mathbf{b}$ are related as indicated above.

Theorem 3.10 ([26, 54]) *The symmetric rank-one QUBO is NP-hard.*

Proof We reduce the SUBSET SUM problem to the symmetric rank-one QUBO. The SUBSET SUM problem can be stated as follows. Given n numbers $\alpha_1, \alpha_2, \ldots, \alpha_n$ and a constant K, we want to determine if there exist a subset $S \subseteq \{1, 2, \ldots, n\}$ such that $\sum_{j \in S} \alpha_j = K$. From an instance of SUBSET SUM, construct an instance of the symmetric rank-one QUBO as follows. Choose $a_i = \alpha_i$, $b_i = -a_i$, and $c_i = 2K\alpha_i$. For $\mathbf{x} \in \{0, 1\}^n$, define $h(\mathbf{x}) = -(\mathbf{a}^T\mathbf{x} - K)^2$. Then $h(\mathbf{x}) \leq 0$ for all $\mathbf{x} \in \{0, 1\}^n$ and $h(\mathbf{x}) = 0$ precisely when the required partition exists. But maximizing $h(\mathbf{x})$ is equivalent to maximizing $f(\mathbf{x}) = -(\mathbf{a}^T\mathbf{x})^2 + 2K(\mathbf{a}^T\mathbf{x}) = (\mathbf{a}^T\mathbf{x})(\mathbf{b}^T\mathbf{x}) + \mathbf{c}^T\mathbf{x}$ and the result follows. □

Again, an analogous NP-hardness result for the Ising QUBO can be obtained when $\mathbf{A}$ is a rank-one matrix. Later we will examine the more general problem of QUBO with rank of $\mathbf{Q}$ being fixed or an appropriate function of n.

Let us now examine another related special case called the *half product* QUBO. In Chap. 1, applications of the half product QUBO in the context of machine scheduling are discussed. Recall that a matrix $\mathbf{Q}$ is called a *half product matrix* if it is upper triangular with diagonal elements zero and there exist vectors $\mathbf{a} = (a_1, a_2, \ldots, a_n)$ and $\mathbf{b} = (b_1, b_2, \ldots, b_n)$ such that $q_{ij} = a_i b_j$ for $i = 1, 2, \ldots, n$ and $j > i$. The *half-product* QUBO is the special case of QUBO where $\mathbf{Q}$ is restricted to a half-product matrix. The half-product QUBO (in minimization form) was introduced by Badics and Boros [5] and independently by Kubiak [64]. Note that we consider the half product QUBO in the maximization form.

Theorem 3.11 ([5]) *The half product QUBO is NP-hard.*

Proof The proof follows by a reduction from the subset sum problem. □

For more complexity results relating to the approximability of QUBO and the Ising QUBO, we refer to the Chap. 8 and the references included there.

3.3 Polynomially Solvable Matrix Structures

In this section we look at special properties of the matrix $\mathbf{Q}$ and the matrix $\mathbf{A}$ to identify polynomially solvable cases of the QUBO and the Ising QUBO. We distinguish between properties which can be specified in terms of simple linear equations and inequalities to be fulfilled by the entries of $\mathbf{Q}$ and also rank-based properties.

3.3.1 *Linear Restrictions on the Cost Matrices*

Sum Matrices A matrix $\mathbf{Q}$ is said to be a *sum matrix* if there exist $a_i, b_i, i = 1, 2, \ldots, n$ such that $q_{ij} = a_i + b_j$ for all $i, j = 1, 2, \ldots, n$.

Theorem 3.12 *If* $\mathbf{Q}$ *is a sum matrix then the QUBO* $(\mathbf{Q}, \mathbf{c})$ *can be solved in* $O(n^2)$ *time.*

Proof When $\mathbf{Q}$ is a sum matrix,

$$\begin{aligned}\mathbf{x}^T\mathbf{Q}\mathbf{x} + \mathbf{c}^T\mathbf{x} &= \sum_{i=1}^{n} a_i x_i \sum_{j=1}^{n} x_j + \sum_{j=1}^{n} b_j x_j \sum_{i=1}^{n} x_i + \mathbf{c}^T\mathbf{x} \\ &= \left(\sum_{i=1}^{n}(a_i + b_i)x_i\right)\left(\sum_{i=1}^{n} x_i\right) + \mathbf{c}^T\mathbf{x} \qquad (3.5)\end{aligned}$$

Now, consider the problem

$$\begin{aligned}\text{P}(\lambda):\quad \text{Maximize} \quad & \lambda \sum_{i=1}^{n}(a_i + b_i)x_i + \mathbf{c}^T\mathbf{x} \\ \text{Subject to:} \quad & \sum_{i=1}^{n} x_i = \lambda, \mathbf{x} \in \{0, 1\}^n .\end{aligned}$$

for some $\lambda \in \{0, 1, \ldots, n\}$. Notice that the objective function of P(λ) can be written as $\sum_{i=1}^{n}[\lambda(a_i + b_i) + c_i]x_i$. Now, for $\lambda \in \{1, \ldots, n\}$, the λth maximum (lexicographically), say (ω, λ), of the ordered pairs $(\lambda(a_i + b_i) + c_i, i), i = 1, 2, \ldots, n$ can be identified in $O(n)$ time. Let $S = \{i : (\lambda(a_i+b_i)+c_i, i) \succcurlyeq (\omega, \lambda)\}$, where $\succcurlyeq$ means lexicographically greater than. Choose $x_j = 1$ if $j \in S$ and $x_j = 0$ otherwise and this solution will be optimal for $P(\lambda)$, $\lambda \neq 0$. For $\lambda = 0$ the zero vector in the optimal solution for $P(\lambda)$. Let $\mathbf{x}^{(\lambda)}$ be an optimal solution to P(λ) with corresponding objective function value $v(\lambda) = \sum_{i=1}^{n}(\lambda(a_i + b_i) + c_i)x_i^{(\lambda)}$, for any $\lambda \in \{0, 1, \ldots, n\}$. Then $\mathbf{x}^{(\bar{\lambda})}$ for $\bar{\lambda} \in argmax\{v(\lambda) \colon \lambda \in \{0, 1, \ldots, n\}\}$ is an optimal solution to the QUBO given in Eq. (3.5) and the result follows. □

A matrix $\mathbf{Q}$ is said to be a *weak sum matrix* if there exist $a_i, b_i, i = 1, 2, \ldots, n$ such that $q_{ij} = a_i + b_j$, for all $i, j = 1, 2, \ldots, n$ and $i \neq j$.

Corollary 3.3 *If* $\mathbf{Q}$ *is a weak sum matrix then the QUBO* $(\mathbf{Q}, \mathbf{c})$ *can be solved in* $O(n^2)$ *time.*

Proof Define the matrix $\mathbf{Q}'$ where $q'_{ij} = a_i + b_j$ for all $i, j = 1, 2, \ldots, n$ and the vector $\mathbf{c}'$ where $c'_i = c_i + q_{ii} - a_i - b_i$. Then

$$\mathbf{x}^T\mathbf{Q}\mathbf{x} + \mathbf{c}^T\mathbf{x} = \mathbf{x}^T\mathbf{Q}'\mathbf{x} + (\mathbf{c}')^T\mathbf{x}$$

for $\mathbf{x} \in \{0, 1\}^n$. But, $\mathbf{Q}'$ is a sum matrix and hence by Theorem 3.12 the QUBO $(\mathbf{Q}', \mathbf{c}')$ can be solved in $O(n^2)$ time and the result follows. □

Analogous results can be obtained for the Ising QUBO. It is not surprising that sum matrices and weak sum matrices lead to polynomially solvable special cases of the QUBO. Such matrices are also studied in the context of other prominent combinatorial optimization problems, e.g. the travelling salesman problem, where they also lead to polynomially solvable special cases.

Product Matrices Another class of specially structured matrices involved in polynomially solvable special cases of hard combinatorial optimization problems are the product matrices. For example, a number of results on polynomially solvable special cases of the well studied quadratic assignment problem (QAP), see for example [23, 25, 69], are related to product matrices. Results on (pseudo) polynomially solvable special cases of the QAP involving product matrices carry over to QUBO as we will explain below.

An $n \times n$ matrix $\mathbf{A} = (a_{ij})$ is called a *product matrix*, if there exist the real numbers $\alpha_1, \ldots, \alpha_n, \beta_1, \ldots, \beta_n$ such that

$$a_{ij} = \alpha_i \beta_j \qquad \text{for } 1 \leq i, j \leq n. \tag{3.6}$$

When $\alpha_i = \beta_i \in \{0, 1\}$ for all $i \in \{1, 2, \ldots, n\}$, $\mathbf{A}$ is called a *symmetric* 0-1 *product matrix*.

The QAP in Koopmans-Beckmann form [62] takes as input two $n \times n$ square matrices $\mathbf{A} = (a_{ij})$ and $\mathbf{B} = (b_{ij})$ with real entries and searches for a permutation π that minimizes the objective function

$$Z_\pi(\mathbf{A}, \mathbf{B}) := \sum_{i=1}^{n}\sum_{j=1}^{n} a_{\pi(i)\pi(j)} b_{ij}, \tag{3.7}$$

where π ranges over the set S_n of all permutations of $\{1, 2, \ldots, n\}$. We denote this QAP instance by $QAP(\mathbf{A}, \mathbf{B})$. In general, the QAP is extremely difficult to solve and hard to approximate, but there are some strongly structured polynomially solvable special cases.

In this context a special case is characterized by certain (combinatorial) properties of the coefficient matrices **A** and **B**. To formalize the concept, let us denote by $\mathcal{M}_P$ the set of square matrices having a certain property P. P could be for example "being a sum matrix" or "being a product matrix" as defined above.

Given two properties P_1 and P_2 the P_1-P_2 special case of the QAP consists of the family of instances $QAP(\mathbf{A}, \mathbf{B})$ with $\mathbf{A} \in \mathcal{M}_{P_1}, \mathbf{B} \in \mathcal{M}_{P_2}$.

Consider now a $\lambda \in \{1, 2, \dots, n\}$ and let $\mathbf{A}^{(\lambda)} = (a_{ij}^{(\lambda)})$ be a 0-1 product matrix given by $a_{ij}^{(\lambda)} = x_i x_j$, for $i, j \in \{1, 2, \dots n\}$, where $\mathbf{x} = (x_1, x_2, \dots, x_n)^T \in \{0, 1\}^n$ with $x_i = 0$, if and only if $i \in \{1, 2, \dots, \lambda\}$. For every $\pi \in S_n$ let $\mathbf{x}^\pi := (x_{\pi(1)}, x_{\pi(2)}, \dots, x_{\pi(n)})^T$ be the vector obtained by permuting **x** according to π. Clearly $\mathbf{x}^\pi \in \{0, 1\}^n$ and the sum of its entries equals λ. Conversely, any vector in $\{0, 1\}^n$ with sum of the entries equal to λ can be written as $\mathbf{x}^\pi$ for some $\pi \in S_n$. Then the objective function of $QAP(\mathbf{A}^{(\lambda)}, B)$ can be written as $Z_\pi(\mathbf{A}^{(\lambda)}, \mathbf{B}) = -f_{\mathbf{Q},\mathbf{c}}(\mathbf{x}^\pi)$, where $\mathbf{Q} = -\mathbf{B}$ and $\mathbf{c} = \mathbf{0}$. Hence $QAP(\mathbf{A}^{(\lambda)}, \mathbf{B})$ is equivalent to

$$\max \left\{ f_{\mathbf{Q},\mathbf{c}}(\mathbf{x}) : \mathbf{x} \in \{0, 1\}^n, \sum_{i=1}^{n} x_i = \lambda \right\},$$

and $QUBO(\mathbf{Q}, \mathbf{0})$ can be written as $\max\limits_{\lambda \in \{0,1,\dots,n\}} \min\limits_{\pi \in S_n} Z_\pi(A^{(\lambda)}, -Q)$. Thus $QUBO(\mathbf{Q}, \mathbf{0})$ can be solved by solving $QAP(A^{(\lambda)}, -Q)$ for all $\lambda \in \{0, 1, \dots, n\}$. Consequently every P_1-P_2 (pseudo) polynomially solvable special case of the QAP, where P_1 is the property of being a symmetric 0-1 product matrix, yields a (pseudo) polynomially solvable special case of the $QUBO(\mathbf{Q}, \mathbf{0})$.

Lemma 3.1 *Let P_1 be the property "being a symmetric* 0-1 *product matrix" and P_2 be some other matrix property such that the P_1-P_2 special case of the QAP is (pseudo) polynomially solvable. Then $QUBO(\mathbf{Q}, \mathbf{0})$ is (pseudo) polynomially solvable over the class of matrices $\mathbf{Q}$ such that $(-\mathbf{Q})$ has the property P_2.*

Pseudo (polynomially) solvable special cases of the QAP involving symmetric 0-1 product matrices have been investigated in [22, 27, 28]. By applying Lemma 3.1 these special cases immediately imply the following results for the QUBO.

Theorem 3.13 *The following special cases of the QUBO* $(\mathbf{Q}, \mathbf{0})$ *are (pseudo) polynomially solvable.*

(i) Block matrices [28].
Let q be a fixed natural number and let $\mathbf{P} = (p_{ij})$ be a $q \times q$ matrix such that $p_{ii} + p_{jj} \geq 2p_{ij}$, for any $i, j \in \{1, 2, \dots, q\}$. Then, QUBO $(\mathbf{Q}, \mathbf{0})$ is solvable in $O(n(q^2 q! + n \log n))$ time if $\mathbf{Q}$ is a block matrix with block pattern $\mathbf{P}$, i.e. (a) there exists a partition of the row and column set $\{1, \dots, n\}$ into q (possibly empty) intervals $I_1, \dots, I_q$ such that for $1 \leq k \leq q-1$ all elements of interval I_k are smaller than all elements of interval I_{k+1} and (b) for all indices i and j with $1 \leq i, j \leq n$ and $i \in I_k$ and $j \in I_\ell$, we have $q_{ij} = p_{k\ell}$.

(ii) *Toeplitz matrices [22].*
Let $\mathbf{Q}$ *be a symmetric* $n \times n$ *Toeplitz matrix with* $q_{ij} = f(|i - j|)$ *for* $i, j \in \{1, 2, \dots, n\}$ *and* $f\colon \{0, 1, \dots, n-1\} \to \mathbb{R}$ *with the properties (a)* $f(1) \geq f(i+1)$ *for* $1 \leq i \leq \lfloor \frac{n}{2} \rfloor - 1$, *and (b)* $f(i) \geq f(n-i)$ *for all* $1 \leq i \leq \lceil \frac{n}{2} \rceil - 1$. *Then* $QUBO(\mathbf{Q}, \mathbf{0})$ *is solvable in polynomial time. More precisely, an optimal solution can be found among the vectors* $\mathbf{x}^{(\lambda)}$, *for* $\lambda \in \{0, 1, \dots, n\}$, *where* $x_i^{(\lambda)} = 0$ *iff* $i \in \{0, 1, \dots, \lceil \frac{n}{2} \rceil\} \cup \{n - \lfloor \frac{n}{2} \rfloor + 1, \dots, n\}$.

(iii) *Distance matrices of one-dimensional point sets [27].*
$QUBO(\mathbf{Q}, \mathbf{0})$ *is solvable in pseudo polynomial time if* $\mathbf{Q}$ *is the distance matrix of a one-dimensional point set.*

In the following we give some examples to illustrate the results presented in Theorem 3.13. For example let $\mathbf{P}$ be a 3×3 matrix as follows

$$\mathbf{P} = \begin{pmatrix} d_1 & a & b \\ a & d_2 & c \\ b & c & d_3 \end{pmatrix},$$

such that d_1, d_2, d_3 are arbitrary reals and the other entries of the matrix fulfill $2a \leq d_1 + d_2$, $2b \leq d_1 + d_3$, $2c \leq d_2 + d_3$. Then, according to Theorem 3.13 (i), $QUBO(\mathbf{Q}, \mathbf{0})$ is solvable in $O(n(54 + n \log n) = O(n^2 \log n)$ time, if $\mathbf{Q}$ is a block matrix with block pattern $\mathbf{P}$, i.e. if there exist natural numbers $n_1, n_2, n_3 \in \mathbb{N}$ with $n = n_1 + n_2 + n_3$ such that the entries of $\mathbf{Q}$ are given as follows:

$$q_{ij} = \begin{cases} d_1 \text{ if } i, j \in \{1, \dots, n_1\}, \\ d_2 \text{ if } i, j \in \{n_1 + 1, \dots, n_1 + n_2\}, \\ d_3 \text{ if } i, j \in \{n_1 + n_2 + 1, \dots, n\} \end{cases}$$

$$q_{ij} = q_{ji} = \begin{cases} a \text{ if } i \in \{1, \dots, n_1\},\ j \in \{n_1 + 1, \dots, n_1 + n_2\}, \\ b \text{ if } i \in \{1, \dots, n_1\},\ j \in \{n_1 + n_2 + 1, \dots, n\}, \\ c \text{ if } i \in \{n_1 + 1, \dots, n_1 + n_2\},\ j \in \{n_1 + n_2 + 1, \dots, n\} \end{cases}.$$

As an illustrative example for Theorem 3.13 (ii) consider a symmetric 8×8 Toeplitz matrix $\mathbf{Q}$ generated by $f\colon \{0, 1, \dots, 7\} \to \mathbb{R}$ given by $f(0) = 2$, $f(1) = 5$, $f(2) = 2$, $f(3) = 1$, $f(4) = 0$, $f(5) = 1$, $f(6) = 0$, $f(7) = 3$ which obviously fulfills the conditions specified in (ii). In this case the results of [22] would imply that an optimal solution of the QUBO $(\mathbf{Q}, \mathbf{0})$ can be found among the the vectors $\{\mathbf{x}^{(i)} : i \in \{0, 1, \dots, 7\}\}$ given as

$$(\mathbf{x}^{(0)})^T = (0, 0, 0, 0, 0, 0, 0, 0), \quad (\mathbf{x}^{(1)})^T = (0, 1, 1, 1, 1, 1, 1, 1),$$

$$(\mathbf{x}^{(2)})^T = (0, 1, 1, 1, 1, 1, 1, 0), \quad (\mathbf{x}^{(3)})^T = (0, 0, 1, 1, 1, 1, 1, 0),$$

$$(\mathbf{x}^{(4)})^T = (0, 0, 1, 1, 1, 1, 0, 0), \quad (\mathbf{x}^{(5)})^T = (0, 0, 0, 1, 1, 1, 0, 0),$$

$$(\mathbf{x}^{(6)})^T = (0,0,0,1,1,0,0,0), \quad (\mathbf{x}^{(7)})^T = (0,0,0,0,1,0,0,0),$$
$$(\mathbf{x}^{(8)})^T = (1,1,1,1,1,1,1,1).$$

This is due to the fact that $\mathbf{x}^{(\lambda)}$ is an optimal solution to the $QAP(\mathbf{A}^{\lambda}, -\mathbf{Q})$, for $\lambda \in \{0, 1, \dots, 8\}$.

To illustrate Theorem 3.13 (iii) consider a set of n colinear points P_i in $\mathbb{R}^k$, $i \in \{1, 2, \dots, n\}$, and let $\mathbf{Q} = (q_{ij})$ be such that $q_{ij} = d_2(P_i, P_j)$, for all $i, j \in \{1, 2, \dots, n\}$, where d_2 is the Euclidean distance in $\mathbb{R}^k$. Then QUBO $(\mathbf{Q}, \mathbf{0})$ can be solved in pseudo-polynomial time by dynamical programming, as implied by the results in [27].

k-Strip Matrices An $n \times n$ matrix is said to be *k-strip matrix* if all of its non-zero entries are located in the first k rows and first k columns. A QUBO when $\mathbf{Q}$ is a k-strip matrix is called a *k-strip QUBO*.

Theorem 3.14 *A k-strip QUBO is polynomially solvable when $k = O(\log n)$ and is NP-hard when $k = O(\sqrt[t]{n})$ for any fixed $t > 0$.*

Proof From the given k-strip matrix, construct the $k \times n$ matrix Q' as

$$q'_{ij} = \begin{cases} q_{ij} & \text{if } 1 \le i, j \le k \\ q_{ij} + q_{ji} & \text{if } i = 1, 2, \dots, k,\ j = k+1, \dots, n. \end{cases}$$

Then, QUBO is equivalent to the pseudo-bilinear program [25]

$$\text{PBL: Maximize } \sum_{i=1}^{k}\sum_{j=1}^{k} q'_{ij} z_i z_j + \sum_{i=1}^{k} c_i z_i + \sum_{i=1}^{k}\sum_{j=k+1}^{n} (q_{ij} + q_{ji}) z_i y_j + \sum_{j=k+1}^{n} c_j y_j$$

$$\text{Subject to: } z_i \in \{0, 1\} \text{ for } i = 1, 2 \dots, k$$
$$y_j \in \{0, 1\} \text{ for } j = k+1, \dots, n.$$

For each $\mathbf{z}^p \in \{0, 1\}^k$ construct the best y_j^p, $j = k+1, \dots, n$ as

$$y_j^p = \begin{cases} 1 & \text{if } c_j + \sum_{i=1}^{k} (q_{ij} + q_{ji}) z_i^p > 0 \\ 0 & \text{otherwise.} \end{cases}$$

Now for each $z^p \in \{0, 1\}^k$, $p = 1, 2, \dots, 2^k$, construct the solution $\mathbf{x}^p = (z_1^p, z_2^p, \dots z_k^p, y_{k+1}^p, \dots, y_n^p)$ for the QUBO. Choose $\mathbf{x}^r \in \{\mathbf{x}^1, \mathbf{x}^2, \dots, \mathbf{x}^{2^k}\}$ such that $f_{\mathbf{Q},\mathbf{c}}(\mathbf{x}^r) = \max\{f_{\mathbf{Q},\mathbf{c}}(\mathbf{x}^i) : i = 1, 2, \dots, 2^k\}$. Then x^r will be an optimal solution to the k-strip QUBO. When $k = O(\log n)$ the enumeration scheme discussed above is polynomially bounded. The second part of the theorem can be proved using arguments similar to those used in the proof of Theorem 10.7 in Chap. 10. □

A matrix is said to be a *permuted k-strip matrix* if there exists a permutation of rows and columns that transforms the matrix to a k-strip matrix. Equivalently, $\mathbf{Q}$ is a permuted k-strip matrix if there exists a set $S \subset \{1, 2, \ldots, n\}$ such that $|S| = k$ and deleting rows and columns of $\mathbf{Q}$ corresponding to the indices in S results in a zero matrix. In view of Theorem 3.14, QUBO can be solved in polynomial time if the cost matrix $\mathbf{Q}$ is a permuted k-strip matrix for a given permutation (or a given index set S as indicated above). An example of a permuted 3-strip matrix is given below. The permutation that interchanges row 1 and 6 and column 1 and 6 yields a 3-strip matrix (equivalently choose $S = \{2, 3, 6\}$).

$$Q = \begin{pmatrix} 0 & -5 & -2 & 0 & 0 & 6 \\ -2 & 1 & 0 & 1 & 3 & 2 \\ 1 & -6 & 5 & 2 & 3 & 1 \\ 0 & 3 & -4 & 0 & 0 & 2 \\ 0 & 5 & 3 & 0 & 0 & 1 \\ -3 & 2 & 3 & -1 & 5 & 1 \end{pmatrix}$$

However, recognizing a permuted k-strip matrix in general is a difficult problem.

Theorem 3.15 *Recognizing a permuted k-strip matrix is strongly NP-hard. When k is fixed, a permuted k-strip matrix can be recognized in polynomial time and when* $k = O(\log n)$ *it can be recognized in quasi-polynomial time.*

Proof Consider a matrix $\mathbf{Q}$ and its associated support graph G. Then, $\mathbf{Q}$ is a k-strip matrix if and only if G has a stable set of size at least $n - k$. Thus, a polynomial time algorithm for recognizing a permuted k-strip matrix can be used to test if a graph contains a stable set of size at least $n - k$. This proves the strong NP-hardness. Now, from the graph G, remove k vertices along with their incident edges and check if the resulting graph is a collection of isolated nodes. If the answer is yes, the matrix is k-strip and it will also identify the permutation (or the set S). There are $\binom{n}{k}$ ways to select k nodes and hence the process is polynomially bounded for fixed k and quasi-polynomial for $k = O(\log n)$. □

Non-negative Off-diagonal Elements Let us now discuss the special case of QUBO when the off-diagonal elements of $\mathbf{Q}$ are non-negative. There are no restrictions on the diagonal elements or the elements of $\mathbf{c}$. In this case, QUBO can be written as the $0 - 1$ linear program (which follows from the Glover-Woolsey linearization discussed in Chap. 1 and the references there in)

$$\begin{aligned} \text{GWO:} \quad & \text{Maximize} \quad \sum_{i=1}^{n} \sum_{j \in P_i, j \neq i} q_{ij} y_{ij} + \sum_{i=1}^{n} (c_i + q_{ii}) x_i \\ & \text{Subject to:} \quad y_{ij} - x_i \leq 0 \text{ for all } j \in P_i, i = 1, 2, \ldots, n \\ & \qquad\qquad\quad y_{ij} - x_j \leq 0 \text{ for all } j \in P_i, i = 1, 2, \ldots, n \\ & \qquad\qquad\quad x_i \in \{0, 1\} \text{ for all } i = 1, 2, \ldots, n, \end{aligned}$$

where $P_i = \{j : q_{ij} > 0\}$. The variable y_{ij} is unrestricted but has an implicit upper bound of 1. Further, in an optimal solution, y_{ij} cannot be negative.

Lemma 3.2 *The coefficient matrix of the LP relaxation of GWO is totally unimodular.*

Proof The coefficient matrix, say **C**, of the LP relaxation of GWO can be written as $\mathbf{C} = \begin{pmatrix} \mathbf{B} \\ \hline \mathbf{I} \end{pmatrix}$ where **B** corresponds to the part of the coefficient matrix for general constraints and the identity matrix **I** corresponds to the upper bound constraints (including the implicit bound on the y_{ij} variables). Thus, **C** is totally unimodular if **B** is totaly unimodular and **B** is totally unimodular if $\mathbf{B}^T$ is totally unimodular. But $\mathbf{B}^T$ has a 1 and a -1 in each column with the remaining entries being zero and hence it is the node-arc incidence matrix of some directed graph. This shows that $\mathbf{B}^T$ is totally unimodular and the result follows. □

Theorem 3.16 *When* **Q** *is a square matrix with non-negative off diagonal elements, then the QUBO* (**Q**, **c**) *can be solved in polynomial time.*

Proof The type of QUBO as discussed in the statement of the theorem can be solved by solving GWO. From Lemma 3.2, GWO can be solved in polynomial time by simply solving its linear programming relaxation. □

Theorem 3.16 was proved earlier by many authors [66, 84, 85]. Similarly, when the off-diagonal elements of the Ising QUBO (**A**, **b**) are non-negative the Ising QUBO can be solved in polynomial time. This follows from the fact that such an Ising QUBO can be reduced to a QUBO with a cost matrix having non-negative off-diagonal elements. Interestingly, when the off-diagonal elements of **Q** are non-positive, the complexity results from Sect. 3.2 shows that the resulting QUBO is NP-hard.

Let us now consider a more efficient algorithm to solve GWO, without using a general purpose linear programming solver. In particular, we show that GWO can be solved as a maximum flow problem (equivalently a minimum (s, t)-cut problem with non-negative edge weights).

For $i = 1, 2, \ldots, n$, let $r_i = \sum_{j=1}^{n} q_{ij}$. Consider the graph $G = (V, E)$ associated with the matrix **Q**, where $V = \{s, t, 1, 2, \ldots, n\}$ and $E = \{(i, j) : q_{ij} \neq 0\} \cup \{(s, j), (j, t) : j = 1, 2, \ldots, n\}$. Define the weights w_{ij} of the edges of G as follows:

$$w_{ij} = \begin{cases} \min\{0, -r_j - c_j\} & \text{if } i = s \\ \min\{0, r_i + c_i\} & \text{if } j = t \\ -q_{ij} & \text{otherwise.} \end{cases}$$

An (s, t)-cut in G is a partition (S, T) of the vertex set V such that $s \in S$ and $t \in T$. The capacity of the (s, t)-cut (S, T) is denoted by $\delta(S, T)$ and is given as the sum of the weights of the edges in the cut, i.e. $\delta(S, T) = \sum_{i \in S, j \in T} w_{ij}$. There is a one-to-one correspondence between (s,t)-cuts in G and $\mathbf{x} \in \{0,1\}^n$. More precisely, for any $\mathbf{x} \in \{0,1\}^n$ there is an (s,t)-cut (S,T) where $S = \{s\} \cup \{j : x_j = 1\}$ and $T = \{t\} \cup \{j : x_j = 0\}$. Likewise for any (s,t)-cut (S,T) in G, we get an $\mathbf{x} \in \{0,1\}^n$ such that $x_j = 1$ if $j \in S, j \neq s$ and $x_j = 0$ if $j \in T, j \neq t$.

Theorem 3.17 ([84, 85]) *Let G be the weighted graph associated with the symmetric matrix* $\mathbf{Q}$ *as defined above. Then, G contains an* (s,t)*-cut of value* $\delta(S,T)$ *if and only if there exists an* $\mathbf{x} \in \{0,1\}^n$ *such that* $\mathbf{x}^T\mathbf{Q}\mathbf{x} + \mathbf{c}^T\mathbf{x} + \sum_{j=1}^n w_{sj} = \delta(S,T)$.

Proof The value $\delta(S,T)$ of the (s,t)-cut is given by

$$
\begin{aligned}
\delta(S,T) &= \sum_{j=1}^n w_{sj}(1-x_j) + \sum_{i=1}^n\sum_{j=1}^n w_{ij}x_i(1-x_j) + \sum_{j=1}^n w_{jt}x_j \\
&= \sum_{j=1}^n w_{sj} - \sum_{j=1}^n w_{sj}x_j - \sum_{i=1}^n\sum_{j=1}^n q_{ij}x_i(1-x_j) + \sum_{j=1}^n w_{jt}x_j \\
&= \sum_{j=1}^n w_{sj} - \sum_{i=1}^n \min\{0, -r_i - c_i\}x_i + \mathbf{x}^T\mathbf{Q}\mathbf{x} - \sum_{i=1}^n r_i x_i + \sum_{i=1}^n \min\{0, r_i + c_i\}x_i \\
&= \sum_{j=1}^n w_{sj} + \mathbf{x}^T\mathbf{Q}\mathbf{x} + \mathbf{c}^T\mathbf{x}.
\end{aligned}
$$

The converse can be proved using the above equations, by working backwards. □

To clarify the construction discussed above, consider the instance $(\mathbf{Q}, \mathbf{c})$ of QUBO where

$$
\mathbf{Q} = \begin{pmatrix} 0 & 2 & 0 & 4 & 5 \\ 2 & 0 & 1 & 0 & 3 \\ 0 & 1 & 0 & 2 & 6 \\ 4 & 3 & 2 & 0 & 4 \\ 5 & 0 & 6 & 4 & 0 \end{pmatrix} \text{ and } \mathbf{c} = \begin{pmatrix} -8 \\ -8 \\ -12 \\ -7 \\ -15 \end{pmatrix}
$$

Now, the vector of the r_i values is given by $\mathbf{r}^T = (11, 6, 9, 13, 15)$. Then, the QUBO $(\mathbf{Q}, \mathbf{c})$ is equivalent solving the maximum $s - t$ cut problem on the graph G in Fig. 3.1.

Take any (s,t)-cut, say $S = \{s, 1, 2\}$. Then $\delta(S,T) = -21$. $\sum_{j=1}^5 w_{sj} = -9$. The incidence vector of S is $\mathbf{x} = (1, 1, 0, 0, 0)$. Then $\mathbf{x}^T\mathbf{Q}\mathbf{x} + \mathbf{c}^T\mathbf{x} + \sum_{j=1}^5 w_{sj} = -21 = \delta(S,T)$.

Thus, the QUBO can be solved by solving the maximum (s,t)-cut problem on G. But, when the off-diagonal elements of $\mathbf{Q}$ are non-negative, the weights w_{ij} of the edges of the graph G constructed above are non-positive. Thus, the maximum (s,t)-cut on G can be identified by solving the minimum (s,t)-cut problem on G

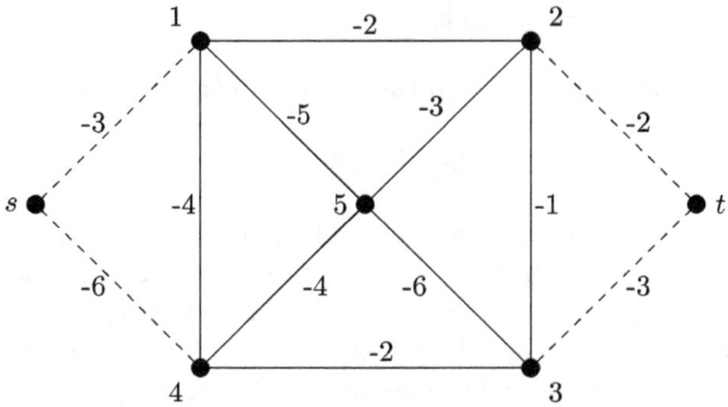

Fig. 3.1 The instance of Maximum (s, t)-cut equivalent to $(\mathbf{Q}, \mathbf{c})$

by taking the negative of the weights on the edges. Thus, the special case of QUBO where the off-diagonal elements of $\mathbf{Q}$ are non-negative can be solved as a minimum cut problem with non-negative edge weights, which is equivalent to the maximum flow problem in a directed graph. The corresponding Ising QUBO can also be solved as a minimum $s - t$-cut problem.,

In Theorem 3.16 we showed that the QUBO $(\mathbf{Q}, \mathbf{c})$ can be solved in polynomial time, if the off diagonal elements of $\mathbf{Q}$ are non-negative. If the off-diagonal elements are non-positive, QUBO $(\mathbf{Q}, \mathbf{c})$ is strongly NP-hard as proved in Sect. 3.2. However, if additional restrictions are imposed, we obtain a polynomially solvable special case of QUBO $(\mathbf{Q}, \mathbf{c})$ with non-positive off diagonal entries of $\mathbf{Q}$.

Theorem 3.18 *If the off diagonal elements of the matrix* $\mathbf{Q}$ *are non-positive and the support graph of* $\mathbf{Q}$ *is an undirected bipartite graph, then the QUBO* $(\mathbf{Q}, \mathbf{c})$ *can be solved in polynomial time.*

This result was obtained by Billionnet [12] and is proved in Chap. 10 in the context of BQUBO.

Pseudo k-Strip Matrices and Further Generalizations Let us now generalize Theorem 3.14 a little bit further. An $n \times n$ matrix $\mathbf{Q}$ is said to be a *pseudo k-strip matrix* if the off-diagonal elements of the matrix obtained from $\mathbf{Q}$ by deleting the first k rows and columns are non-negative. An instance $(\mathbf{Q}, \mathbf{c})$ of QUBO is called a *pseudo k-strip QUBO* if $\mathbf{Q}$ is a pseudo k-strip matrix.

Theorem 3.19 *The pseudo k-strip QUBO is solvable in polynomial time if* $k = O(\log n)$ *and NP-hard if* $k = O(\sqrt[t]{n})$ *for any fixed* $t > 0$.

Proof For a given $\mathbf{z} \in \{0, 1\}^k$, define the $(n-k)$-vector $\mathbf{d}^z = (d_1^z, d_2^z, \ldots, d_{n-k}^z)$ where $d_j^z = c_{k+j} + q_{k+j,k+j} + \sum_{i=1}^{k}(q_{i(k+j)} + q_{(k+j)i})z_i$, $j = 1, 2, \ldots, (n-k)$. Also, define the $(n-k) \times (n-k)$ matrix Q' such that $q'_{ij} = q_{(k+i)(k+j)}$ for $i, j = 1, 2, \ldots, (n-k), i \neq j$ and 0 if $i = j$. Let $\mathbf{y}^z$ an optimal solution to the reduced

QUBO $(\mathbf{Q}', \mathbf{d})$ of size $n - k$ given by:

$$QP'(\mathbf{z}): \quad \text{Maximize} \quad f(\mathbf{y}) = \mathbf{y}^T\mathbf{Q}'\mathbf{y} + \mathbf{d}^z\mathbf{y}$$
$$\text{Subject to:} \quad \mathbf{y} \in \{0, 1\}^{n-k},$$

Then, the best solution to the given QUBO $(\mathbf{Q}, \mathbf{c})$ subject to the condition that $x_i = z_i, i = 1, 2, \ldots, k$ is $\mathbf{x} = (\mathbf{z}, \mathbf{y}^z)$ with the corresponding objective function value $f(\mathbf{y}^z) + \sum_{i=1}^{k}\sum_{j=1}^{k} q_{ij}z_iz_j + \sum_{j=1}^{k} c_jz_j$. Since Q is a pseudo k-strip matrix, $QP'(\mathbf{z})$ is of the type GWO and hence can be solved in polynomial time by linear programming. Choose all possible values of $\mathbf{z} \in \{0, 1\}^k$, compute the corresponding best solution $\mathbf{x} = (\mathbf{z}, \mathbf{y}^z)$ and select the overall best solution so obtained, which is an optimal solution to the given QUBO $(\mathbf{Q}, \mathbf{c})$. When, $k = O(\log n)$ this method is polynomially bounded. The second part of the theorem follows from Theorem 3.14. □

The above theorem also extends to the Ising QUBO. As in the case of the k-strip matrix, we can define a *permuted pseudo k-strip matrix*. Let $S \subseteq \{1, 2, \ldots, n\}$ be a subset of the index set for rows (columns) of $\mathbf{Q}$ with $|S| = k$. Then, $\mathbf{Q}$ is a *permuted pseudo k-strip matrix* if the matrix obtained from $\mathbf{Q}$ by deleting rows and columns corresponding to S have off-diagonal elements non-negative. As established in Theorem 3.15 for k-strip matrices, recognizing a permuted pseudo k-strip matrix is strongly NP-hard in general but when k is fixed it is polynomially bounded. The QUBO and the Ising QUBO can be solved in polynomial time when the associated cost matrix is a permuted pseudo k-strip matrix when k is fixed.

Motivated by the examples of the k-strip matrix, the pseudo k-strip matrix and their permuted versions, we can define a more general class of matrices $\mathbf{Q}$ so that the corresponding QUBO $(\mathbf{Q}, \mathbf{c})$ can be solved in polynomial time. Let $\mathcal{P}$ be a prescribed matrix property. Then, the matrix $\mathbf{Q}$ is said to be *$(\mathcal{P}, k)$-reducible* if there exists $S \subset \{1, 2, \ldots, n\}$ such that $|S| = k$ and the matrix $\mathbf{Q}'$ obtained by deleting the rows and columns of $\mathbf{Q}$ corresponding to S satisfies $\mathcal{P}$. If $\mathcal{P}$ is such that the instance $(\mathbf{Q}', \mathbf{d})$ of the QUBO is solvable in polynomial time for any vector $\mathbf{d}$ then the QUBO $(\mathbf{Q}, \mathbf{c})$ can be solved in polynomial time whenever $k = O(\log n)$ and S is given. This observation provides additional enhancements to the polynomially (pseudo-polynomially) solvable cases of QUBO that we discuss hereafter. These results also extend to the case of the Ising QUBO.

3.3.2 Low Rank Cost Matrices

From Theorem 3.10, we know that the QUBO $(\mathbf{Q}, \mathbf{c})$ is NP-hard even if $\mathbf{Q}$ is symmetric and its rank is equal to one. If the rank of $\mathbf{Q}$ is low and $\mathbf{Q}$ satisfies some particular additional properties, we can get polynomially solvable special cases of the QUBO $(\mathbf{Q}, \mathbf{c})$. The matrix $\mathbf{Q}$ is of rank r if and only if there exist matrices $\mathbf{U}, \mathbf{V}$, each of size $r \times n$ such that $\mathbf{Q} = \mathbf{U}^T\mathbf{V}$. Equivalently, $\mathbf{Q}$ is of rank r if and only if $q_{ij} = \sum_{k=1}^{r} u_{ki} v_{kj}$ for all i and j. The matrices $\mathbf{U}$ and $\mathbf{V}$ can be identified from the reduced echelon form of $\mathbf{Q}$ or by using the singular value decomposition. There are many equivalent representations of QUBO which affect the rank of $\mathbf{Q}$. Through out the rest of this section, we assume that the rank factorization $\mathbf{Q} = \mathbf{U}^T\mathbf{V}$ is given.

There are a number of special cases of QUBO with a given rank factorization of $\mathbf{Q}$ investigated in the literature. In the following discussion we will distinguish the cases $\mathbf{c} = \mathbf{0}$ and $\mathbf{c} \neq \mathbf{0}$.

3.3.2.1 The Special Case c = 0

Here we consider $QUBO(\mathbf{Q}, \mathbf{0})$ where $\mathbf{Q}$ has rank r and a rank factorization is given. First notice that if the symmetric matrix $\mathbf{Q}$ has rank one, the problem is solvable in polynomial time. Indeed, for a symmetric matrix $\mathbf{Q} = (q_{ij})$ of rank one either (a) $q_{ij} = \alpha_i \alpha_j$ for all i, j or (b) $q_{ij} = -\alpha_i \alpha_j$ for all i, j hold, $i, j \in \{1, 2, \ldots, n\}$, $n \in \mathbb{N}$. In Case (a) we have $\mathbf{x}^T\mathbf{Q}\mathbf{x} = \sum_{i,j=1}^{n} \alpha_i \alpha_j x_i x_j = (\sum_{i=1}^{n} \alpha_i x_i)^2$ and $\max\{\mathbf{x}^T\mathbf{Q}\mathbf{x} \colon \mathbf{x} \in \{0, 1\}^n\}$ is equivalent to $\max\{|\sum_{i=1}^{n} \alpha_i x_i| \colon \mathbf{x} \in \{0, 1\}^n\}$. Notice that an optimal solution of the later problem can be found among $\{\underline{\mathbf{x}}, \bar{\mathbf{x}}\}$, where $\underline{\mathbf{x}} = (\underline{x}_i)$ and $\bar{\mathbf{x}} = (\bar{x}_i)$ are given as

$$\underline{x}_i = \begin{cases} 1 \text{ if } \alpha_i < 0 \\ 0 \text{ otherwise} \end{cases} \qquad \bar{x}_i = \begin{cases} 1 \text{ if } \alpha_i > 0 \\ 0 \text{ otherwise} \end{cases}$$

This is due to the inequalities below which trivially hold for any $\mathbf{x} \in \{0, 1\}^n$

$$\sum_{i=1}^{n} \alpha_i \underline{x}_i = \sum_{i \in I^-} \alpha_i \le \sum_{i=1}^{n} \alpha_i x_i = \sum_{i \in I^+} \alpha_i x_i + \sum_{i \in I^-} \alpha_i x_i \le \sum_{i \in I^+} \alpha_i = \sum_{i=1}^{n} \alpha_i \bar{x}_i$$

where $I^+ := \{i \in \{1, 2, \ldots, n\} \colon \alpha_i > 0\}$ and $I^- := \{i \in \{1, 2, \ldots, n\} \colon \alpha_i < 0\}$. The inequalities above imply that

$$\left| \sum_{i=1}^{n} \alpha_i x_i \right| \le \max \left\{ \left| \sum_{i=1}^{n} \alpha_i \underline{x}_i \right|, \left| \sum_{i=1}^{n} \alpha_i \bar{x}_i \right| \right\} .$$

Thus, $\underline{\mathbf{x}} = (\underline{x}_i)$ or $\bar{\mathbf{x}} = (\bar{x}_i)$ is the optimal solution of $QUBO(\mathbf{Q}, \mathbf{0})$ in this case.

Case (b) is trivial because $\mathbf{x}^T\mathbf{Q}\mathbf{x} = -\sum_{i,j=1}^{n} \alpha_i \alpha_j x_i x_j = -(\sum_{i=1}^{n} \alpha_i x_i)^2$ in this case and $\max\{\mathbf{x}^T\mathbf{Q}\mathbf{x} \colon \mathbf{x} \in \{0, 1\}^n\} = 0$ and is obtained for $\mathbf{x} = \mathbf{0}$. If the rank is larger than 1 the complexity of the problem depends on other factors. For example the (positive) definiteness of $\mathbf{Q}$. Hammer et al. [54] showed that $QUBO(\mathbf{Q}, \mathbf{0})$ remains NP-hard even if $\mathbf{Q}$ is indefinite and is of rank 2. Hammer et al. [54] gave an $O(n \log n)$ algorithm to solve the continuous relaxation of this problem. Moreover, they characterized the cases where an optimal solution of the continuous relaxation is binary and thus constitutes an optimal solution of the QUBO.

Allemand et al. [3] considered the case where $\mathbf{Q}$ is positive semidefinite and has a constant rank r. They proposed an $O(n^{r-1})$ algorithm to solve the problem when rank of $\mathbf{Q}$ is $r \geq 3$ and an $O(n^2)$ algorithm when $r \leq 2$. Thus, in this case the QUBO $(\mathbf{Q}, \mathbf{0})$ can be solved in polynomial time when the rank of $\mathbf{Q}$ is fixed. The algorithm of Allemand et al. [3] involves the enumeration of the extreme points of a special polytope called *zonotope* which will be defined later. Ferrez et al. [38] proposed an improved method for enumerating the extreme points of the zonotope which is also relevant for the practical implementation of the method of Allemand et al. [3].

Let us take a closer look at the approach of Allemand et al. [3]. We assume that $\mathbf{Q} = \mathbf{V}^T\mathbf{V}$, where $\mathbf{V}$ is a given $r \times n$ matrix such that its rows $\mathbf{v}^{(i)T}$ are the appropriately scaled eigenvectors of $\mathbf{Q}$. Then the QUBO $(\mathbf{Q}, \mathbf{0})$ can equivalently be formulated as

$$\max f(\mathbf{x}) = \mathbf{x}^T\mathbf{V}^T\mathbf{V}\mathbf{x} = \sum_{i=1}^{r} \langle \mathbf{v}^{(i)}, \mathbf{x}\rangle^2 \tag{3.8}$$
$$\text{subject to} \quad \mathbf{x} \in \{0, 1\}^n ,$$

where $\langle \mathbf{a}, \mathbf{b}\rangle$ denotes the scalar product of the vectors $\mathbf{a}$ and $\mathbf{b}$. Consider the linear mapping $\mathbf{z} = \mathbf{V}\mathbf{x}$ of $\mathbb{R}^n$ to $\mathbb{R}^r$. Let Z be the image of the hypercube $[0, 1]^n$ under this mapping. Clearly, Z is a convex polytope which is generally called a *zonotope*. Moreover, for every extreme point $\bar{\mathbf{z}}$ of Z there exist an extreme point $\bar{\mathbf{x}}$ of $[0, 1]^n$, i.e. an $\bar{\mathbf{x}} \in \{0, 1\}^n$, such that $\bar{\mathbf{z}} = \mathbf{V}\bar{\mathbf{x}}$. For the optimal objective function value f^* of the problem (3.8), we get

$$f^* = \max\left\{\sum_{i=1}^{r} \langle \mathbf{v}^{(i)}, \mathbf{x}\rangle^2 : \mathbf{x} \in \{0, 1\}^n\right\} = \max\left\{\sum_{i=1}^{r} \langle \mathbf{v}^{(i)}, \mathbf{x}\rangle^2 : \mathbf{x} \in [0, 1]^n\right\}$$
$$= \max\left\{\sum_{i=1}^{r} z_i^2 : \mathbf{z} \in Z\right\}, \tag{3.9}$$

where the second equality is due to the convexity of the function $\sum_{i=1}^{r} \langle \mathbf{v}_i, \mathbf{x}\rangle^2$ which holds because $\mathbf{Q}$ is assumed to be positive semidefinite. Clearly, the maximum of the convex function $\sum_{i=1}^{r} z_i^2$ over the zonotope Z is attained at some extreme point of Z. Thus, this special case of the QUBO $(\mathbf{Q}, \mathbf{0})$ can be formulated as an enumeration problem over the extreme points of Z. Since Z is a polytope in $\mathbb{R}^r$ it has $O(n^{r-1})$ extreme points. Moreover, for every $\mathbf{z} \in Z$ we have $\mathbf{z} = \mathbf{V}\mathbf{x} = \sum_{i=1}^{n} x_i \mathbf{v}^{(i)}$ for some $\mathbf{x} \in [0, 1]^n$, where $\mathbf{v}^{(i)}$, $i \in \{1, 2, \ldots, n\}$, are the columns of $\mathbf{V}$. Thus Z is the Minkowski sum of the closed line segments $[\mathbf{0}, \mathbf{v}^{(i)}]$ for $i \in \{1, 2, \ldots, n\}$. From the duality theory in discrete geometry it is known that the set S of the extreme points of Z together with the corresponding points $\mathbf{x} \in \{0, 1\}^n$

with $\mathbf{z} = \mathbf{V}\mathbf{x}$ for every $\mathbf{z} \in S$ can be computed in $O(n^{r-1})$ time for $r \geq 3$ and in $O(n^2)$ time for $r = 2$, see e.g. [35].

3.3.2.2 The General Case: $\mathbf{c} \in \mathbb{R}^n$

Now we drop the assumption $\mathbf{c} = 0$ and consider QUBO $(\mathbf{Q}, \mathbf{c})$, where $\mathbf{Q}$ has rank r and a rank factorization of $\mathbf{Q}$ is given.

If $\mathbf{Q}$ is a non-negative matrix, we have already shown that QUBO $(\mathbf{Q}, \mathbf{c})$ can be solved in polynomial time, even if no assumptions are made on the rank of $\mathbf{Q}$, see Theorems 3.17. One way to partially relax the non-negativity assumption on the elements of $\mathbf{Q}$ is identified in Theorem 3.19. Another line of reasoning to achieve computational advantage is by restricting the rank of $\mathbf{Q}$. If we assume a fixed rank of $\mathbf{Q}$, the requirements on the non-negativity of the entries of $\mathbf{Q}$ can be weakened in a number of ways and lead to several polynomially solvable special cases of the low rank QUBO as described below.

Q Has Non-negative Diagonal Entries The second equality in (3.9) shows that QUBO and its continuous version are equivalent, if $\mathbf{Q}$ is positive semidefinite. Thus, the algorithm of Allemand et al. [3] also solves the continuous version of QUBO in polynomial time, when the rank of $\mathbf{Q}$ is fixed. By extending and generalizing the work of Allemand et al. [3], Hladík et al. [56] considered the continuous quadratic optimization problem with bounded variables defined as follows

$$\begin{aligned} \text{BQP: Maximize } & \mathbf{x}^T\mathbf{Q}\mathbf{x} + \mathbf{c}^T\mathbf{x} \\ \text{subject to: } & x_i \in [\ell_i, u_i], i = 1, 2, \ldots, n, \end{aligned}$$

where ℓ_i and u_i are rational numbers for $i = 1, 2, \ldots, n$. The bound constraints allow the flexibility to fix the value of a variable disclosing the possibility of augmenting $\mathbf{Q}$ using $\mathbf{c}$ as described below. Consider the $(n + 1) \times (n + 1)$ matrix

$$\mathbf{Q}' = \left[\begin{array}{c|c} \mathbf{Q} & \frac{1}{2}\mathbf{c} \\ \hline \frac{1}{2}\mathbf{c}^T & 0 \end{array}\right] \tag{3.10}$$

Define $\ell_{n+1} = u_{n+1} = 1$. Note that if the rank of $\mathbf{Q}$ is r, then rank of $\mathbf{Q}' \leq r + 2$. Now, BQP can be written as

$$\begin{aligned} \text{BQP1: Maximize } & \mathbf{x}^T\mathbf{Q}'\mathbf{x} \\ \text{subject to: } & x_i \in [\ell_i, u_i], i = 1, 2, \ldots, n, n + 1. \end{aligned}$$

It is easy to verify that BQP and BQP1 are equivalent. To solve BQP, Hladík et al. [56] essentially solved BQP1 and obtained the following result.

Theorem 3.20 ([56]) *BQP can be solved in* $O(n^{2r+3}\psi(n, L))$ *time, where* r *is the rank of* $\mathbf{Q}$ *and* $\psi(n, L)$ *is the complexity of solving a linear program with* n *variables,* n *constraints and input size* L.

Theorem 3.20 has interesting consequences on the polynomial solvability of QUBO.

Theorem 3.21 *The QUBO* $(\mathbf{Q}, \mathbf{c})$ *can be solved in polynomial time if the diagonal elements of* $\mathbf{Q}$ *are non-negative and the rank of* $\mathbf{Q}$ *is fixed.*

Proof Choose $\ell_i = 0$ and $u_i = 1$ for $i = 1, 2, \ldots, n$. Since the diagonal elements of $\mathbf{Q}$ are non-negative, and by the choice of ℓ_i and u_i, there exists an optimal solution to BQP which is at an extreme point of the hypercube $[0, 1]^n$ (see Chap. 8 and the references there in). Now, solve the continuous optimization problem BQP. If the resulting solution is binary, we have an optimal solution to the QUBO $(\mathbf{Q}, \mathbf{c})$. If there are fractional components, apply a simple polynomial time rounding algorithm to convert this to a 0-1 solution without worsening the objective function value. (For example apply the [0, 1]-rounding algorithm discussed in Chap. 8.) Since the rank of $\mathbf{Q}$ is fixed, the result follows from Theorem 3.20. □

Note that in Theorem 3.21 we replaced the assumption on positive semidefiniteness of $\mathbf{Q}$ by a weaker one, namely the non-negativity of the diagonal entries of $\mathbf{Q}$. These results can be further extended to some special classes of full rank matrices.

A matrix $\mathbf{Q}$ is said to be of *pseudo rank* r if there exists a diagonal matrix $\mathbf{D}$ such that $\mathbf{Q} + \mathbf{D}$ is of rank r. Matrices with pseudo rank r are precisely those matrices $\mathbf{Q}$ where $q_{ij} = \sum_{k=1}^{r} u_{ki} v_{kj}$ hold for all $i, j \in \{1, 2, \ldots, n\}$ with $i \neq j$. Note that a matrix with pseudo rank r could have rank up to n. Our next corollary shows that a special case of $QUBO(\mathbf{Q}, \mathbf{c})$ with $\mathbf{Q}$ having pseudo rank r can be solved in polynomial time when r is fixed.

Corollary 3.4 *If* $\mathbf{D}$ *is a diagonal matrix such that* $\mathbf{Q} + \mathbf{D}$ *is of rank* r *with non-negative diagonal entries, then the QUBO* $(\mathbf{Q}, \mathbf{c})$ *can be solved in polynomial time when* r *is fixed.*

Proof The QUBO $(\mathbf{Q}, \mathbf{c})$ is equivalent to the QUBO $(\mathbf{Q}+\mathbf{D}, \mathbf{c} - diagv(\mathbf{D}))$, where $diagv(\mathbf{D})$ is the n-dimensional vector formed by the diagonal entries of $\mathbf{D}$ such that the ith entry of $diagv(\mathbf{D})$ is d_{ii}. The result follows by applying Theorem 3.21 to the QUBO $(\mathbf{Q} + \mathbf{D}, \mathbf{c} - diagv(\mathbf{D}))$. □

As an example of a matrix of the type discussed in Corollary 3.4, consider a matrix $\mathbf{Q}'$ of rank r such that $q'_{ii} \geq 0$ for all i. Let $\mathbf{Q}$ be the matrix obtained by replacing the diagonal elements of $\mathbf{Q}'$ with values $\alpha_i, i = 1, 2, \ldots, n$, so that rank of $\mathbf{Q}$ is n. Now consider the QUBO $(\mathbf{Q}, \mathbf{c})$. Since the rank of $\mathbf{Q}$ is n, Theorem 3.21 does not help to solve the QUBO $(\mathbf{Q}, \mathbf{c})$ in polynomial time. However, for fixed r Corollary 3.4 allows to solve the QUBO $(\mathbf{Q}, \mathbf{c})$ in polynomial time by using the diagonal matrix $\mathbf{D}$ with $d_{ii} = q'_{ii} - \alpha_i$, for $i \in \{1, 2, \ldots, n\}$.

The polynomially solvable cases discussed here for QUBO $(\mathbf{Q}, \mathbf{c})$ extend to the case of the Ising QUBO since the transformation between the QUBO and the

Ising QUBO preserves the rank of the cost matrix. For the Ising QUBO $(\mathbf{A}, \mathbf{0})$ (in the minimization form) Ben-Ameur and Neto [11] proposed a polynomial time algorithm when the rank of **A** is fixed and the number of positive entries on the diagonal of **A** is $O(\log n)$. This offers yet another special case of the low rank QUBO solvable in polynomial time.

Non-negativity Constraints Imposed to the Entries of Q Some other polynomially solvable special cases of the QUBO $(\mathbf{Q}, \mathbf{c})$ with a given rank factorization of **Q** and of a fixed rank r arise if more complex non-negativity restrictions are imposed to the entries of **Q**, see Çela et al. [26]. The non-negativity restrictions are expressed by means of graph models or more generally, hypergraph models. In particular, let $\mathcal{H}$ be a family of hypergraphs H with vertex set $V(H) = \{1, 2, \ldots, n\}$, where n is the size of **Q**, and the edge set $E(H)$ fulfilling the following three properties:

(i) The cardinality of all edges $F \in E(H)$ is bounded by a constant (independent on n).
(ii) The cardinality of the largest stable set in H is $O(logn)$.
(iii) The number of maximal stable sets in H is polynomial in n.

Then, the following theorem holds:

Theorem 3.22 (Çela et al. [26]) *The QUBO* $(\mathbf{Q}, \mathbf{c})$, *where* **Q** *is of rank* r *with a given rank factorization, is solvable in polynomial time if there is a hypergraph* $H \in \mathcal{H}$ *such that* $\sum_{i,j \in F} q_{ij} > 0$ *for all* $F \in E(H)$.

Consider now a family $\mathcal{G}$ of (loop-free) graphs G with vertex set $V(G) = \{1, 2, \ldots, n\}$, where n is the size of **Q**, and edge set $E(G)$ fulfilling the following property:

(iv) The number of stable sets in G is polynomial in n.

The following theorem holds:

Theorem 3.23 (Çela et al. [26]) *The* $QUBO(\mathbf{Q}, \mathbf{c})$, *where* **Q** *is of rank* r *with a given rank factorization, is solvable in polynomial time if there is a (loop-free) graph* $G \in \mathcal{G}$ *such that* $q_{ii} + q_{jj} - 2|q_{ij}| > 0$ *for all* $\{i, j\} \in E(G)$.

Notice that the condition (iv) is equivalent to the following condition

(iv') The cliques in the complement G^c of G can be enumerated in polynomial time.

Several graphs classes fulfilling (iv') have been investigated in the literature. A prominent example is the class of graphs with bounded treewidth [90] (including series-parallel graphs, and also trees as special series-parallel graphs). Also the planar graphs and more generally, the graphs with bounded thickness, have a polynomial number of cliques. *The thickness of a graph G* is defined as the minimum number of planar graphs whose union yields G. It is known that the size of a maximum clique in a graph with thickness t is not more than $6t - 2$, see [10]. By considering the complementary classes of the classes mentioned above we get polynomially solvable special cases of the low-rank QUBO in the sense of Theorem 3.23.

We are not aware of non-trivial classes of (proper) hypergraphs fulfilling the properties (i)–(iii) above. (Hypergraphs which are not graphs, i.e. which have at least one edge of cardinality more than 2, are called proper hypergraphs.) A class of hypergraphs which trivially fulfill properties (i)–(iii) are the complete p-hypergraphs, i.e. the hypergraphs containing all possible edges of size p for some constant $p \in \mathbb{N}$.

Notice that the proofs of the above theorems are based on a general algorithm which solves the following equivalent representation of QUBO $(\mathbf{Q}, \mathbf{c})$

$$\min_{x \in \{0,1\}^n} f(x) = \langle \mathbf{c}'\mathbf{x} \rangle + \sum_{j=1}^{d} \lambda_j (\beta_j + \langle \mathbf{u}^{(j)}, \mathbf{x} \rangle)^2 . \tag{3.11}$$

The time complexity analysis reveals that the algorithm runs in polynomial time in the special case when the conditions of Theorems 3.22 and 3.23 are fulfilled. Moreover, notice that the representation in (3.11) is not unique and that different values for $\mathbf{c}'$ and d influence the time complexity of the above mentioned algorithm. For example, such a representation is obtained by setting $d := r$, $\mathbf{c}' := \mathbf{c}$, $\beta_j := 0$ for all $j \in \{1, 2, \ldots, r\}$, λ_j to be the eigenvalues and $\mathbf{u}^{(j)}$ to be the eigenvectors of $\mathbf{Q}$, for $j \in \{1, 2, \ldots, r\}$. In this setting irrational values in the quantities λ_j and $\mathbf{u}^{(j)}$ may arise even if $\mathbf{Q}$ and $\mathbf{c}$ are rational. An alternative approach to avoid irrational entries would be to compute the LDU-decomposition of $\mathbf{Q}$ resulting in $\mathbf{Q} = \mathbf{LDL}^T$, where $\mathbf{L}$ is a lower triangular matrix and $\mathbf{D}$ is a diagonal matrix of rank r.

Later in Sect. 3.5 we discuss pseudo-polynomial algorithms and related results for QUBO when rank of $\mathbf{Q}$ is fixed. Many of the complexity results discussed in this section and the section to follow involve terms of the type n^r as a factor along with possibly other polynomial factors. When r is fixed, such algorithms are polynomially bounded. When $r = O(\log n)$, $O(\log \log n)$ etc., we still get non-exponential growth with slow growing complexity expressions. In such cases, our algorithms are not exactly polynomial time but they are quasi-polynomial time.

3.4 Polynomially Solvable Graph Structures

Let us now discuss some polynomially solvable special cases of QUBO by exploiting the properties of the support graph of $\mathbf{Q}$. The support graph offers significant information of the relative difficulty of the QUBO and the Ising QUBO. In Sect. 3.2 we have seen that even when the support graph is very sparse, the complexity of QUBO remains unaltered. At the other extreme, if the support graph is diagonal, QUBO can be solve by a greedy algorithm since in this case the problem reduces to the optimization of an unconstrained linear function in binary variables. Let us now explore the linkages between the polynomial solvability of the QUBO and the

Ising QUBO on one side and the solvability of some fundamental graph theoretic optimization problems on the other.

3.4.1 *QUBO and the Maximum Cut Problem*

Theorem 3.5 implies that the Ising QUBO $(\mathbf{A}, \mathbf{0})$ can be solved in polynomial time if and only if the maximum weight cut problem on the support graph of $\mathbf{A}$ can be solved in polynomial time. There are many classes of graphs on which the maximum weight cut problem can be solved in polynomial time. For example,

Theorem 3.24 *The maximum weight cut problem is polynomially solvable on the following classes of graphs.*
1. *Planar graphs [67, 87]*
2. *Graphs of bounded bandwidth [65]*
3. *Graphs of bounded treewidth [16, 32]*
4. *Graphs of bounded genus [41].*

Thus, as an immediate consequence of Theorems 3.24 and 3.5 we have

Corollary 3.5 *If the support graph G of $\mathbf{A}$ is planer then the Ising QUBO $(\mathbf{A}, \mathbf{0})$ can be solved in polynomial time. Further, if the bandwidth, the treewidth, or the genus of G bounded by some fixed k then the Ising QUBO $(\mathbf{A}, \mathbf{0})$ can be solved in polynomial time.*

The special case of QUBO when the support graph has bandwidth k was studied by various authors when $k = 3, 5, 7$ and for a fixed value of k [51, 66, 76]. Note that the Ising QUBO $(\mathbf{A}, \mathbf{b})$ is equivalent to the Ising QUBO $(\mathbf{A}', \mathbf{0})$, where

$$\mathbf{A}' = \left[\begin{array}{c|c} A' & \frac{1}{2}\mathbf{b} \\ \hline \frac{1}{2}\mathbf{b}^T & 0 \end{array}\right] \tag{3.12}$$

Thus, without loss of generality, we can assume that the Ising QUBO is given in the form $(\mathbf{A}, \mathbf{0})$, where $\mathbf{0}$ is the zero vector in $\mathbb{R}^n$. Notice that the transformation from $(\mathbf{A}, \mathbf{b})$ to $(\mathbf{A}', \mathbf{0})$ does impact the structure of the support graph of $\mathbf{A}'$ and the latter might not inherit all the "nice" properties of the support graph of $\mathbf{A}$. However, if additional conditions are imposed on the elements of $\mathbf{A}$ and $\mathbf{b}$ or further restrictions on the support graph of $\mathbf{A}$ are imposed, the "nice" structure of the support graph of $\mathbf{A}$ can be guaranteed for the support graph of $\mathbf{A}'$. For example, if the support graph of $\mathbf{A}$ is outerplanar, then the support graph of $\mathbf{A}'$ will be planar. This is because, for an outerplanar graph, there exists a planar embedding where all vertices are on the outer face. Hence, if the support graph of $\mathbf{A}$ is outerplanar, the Ising QUBO $(\mathbf{A}, \mathbf{b})$ (and consequently also the corresponding QUBO) can be solved in polynomial time. Polynomially solvable special cases of the QUBO $(\mathbf{Q}, \mathbf{c})$ can be obtained by applying the transformation between the QUBO and the Ising QUBO discussed earlier in this chapter or from the fact that the QUBO $(\mathbf{Q}, \mathbf{c})$ of size n can be formulated as a maximum weight cut problem on a graph on $n + 1$ nodes (see Chap. 1). Thus, if the maximum weight cut problem on that graph is solvable in polynomial time, the QUBO $(\mathbf{Q}, \mathbf{c})$ can be solved polynomial time.

The first polynomial time algorithm for the weighted maximum cut problem on planar graphs was obtained by Hadlock [53] in the case of non-negative weights. (See Orlova and Dorfman [75] for a branch and bound type algorithm on that problem.) Thus, for our purpose of solving QUBO in polynomial time, these algorithms are not of much use. Shih et al. [87] and Liers and Pardella [67] proposed $O(n^{3/2} \log n)$ algorithms to solve the weighted maximum cut problem on planar graphs, where arbitrary weights are allowed. Both of these algorithms can be adapted to solve the Ising QUBO $(\mathbf{A}, \mathbf{0})$ when the support graph of $\mathbf{A}$ is planar.

Let us now discuss very briefly the algorithm of Liers and Pardella [67] to solve the maximum cut problem on a planar graph. Let $G = (V, E)$ be a planar graph and we assume that a planar embedding of G is also given. Moreover the edge weights w_{ij} for each edge $\{i, j\} \in E$ are given and no assumptions on the sign of w_{ij} are made. Let G_D be the dual of G with respect to the given planar embedding. The weight of an edge e in G_D is the weight of the edge $\{i, j\}$ in G that e crosses. First a pre-processing step is performed so as to 'balance' the graph G_D. To this end each node v of G_D having degree more than 4 is split into $\lfloor (d(v) - 1)/2 \rfloor$ nodes connected by new edges of weight zero to form a path P_v. Here, $d(v)$ is the degree of the dual node v. Now allocate the $d(v)$ edges incident on v to unique nodes in P_v so that the degree of each node is at most 4. Let $G_T = (V_T, E_T)$ be the transformed dual graph. It can be verified that the degree of any node in G_T is either 3 or 4 (exceptions can be handled easily).

Now, replace each node v of G_T by a K_4, the complete graph on 4 nodes $\{v_1, v_2, v_3, v_4\}$. Note that degree of node v in G_T is 3 or 4. Suppose degree of v is 4 and let $\{i_1, v\}, \{i_2, v\}, \{i_3, v\}, \{i_4, v\}$ be the edges in G_T incident on v. Connect i_j to v_j by an edge of weight the same as that of $\{i_j, v\}$ for $j = 1, 2, 3, 4$. Without loss of generality assume that v_1, v_2, v_3, v_4 are labelled such that this operation leaves a planar embedding. If the degree of v is three, we connect the three nodes adjacent to v in G_T to three nodes of the K_4 in an analogous way. Nodes with self loops lead to multi-arcs in K_4. (See [67] for further details and exceptions.) The weights of the edges of the complete graphs K_4 introduced as described above are set to zero. The resulting expanded planar graph is denoted by G_E.

Now, find a maximum weight perfect matching M in G_E. Then shrink back all complete graphs K_4 that were created along with all split nodes, if any such nodes were introduced, while keeping track of the matched edges. Consider the subgraph induced by the matching edges, after the shrinking operation. This graph is a maximum weight Eulerian graph in the dual which yields an optimal cut in the primal.

Theorem 3.25 ([67]) *The algorithm discussed above correctly computes a maximum weight cut in a planar graph G in $O(n^{3/2} \log n)$ time, where n is the number of edges in G.*

A graph G is called 1-planar if it can be drawn in the plane in such a way that at most one crossing occurs per edge [33]. Dahn et al. [33] used the algorithm of Liers and Pardella [67] to solve the maximum weight cut problem on a 1-planar graph on n nodes and k edge crossings in $O(3^k(n^{3/2}\log n)$ time. Thus, when $k = O(\log n)$ this algorithm is polynomially bounded. Chimani et al. [30] and Kobayashi et al. [60] generalized this further to graphs that can be drawn in the plane with at most k crossings, relaxing the assumption of 1-planarity, and proposed an $O(2^k p(n+k))$ algorithm where p is the polynomial running time for solving the maximum weight cut problem on a planar graph. These algorithms for the maximum weight cut problem on these special graphs immediately lead to polynomial time algorithms for QUBO $(\mathbf{Q}, \mathbf{c})$ or the Ising QUBO $(\mathbf{A}, \mathbf{b})$ if the graphs of the associated equivalent maximum cut problems have the structure discussed above.

The polynomial time algorithm for solving the maximum weight cut problem in planar graphs can be used to solve the maximum cut problem on graphs not contractible to K_5 [7]. Note that this class of graphs include planar graphs and hence is yet another generalization of planar graphs. Barahona [7] presented a description of the cut polytope along with polynomial time separation algorithms on graphs not contractible to K_5. He also proposed a polynomial time combinatorial algorithm to solve the maximum weight cut problem on such graphs. This leads to yet another polynomially solvable special case of the QUBO and the Ising QUBO.

Polynomially solvable special cases of the maximum weight cut problem on special graphs with additional restrictions on edge weights are also studied in literature. Some results of this kind are summarized in the following theorem.

Theorem 3.26 *The maximum weight cut problem is polynomially solvable on the following classes of graphs.*

1. *Weakly bipartite with positive edge weights [48]*
2. *Graphs having no long odd cycles with positive edge weights [47]*
3. *Line graphs [52]*
4. *Cographs [16].*

The graph structures identified in Theorems 3.26 also lead to polynomial solvable special cases of the Ising QUBO $(\mathbf{A}, \mathbf{b})$ and QUBO $(\mathbf{Q}, \mathbf{c})$ with corresponding assumptions on the related support graphs and on the elements of $\mathbf{A}$ and $\mathbf{b}$ or on the elements of $\mathbf{Q}$ and $\mathbf{c}$.

Another family of polynomially solvable special cases of the maximum weight cut problem (MWCP) was investigated by McCormick et al. [71] who classify easy and hard instances of the problem based solely on the sign of the edge weights while making assumptions neither on the magnitude of the edge weights nor on the structure of the graph. More precisely, for an instance (G, w) of the MWCP, denote by E^- (E^+) the set of edges with negative (positive) weights in G, i.e. $E^- = \{\{i,j\} \in E \colon w_{ij} < 0\}$ and $E^+ = \{\{i,j\} \in E \colon w_{ij} > 0\}$. The solution of the MWCP is trivial if $E^+ = \emptyset$, since the empty cut $\delta(\emptyset) = \delta(V)$ is optimal in this case. In this case, even finding a non-trivial cut of maximum weight, i.e. a maximum weight cut $\delta(S)$ for $\emptyset \neq S \subsetneq V$, is polynomially solvable e.g. by the Gomory-Hu algorithm [46]. McCormick et al. [71] identified a borderline between polynomially solvable and NP-hard special cases of the MWCP and investigated to what extent the assumption $E^+ = \emptyset$ can be relaxed while the MWCP remains polynomially solvable. They consider the subgraph $G_w^+ = (V, E^+)$ of G, where V is the set of vertices in G and E^+ is as defined above. Then the pair (G, w) is said to be nearly negative if the cover number $c(G_w^+)$ of G_w^+ fulfills $c(G_w^+) = O(\log |V|^k)$ for some fixed integer $k > 0$. Also the matching number $m(G_w^+)$ of the subgraph G_w^+, i.e. the largest cardinality of a matching in G_w^+, turns out to be an important parameter when dealing with the complexity of the MWCP, as expressed in the following theorem.

Theorem 3.27 (McCormick et al. [71]) *For any fixed integer $k > 0$, the maximum weight cut problem with input (G, w) is solvable in polynomial time if one of the following conditions is fulfilled:*

(a) the cover number of G_w^+ fulfills $c(G_w^+) = O(\log |V|^k)$, where V is the set of vertices of G,
(b) the matching number of G_w^+ fulfills $m(G_w^+) = O(\log |V|^k)$, where V is the set of vertices of G.

Further, MWCP is strongly NP-hard if one of the following conditions is fulfilled:

(c) $m(G_w^+) = \Omega(|V|^{1/k})$, for some fixed integer $k > 0$,
(d) V is partitioned into two subsets U and W such that all edges with both endpoints in U or in W have non-positive weights, while all other edges have arbitrary weights.

Notice that the statement (d) above implies that MWCP is NP-hard on bipartite graphs with arbitrary edge weights, complementing the first statement of Theorem 3.26.

As a straightforward consequence of Theorem 3.27 we obtain the following corollary.

Corollary 3.6 *Let $\mathbf{A}$ be a symmetric $n \times n$ matrix and let G be the support graph of A. Moreover, let $G_w^+ = (\{1, 2, \ldots, n\}, E^+)$, where $E^+ := \{\{i, j\}: a_{ij} > 0\}$. For any fixed integer $k > 0$ the Ising $QUBO(\mathbf{A}, \mathbf{0})$ is polynomially solvable if $c(G_w^+) = O(\log n^k)$ or $m(G_w^+) = O(\log n^k)$, where $c(G_w^+)$ and $m(G_w^+)$ are the cover number and the matching number of G_w^+, respectively.*

The Ising $QUBO(\mathbf{A}, \mathbf{0})$ is strongly NP-hard if (i) $m(G_w^+) = \Omega(n^{1/k})$, for some fixed integer $k > 0$, or (ii) if there is a partition $U \dot\cup W = \{1, 2, \ldots, n\}$ such that $a_{ij} \le 0$ holds for all $i \ne j$ with $i, j \in U$ or $i, j \in W$.

In particular the above corollary implies that for any fixed positive integer k the Ising QUBO $(\mathbf{A}, \mathbf{0})$ is polynomially solvable if there exists a subset $I \subset \{1, 2, \ldots, n\}$ with $|I| = O(\log n^k)$, such that $a_{ij} < 0$ holds for all pairs (i, j) with $i \notin I$ and $j \notin I$. In this case $c(G_w^+) = O(\log n^k)$ would hold.

Some other polynomially solvable special cases of the maximum weight cut problem (and hence QUBO and the Ising QUBO) include series parallel graphs [8], gauge graphs [4], and graphs determined by combinatorial circuit of AND, OR,

NAND, NOR, and NOT logical gates [29]. Bilu et al. [13], using the notion of stability of instances, identified polynomially solvable instances of the maximum weight cut problem which are also relevant to QUBO.

3.4.2 Stable Sets and Cliques

Let us now explore polynomially solvable special cases of the QUBO derived from polynomially solvable special cases of the maximum weight stable set problem (MWSSP). Although MWSSP is strongly NP-hard, many special cases of it are solvable in polynomial time. The next theorem indicates some examples of graphs where the MWSSP is solvable in polynomial time.

Theorem 3.28 *The maximum weight stable set problem is solvable in polynomial time for the following classes of graphs.*

1. *Perfect graphs [49]*
2. *t-perfect graphs [50]*
3. *Claw-free graphs [37, 74]*
4. *Outerplanar graphs [15]*
5. *Graphs of bounded treewidth [17]*
6. *Planar graphs of bounded diameter [36]*

It is well known that any QUBO $(\mathbf{Q}, \mathbf{c})$ can be formulated as a maximum weight stable set problem (MWSSP). (See Chap. 1 and the references given there.) Thus, if the MWSSP is solvable on the corresponding graph generated from the equivalent QUBO $(\mathbf{Q}, \mathbf{c})$, then the QUBO $(\mathbf{Q}, \mathbf{c})$ can be solved in polynomial time. However the structure of this associated graph is significantly different from the support graph of $\mathbf{Q}$. Let us now derive conditions on the elements of $\mathbf{Q}$ so that the QUBO $(\mathbf{Q}, \mathbf{c})$ can be solved in polynomial time when the MWSSP can be solved in polynomial time on the support graph of $\mathbf{Q}$. Without loss of generality assume that $\mathbf{Q}$ is symmetric and $\mathbf{c}$ is the zero vector.

Theorem 3.29 *Let $G = (V, E)$ be the support graph of $\mathbf{Q}$. If the off-diagonal elements of $\mathbf{Q}$ are non-positive and $\max\{q_{ii}, q_{jj}\} \leq |q_{ij} + q_{ji}|$ for all $i, j \in \{1, 2, \ldots, n\}$, where n is the size of $\mathbf{Q}$, then an optimal solution to the QUBO $(\mathbf{Q}, \mathbf{c})$ can be recovered from an optimal solution to the MWSSP on G with vertex weights $w_i = q_{ii}$, for all $i \in \{1, 2, \ldots, n\}$.*

Proof Let S be an optimal solution to the MWSSP problem on G with optimal objective function value Z^*. Construct the solution $\mathbf{x}^0 \in \{0, 1\}^n$ such that $x_i^0 = 1$ if $i \in S$ and $x_i^0 = 0$ if $i \notin S$. Then $Z^* = \sum_{i \in S} q_{ii}$ and $(\mathbf{x}^0)^T \mathbf{Q} \mathbf{x}^0 = \sum_{i \in S} q_{ii} + \sum_{i,j \in S, i \neq j} (q_{ij} + q_{ji}) = \sum_{i \in S} q_{ii}$. Note that since S is an independent set in G and G is the support graph of $\mathbf{Q}$, for $i, j \in S$, $q_{ij} = q_{ji} = 0$ and hence $\sum_{i,j \in S, i \neq j} (q_{ij} + q_{ji}) = 0$. Thus, $(\mathbf{x}^0)^T \mathbf{Q} \mathbf{x}^0 = Z^*$. Now we show that there exists an optimal solution $\mathbf{x}^*$ of the QUBO $(\mathbf{Q}, \mathbf{c})$ such that if $q_{ij} \neq 0$, then at least one of x_i^* or x_j^* is zero.

Choose any optimal solution $\mathbf{x}^*$ of the QUBO $(\mathbf{Q}, \mathbf{c})$. Assume that $x_i^* = x_j^* = 1$ and $q_{ij} \neq 0$. By assumption $q_{ij} = q_{ji} < 0$, and $\max\{q_{ii}, q_{jj}\} \leq |q_{ij} + q_{ij}|$, and all other entries of Q are no more than zero. Without loss of generality assume that $q_{ii} \leq q_{jj}$. Then by setting $x_i^* = 0$ gives an improved solution, violating the optimality of $\mathbf{x}^*$, unless $q_{ii} = q_{jj} = |q_{ij} + q_{ji}|$. In the latter case, setting $x_i^* = 0$ yields an alternative optimal solution to QUBO but with the number of components of weight one is reduced by one. Continuing in this way, after at most n steps, we get an optimal solution to the QUBO which corresponds to a stable set in G. Since $\mathbf{x}^0$ corresponds to the maximum weight stable set in G, $\mathbf{x}^0$ must be an optimal solution to the QUBO $(\mathbf{Q}, \mathbf{c})$ and the result follows. □

Theorem 3.29 allows us to solve QUBO instances satisfying the conditions imposed on the matrix $\mathbf{Q}$ as a maximum weight stable set problem. The theorem is essentially a restatement of the QUBO formulation of MWSSP given in [81] with slightly relaxed conditions on the elements of $\mathbf{Q}$. The clique version of Theorem 3.29 can be stated as follows.

Theorem 3.30 *Let $G = (V, E)$ be the complement of the support graph of $\mathbf{Q}$. If the off-diagonal elements of $\mathbf{Q}$ are non-positive and $\max\{q_{ii}, q_{jj}\} \leq |q_{ij} + q_{ji}|$ for all $i, j \in \{1, 2, \ldots, n\}$, where n is the size of $\mathbf{Q}$, then an optimal solution to the QUBO $(\mathbf{Q}, \mathbf{c})$ can be recovered from an optimal solution to the MWClP on the complement of G with vertex weights $w_i = q_{ii}$, for all $i \in \{1, 2, \ldots, n\}$.*

The maximum clique problem can be solved in polynomial time on the complements of the graphs listed in Theorem 3.28. The theorem below summarizes additional classes of graphs on which the maximum weight clique problem can be solved in polynomial time.

Theorem 3.31 *The maximum weight (stable set) maximum clique problem is solvable polynomial time in the following classes of graphs.*

1. *Perfect graphs [49]*
2. *Planar graphs (trivial, since the maximum clique has at most 4 vertices)*
3. *Circle graphs [44]*
4. *Unit disk graph (with a known geometric representation) [31]*

Note that the complement of a perfect graph is also perfect. Thus, on this class of graphs both MWSSP and MWClP are solvable in polynomial time.

Let us now briefly consider the linear programming formulation of the maximum weight stable set problem on some specific graphs. On a bipartite graph $G = (V, E)$ and weight function w_i for $i \in V$, the maximum weight stable set problem can be solved as the linear program [50]

$$\begin{aligned} \text{Maximize} \quad & \sum_{i \in V} w_i x_i \\ \text{Subject to:} \quad & x_i + x_j \leq 1 \text{ for all } \{i, j\} \in E \\ & x_i \geq 0, i \in V. \end{aligned}$$

On a t-perfect graph the stable set polytope is obtained by just adding the odd circuit inequalities [50] to the linear program above. Further, the separation problem associated with the odd circuit inequalities can be solved in polynomial time. Thus the MWSSP can be solved in polynomial time as the linear program

$$\begin{aligned} \text{Maximize} \quad & \sum_{i \in V} w_i x_i \\ \text{Subject to:} \quad & x_i + x_j \le 1 \text{ for all } \{i, j\} \in E \\ & \sum_{i \in V(C)} x_i \le \frac{1}{2}(|V(C)| - 1) \text{ for all } C \in \mathcal{C} \\ & x_i \ge 0, i \in V, \end{aligned}$$

where $\mathcal{C}$ is the collection of all odd circuits in G and $V(C)$ is the vertex set of $C \in \mathcal{C}$. Bipartite graphs, nearly bipartite graphs (i.e. there exists a vertex whose removal yields a bipartite graph), series parallel graphs, etc. are example of t-perfect graphs.

When G is a perfect graph, a linear representation of the stable set polytope is obtained by replacing the odd circuit inequalities by the clique inequalities [50]. Since the separation problem for the clique inequalities can be solved in polynomial time, the MWSSP on perfect graphs can be solved in polynomial time as the following linear program

$$\begin{aligned} \text{Maximize} \quad & \sum_{i \in V} w_i x_i \\ \text{Subject to:} \quad & \sum_{i \in V(L)} x_i \le 1 \text{ for all cliques } L \text{ of } G, \\ & x_i \ge 0, i \in V. \end{aligned}$$

In the above linear program the inequalities $x_i + x_j \le 1$ for $\{i, j\} \in E$ are already included in the clique inequalities since they arise from cliques of cardinality 2. A large number of specially structured graphs belong to the family of perfect graphs, e.g. bipartite graphs and their complements, interval graphs and their complements, comparability graphs and their complements, among others, see [50]. Thus, linear programming is one way to solve many special cases of the MWSSP. However, there are better, more specialized algorithms exploiting specific properties are available, as discussed earlier.

Graph decomposition is another popular approach in obtaining polynomial time algorithms for specially structured MWSSP. Let us illustrate this using a specific decomposition process called clique separation. Let $G = (V, E)$ be a connected graph and $S \subseteq V$. Then S is called a *clique separator* in G if the subgraph of G induced by S is a clique and the graph $G - S$ is disconnected. A graph is called an *atom* if it does not contain a clique separator. Tarjan [89] showed that any graph on n nodes and m edges, which is not an atom, can be decomposed into atoms in $O(mn)$ time. Let $\mathcal{H}$ be a hereditary class of graphs. Tarjan [89] showed that if the MWSSP can be solved in $O(\xi(n))$ time on atoms in $\mathcal{H}$, then the MWISP can be solved in $O(mn + n^2\xi(n))$ time on any graph in $\mathcal{H}$. This time complexity was improved by Brandstädt et al. [20] to $O(mn + n\xi(n))$. Exploiting this divide-and-conquer approach, polynomial time algorithms for the MWSSP have been developed by many authors for specially structured graphs. These include (hole, co-chair)-free graphs [21], (hole, paraglider)-free graphs [20], (hole, dart)-free graphs [9] etc. Obviously, these lead to polynomial time algorithms for QUBO and the Ising QUBO.

The generalized vertex cover problem [55, 57, 58, 77] is another problem closely related to the MWSSP and QUBO. Hassin and Levin [55] and Pandey and Punnen [77] identified polynomially solvable cases of this problem which directly yield additional polynomially solvable cases for QUBO.

3.5 Pseudopolynomial and Subexponential Algorithms

Since the QUBO and the Ising QUBO are strongly NP-hard, existence of pseudopolynomial time algorithms for the problems are ruled out, unless P=NP. In this section, we explore pseudopolynomial and subexponential algorithms for special cases of these problems.

3.5.1 Low Rank Cost Matrices

QUBO $(\mathbf{Q}, \mathbf{c})$ with a rational matrix $\mathbf{Q}$ of given rank r and a rational vector $\mathbf{c}$ can be solved in pseudo polynomial time, see [26]. In the following we briefly review their approach.

Let $\mathbf{Q}$ be a symmetric rational $n \times n$ matrix of rank r. We can assume without loss of generality that the r leading principal minors of $\mathbf{Q}$ are nonzero (otherwise both rows and columns of $\mathbf{Q}$ could be permuted by an appropriately chosen permutation). Consider the LDU-representation of such a symmetric matrix $\mathbf{Q}$. That is, $\mathbf{Q} = \mathbf{LDL}^T$, where $\mathbf{D}$ is a rational diagonal $n \times n$ matrix with $d_{ii} = 0$ for all $i \geq r + 1$ and $\mathbf{L}$ is a rational $n \times n$ lower triangular matrix with diagonal entries equal to 1 (see e.g. [70]). Denote by $\boldsymbol{\ell}^{(i)}$ the columns of $\mathbf{L}$, $i \in \{1, 2, \ldots, n\}$. The following equalities hold

$$\mathbf{x}^T\mathbf{Q}\mathbf{x} = \mathbf{x}^T\mathbf{LDL}^T\mathbf{x} = \mathbf{y}^T\mathbf{D}\mathbf{y} = \sum_{i=1}^{n} d_{ii}y_i^2 = \sum_{i=1}^{r} d_{ii}y_i^2 ,$$

with $\mathbf{y} = (y_1, y_2, \ldots, y_n)^T = \mathbf{L}^T\mathbf{x}$ and $y_i = \langle \boldsymbol{\ell}^{(i)}, \mathbf{x}\rangle$. Then the objective function f of QUBO $(\mathbf{Q}, \mathbf{c})$ with a rational input as above is given as follows for $\mathbf{x} \in \{0, 1\}^n$ the

$$f(\mathbf{x}) = \mathbf{c}^T\mathbf{x} + \sum_{i=1}^{r} d_{ii}\langle \boldsymbol{\ell}^{(i)}, \mathbf{x}\rangle^2. \tag{3.13}$$

By applying a dynamic programming approach of Papadimitriou [78] the following lemma is proved in [26].

Lemma 3.3 (Çela et al. [26]) *Let $\mathbf{K} = (k_{ij})$ be an integral $m \times n$ matrix and let $\mathbf{b}$ be an integral m-dimensional vector. The problem of deciding whether there exists a vector $\mathbf{x} \in \{0, 1\}^n$ such that $\mathbf{K}\mathbf{x} = \mathbf{b}$ holds can be solved in $\varnothing(mn^{m+1}\kappa^m)$ time where $\kappa := \max\{|k_{ij}|: i \in \{1, 2, \ldots, m\}, j \in \{1, 2, \ldots, n\}\}$.*

The above lemma implies then the following result.

Theorem 3.32 (Çela et al. [26]) *The $QUBO(\mathbf{Q}, \mathbf{c})$ with an integer $n \times n$ symmetric matrix $\mathbf{Q}$ of rank r and an integer n-dimensional vector $\mathbf{c}$ can be solved in $O(rU^{2r+2}n^{2r+3})$ time, where $U := 2\max\left\{|\ell_j^{(i)}|, |c_j|: j \in \{1, 2, \ldots, n\}, i \in \{1, 2, \ldots, r\}\right\}$.*

Proof Consider the equivalent formulation of the QUBO $(\mathbf{Q}, \mathbf{c})$ as in (3.13). Clearly, we can assume w.l.o.g. that for $i \in \{1, 2, \ldots, r\}$ the vectors $\ell^{(i)}$ and the entries d_{ii} as well as the vector $\mathbf{c}$ are integral (e.g. by scaling with the smallest common multiple of all rational inputs.) □

Denote $\boldsymbol{\ell}^{(0)} := \mathbf{c}$. Then $-nU/2 \leq \langle \boldsymbol{\ell}^{(i)}, \mathbf{x}\rangle \leq nU/2$ holds for all $i \in \{0, 1, \ldots, n\}$ and $\mathbf{x} \in \{0, 1\}^n$. Consider now an integral vector $\mathbf{v} \in [-nU/2, nU/2]^{r+1}$ and the following parametric optimization problem associated with it

$$\min \quad g_{\mathbf{v}}(\mathbf{x}) := \quad v_0 + \sum_{i=1}^{r} d_{ii}v_i^2 \tag{3.14}$$

subject to

$$\langle \boldsymbol{\ell}^{(i)}, \mathbf{x}\rangle = v_i\,,\ i \in \{0, 1, \ldots, r\}$$

$$x_i \in \{0, 1\},\quad i \in \{0, 1, \ldots, r\}$$

By applying Lemma 3.3 the feasibility problem defined by the constraints of (3.14) can be solved in $O(rn^{r+2}U^{r+1})$ time for each choice of $\mathbf{v}$. The optimal value of (3.14) is the minimum of $v_0 + \sum_{i=1}^{r} d_{ii}v_i^2$ over all vectors $\mathbf{v}$ which correspond to a feasible problem. Since there are at most $(nU + 1)^{r+1} = O((nU)^{r+1})$ such vectors the problem in (3.13) can be solved in $O(rn^{2r+3}U^{2r+2})$ time.

The pseudopolynomial algorithm discussed above extends to cases of QUBO where $\mathbf{Q}$ has pseudo rank r. See Sect. 3.3.2.2 for details on the notion of pseudo rank of a matrix.

3.5.2 The Half-Product QUBO

Some of the machine scheduling problems discussed in literature have a natural QUBO structure. These applications are mentioned in Chap. 1 and it include minimizing completion time variance on a single machine, minimizing total completion time on two machines, among others, see [5, 59, 64]. Pseudopolynomial algorithms to solve these problems indirectly lead to pseudopolynomial time algorithms to solve the QUBO with a certain special structure. In fact these problems belong to the class of the half-product QUBO [5, 64] which is shown to be NP-hard in Sect. 3.2.

In a half-product QUBO, the matrix $\mathbf{Q}$ is upper triangular with diagonal elements equal to zero and $q_{ij} = a_i b_j$ for $1 \le i < j \le n$, where $\mathbf{a} = (a_1, a_2, \ldots, a_n)$ and $\mathbf{b} = (b_1, b_2, \ldots, b_n)$. Thus, a half-product QUBO can be written as

$$\begin{aligned} &\text{Maximize} \quad \sum_{i=1}^{n-1} a_i x_i \left(\sum_{j=i+1}^{n} b_j x_j \right) + \mathbf{c}^T \mathbf{x} \\ &\text{Subject to:} \quad \mathbf{x} \in \{0, 1\}^n, \end{aligned}$$

The half product QUBO is related to the fixed rank QUBO but note that the rank of the matrix $\mathbf{Q}$ of size n in a half-product QUBO can be up to $n - 1$. A pseudopolynomial dynamic programming algorithm for this problem is given in [59]. The pseudo-polynomial algorithm discussed in Sect. 3.5.1 for the case of fixed rank matrices can also be modified to obtain a pseudopolynomial algorithm for the half-product QUBO.

3.5.3 Subexponential Algorithms

Let us now consider the QUBO $(\mathbf{Q}, \mathbf{c})$ where the support graph of $\mathbf{Q}$ is $\eta(n)$-separable [14, 68]. A graph $G = (V, E)$ on n nodes is $\eta(n)$-separable if the following conditions are satisfied.

1. The vertex set V can be partitioned into sets A, B and C such that no vertex in A is adjacent to any vertex in B.
2. There exist constants $\alpha < 1$ and $\beta > 0$ such that $|A| \le \alpha n$, $|B| \le \alpha n$ and $|C| \le \beta \eta(n)$.
3. The subgraph of G induced by A is $\eta(n_1)$-separable and the subgraph of G induced by B is $\eta(n_2)$-separable, where $n_1 = |A|$ and $n_2 = |B|$.

Lipton and Tarjan [68] showed that a planar graph on n nodes is $\sqrt{n}$-separable. Examples of some other graphs that are $\sqrt{n}$-separable include the grid graph [45, 68] and the one-tape Turing machine graph [68, 82, 83]. The $\eta(n)$-separable property of graphs offers the possibility of developing divide and conquer type algorithms for some graph theoretic optimization problems. Exploiting the $\eta(n)$-separability property Pardalos and Jha [80] developed an exponential time algorithm for QUBO where the exponentiality depends on the parameter $\eta(n)$. In particular, they showed that

Theorem 3.33 ([80]) *If the support graph $G = (V, E)$ of the matrix $\mathbf{Q}$ is $\eta(n)$-separable then the QUBO $(\mathbf{Q}, \mathbf{c})$ can be solved in $O\left(\log n\,(p(n) + t(n))\,2^{k\eta(n)\log n}\right)$ time, where $k > 0$ is a constant, $p(n)$ is the complexity of finding an $\eta(n)$-separator of G, $t(n) \leq \min\{(n-1)\beta\eta(n), |E|\}$, and β is the constant used in the definition of the separable graph (as above).*

It may be noted that the QUBO $(\mathbf{Q}, \mathbf{c})$ is NP-hard even if the support graph of $\mathbf{Q}$ is planar. But for planar graphs, a $\sqrt{n}$-separator can be identified in linear time [68] and hence $p(n) = O(n)$. Thus, from Theorem 3.33, the QUBO $(\mathbf{Q}, \mathbf{c})$ on planar graphs can be solved in $O\left(a\sqrt{n}\log n 2^{k\sqrt{n}\log n}\right)$ time, which is sub-exponential complexity.

The complexity result discussed in Theorem 3.33 also extends to the Ising QUBO. Earlier, we discussed various polynomial time algorithms for solving the QUBO and Ising QUBO where the polynomiality depends on a parameter k which is either fixed or takes values $O\log n)$. When $k = On^{\epsilon}$ for $\epsilon > 0$ these algorithms becomes subexponential for solving QUBO and the BQUBO.

3.6 Conclusions

In this chapter we presented a review on the computational complexity of the QUBO and the Ising QUBO and discussed various special cases of the problems that can be solved in polynomial time, in pseudopolynomial time or in subexponential time. For some special cases detailed proofs are provided while for other special cases appropriate references along with partial results are given. Polynomially solvable cases can also be derived using duality theory and semidefinite programming formulation of the Ising QUBO and QUBO [66, 88] but this topic is not covered here. The algorithms for the tractable special cases discussed here exploit properties of the associated support graphs or properties of the involved cost matrices or both. We want to point out that there is a significant gap between known tractable special cases and special cases of the QUBO known to be NP-hard. The development of general purpose tools to reduce this gap is a challenging and interesting issue for further research.

References

1. A.A. Ageev, A.V. Kel̆manov, A.V. Pyatkin, NP-hardness of the Euclidean max-cut problem. Dokl. Math. **89**, 343–345 (2014)
2. V. Alekseev, The effect of local constraints on the complexity of determination of the graph independence number. Combin. Algebraic Methods Appl. Math., 3–13 (1982) (in Russian)
3. K. Allemand, K. Fukuda, T. Liebling, E. Steiner, A polynomial case of unconstrained zero-one quadratic optimization. Math. Program. **91**, 49–52 (2001)
4. J.C. Anglés, D' Auric, M. Preissmann, A, Sebó, Optimal cuts in graphs and statistical mechanics. Math. Comput. Model. **26**, 1–11 (1997)
5. T. Badics, E. Boros, Minimization of half-products. Math. Oper. Res. **23**, 649–660 (1998)
6. F. Barahona, On the computational complexity of Ising spin glass models. J. Phys. A Math. Gen. **15**, 3241–3253 (1982)
7. F. Barahona, The max-cut problem in graphs not contractible to K_5. Oper. Res. Lett. **2**, 107–111 (1983)
8. F. Barahona, A solvable case of quadratic 0-1 programming. Discrete Appl. Math. **13**, 23–26 (1986)
9. M. Basavaraju, L.S. Chandran, T. Karthick, Maximum weight independent sets in hole- and dart-free graphs. Discrete Appl. Math. **160**, 2364–2369 (2012)
10. L.W. Beineke, Biplanar graphs: a survey. Comput. Math. Appl. **34**, 1–8 (1997)
11. W. Ben-Ameur, J. Neto, A polynomial-time recursive algorithm for some unconstrained quadratic optimization problems. Discrete Appl. Math. **159**, 1689–1698 (2011)
12. A. Billionnet, Solving a cut problem in bipartite graphs by linear programming: Application to a forest management problem. Appl. Math. Model. **34**, 1042–1050 (2010)
13. Y. Bilu, A. Daniely, N. Linial, M. Saks, On the practically interesting instances of MAXCUT, in *30th International Symposium on Theoretical Aspects of Computer Science (STACS 2013)* (2013), pp. 526–537
14. D.K. Blandford, G.E. Blelloch, I.A. Kash, Compact representations of separable graphs, in *SODA '03: Proceedings of the Fourteenth Annual ACM-SIAM Symposium on Discrete algorithms*, pp. 679–688 (2003)
15. H.L. Bodlaender, A Partial k-arboretum of graphs with bounded treewidth. Theor. Comput. Sci. **209**, 1–45 (1988)
16. H.L. Bodlaender, K. Jansen, On the complexity of the maximum cut problem. Nordic J. Comput. **7**, 14–31 (2000)
17. H.L. Bodlaender, A.M.C.A. Koster, Combinatorial optimization on graphs of bounded treewidth. Comput. J. **51**, 255–269 (2008)
18. I.M. Bomze, M. Budinich, P.M. Pardalos, M. Pelillo, The maximum clique problem, in *Handbook of Combinatorial Optimization*, ed. by D.Z. Du, P.M. Pardalos (Kluwer Academic Publishers, 1999), pp. 1–74
19. M. Bonamy, E. Bonnet, N. Bousquet, P. Charbit, S. Thomassé, EPTAS for max clique on disks and unit balls, in *Proceedings of the 59th IEEE Symposium on Foundations of Computer Science* (FOCS'18) (2018), pp. 568–579
20. A. Brandstädt, C.T. Hoáng, On clique separators, nearly chordal graphs, and the maximum weight stable set problem. Theor. Comput. Sci. **389**, 295–306 (2007)
21. A. Brandstädt, V. Giakoumakis, F. Maffray, Clique separator decomposition of hole-free and diamond-free graphs and algorithmic consequences. Discrete Appl. Math. **160**, 471–478 (2012)
22. R.E. Burkard, E. Çela, G. Rote, G.J. Woeginger, The quadratic assignment problem with a monotone anti-Monge and a symmetric Toeplitz matrix: Easy and hard cases. Math. Program. B **82**, 125–158 (1998)
23. R.E. Burkard, M. Dell'Amico, S. Martello, *Assignment Problems* (SIAM, Philadelphia, 2009)
24. S. Cabello, J. Cardinal, S. Langerman, The clique problem in ray intersection graphs. Discrete Comput. Geom. **50**, 771–783 (2013)

25. E. Çela, *The Quadratic Assignment Problem: Theory and Algorithms* (Kluwer Academic Publishers, Dordrecht, 1998)
26. E. Çela, B. Klinz, C. Meyer, Polynomially solvable cases of the constant rank unconstrained quadratic 0-1 programming problem. J. Combin. Optim. **12**, 187–215 (2006)
27. E. Çela, N.S. Schmuck, S. Wimer, G.J. Woeginger, The Wiener maximum quadratic assignment problem. Discrete Optim. **8**, 411–416 (2011)
28. E. Çela, V. Deineko, G.J. Woeginger, Well-solvable cases of the QAP with block-structured matrices. Discrete Appl. Math. **186**, 56–65 (2015)
29. S.T. Chakradhar, M.L. Bushnell, A solvable class of quadratic 0-1 programming. Discrete Appl. Math. **36**, 233–251 (1992)
30. M. Chimani, C. Dahn, M. Juhnke-Kubitzke, N.M. Kriege, P. Mutzel, A. Nover, Maximum cut parameterized by crossing number. J. Graph Algorithms Appl. **24**, 155–170 (2020)
31. B.N. Clark, C.J. Colbourn, D.S. Johnson, Unit disk graphs. Discrete Math. **86**, 165–177 (1990)
32. Y. Crama, P. Hansen, B. Jaumard, The basic algorithm for pseudo-Boolean programming revisited. Discrete Appl. Math. **29**, 171–185 (1990)
33. C. Dahn, N.M. Kriege, P. Mutzel, J. Schilling, Fixed-parameter algorithms for the weighted Max-Cut problem on embedded 1-planar graphs. Theor. Comput. Sci. **852**, 172–184 (2021)
34. J. Díaz, M. Kaminski, MAX-CUT and MAX-BISECTION are NP-hard on unit disk graphs. Theor. Comput. Sci. **377**, 271–276 (2007)
35. H. Edelsbrunner, *Algorithms in Combinatorial Geometry* (Springer, 1987)
36. D. Eppstein, Diameter and treewidth in minor-closed graph families. Algorithmica **27**, 275–291 (2000)
37. Y. Faenza, G. Oriolo, G. Stauffer, Solving the weighted stable set problem in claw-Free graphs via decomposition. J. ACM **61**, 1–41 (2014)
38. J.A. Ferrez, K. Fukuda, Th.M. Liebling, Solving the fixed rank convex quadratic maximization in binary variables by a parallel zonotope construction algorithm. Eur. J. Oper. Res. **166**, 35–50 (2005)
39. H. Fleischner, G. Sabidussi, V.I. Sarvanov, Maximum independent sets in 3- and 4-regular Hamiltonian graphs. Discrete Math. **310**, 2742–2749 (2010)
40. M.C. Francis, D. Gonçalves, P. Ochem, The maximum clique problem in multiple interval graphs. Algorithmica **71**, 812–836 (2015)
41. A. Galluccio, M. Loebl, J. Vondrák, Optimization via enumeration: A new algorithm for the MAX-CUT problem. Math. Program. A **90**, 273–290 (2001)
42. M.R. Garey, D.S. Johnson, *Computers and Intractability: A Guide to the Theory of NP-Completeness* (W. H. Freeman and Co, 1979)
43. M.R. Garey, D.S. Johnson, L. Stockmeyer. Some simplified NP-complete graph problems. Theor. Comput. Sci. **1**, 237–267 (1976)
44. F. Gavril, Algorithms for a maximum clique and a maximum independent set of a circle graph. Networks **3**, 261–273 (1973)
45. J.A. George, Nested dissection of a regular finite element mesh. SIAM J. Numer. Anal. **10**, 345–367 (1973)
46. R.E. Gomory, T.C. Hu, Multi-terminal network flow. SIAM J. Appl. Math. **9**, 551–570 (1961)
47. M. Grötschel, G.L. Nemhauser, A polynomial algorithm for the max-cut problem on graphs without long odd cycles. Math. Program. **29**, 28–40 (1984)
48. M. Grötschel, W.R. Pulleyblank, Weakly bipartite graphs and the max-cut problem. Oper. Res. Lett. **1**, 23–27 (1981)
49. M. Grötschel, L. Lovász, A. Schrijver, Polynomial algorithms for Perfect Graphs. Annals Discrete Math. **21**, 325-356 (1984)
50. M. Grötschel, L. Lovász, A.J. Schrijver, *Geometric Algorithms and Combinatorial Optimization* (Wiley, New York, 1988)
51. S. Gu, R. Cui, J. Peng, Polynomial time solvable algorithms to a class of unconstrained and linearly constrained binary quadratic programming problems. Neurocomputing **198**, 171–179 (2016)

52. V. Guruswami, Maximum cut on line and total graphs. Discrete Appl. Math. **92**, 217–221 (1999)
53. F. Hadlock, Finding a maximum cut of a planar graph in polynomial time. SIAM J. Comput. **4**, 221–225 (1975)
54. P.L. Hammer, P. Hansen, P.M. Pardalos, D.J. Rader Jr, Maximizing the product of two linear functions in 0-1 Variables. Optimization **51**, 511–537 (2002)
55. R. Hassin, A. Levin, The minimum generalized vertex cover problem, in *European Symposium on Algorithms* (2003), pp. 289–300
56. M. Hladík, M. Cerný, M. Rada, A new polynomially solvable class of quadratic optimization problems with box constraints. Optim. Lett. **15**, 2331–2341 (2021)
57. D.S. Hochbaum, Solving integer programs over monotone inequalities in three variables: A frame work for half integrality and good approximations. Eur. J. Oper. Res. **140**, 291–321 (2002)
58. D.S. Hochbaum, A. Pathria, Forest harvesting and minimum cuts: A new approach to handling spatial constraints. Forest Sci. **43**, 544–554 (1997)
59. B. Jurisch, W. Kubiak, J. Józefowska, Algorithms for minclique scheduling problems. Discrete Appl. Math. **72**, 115–139 (1997)
60. Y. Kobayashi, Y. Kobayashi, S. Miyazaki, S. Tamaki, An improved fixed-parameter algorithm for Max-Cut parameterized by crossing number, in *Combinatorial Algorithms - 30th International Workshop, IWOCA 2019* (2019), pp. 327–338
61. G. Kochenberger, J.K. Hao, F. Glover, M. Lewis, Z. Lu, H. Wang, Y. Wang, The unconstrained binary quadratic programming problem: A survey. J. Combin. Optim. **28**, 58–81 (2014)
62. T.C. Koopmans, M.J. Beckmann, Assignment problems and the location of economic activities. Econometrica **25**, 53–76 (1957)
63. J. Kratochvíl, J. Nešetřil, Independent set and clique problems in intersection-defined classes of graphs. Comment. Math. Univ. Carolinae **31**, 85–93 (1990)
64. W. Kubiak, New results on the completion time variance minimization. Discrete Appl. Math. **58**, 157–168 (1995)
65. M. Laurent, The max-cut problem, in *Annotated Bibliographies in Combinatorial Optimization*, ed. by M. Dell'Amico, F. Maffioli and S. Martello (Wiley, Chichester, 1997)
66. D. Li, X. Sun, S. Gu, J. Gao, C. Liu, Polynomially solvable cases of binary quadratic programs, in *Optimization and Optimal Control*, ed. by A. Chinchuluun et al., Springer Optimization and Its Applications, vol. 39 (2010)
67. F. Liers, G. Pardella, Partitioning planar graphs: a fast combinatorial approach for max-cut. Comput. Optim. Appl. **51**, 323–344 (2012)
68. R.J. Lipton, R.E. Tarjan, A planar separator theorem. SIAM J. Appl. Math. **36**, 177–189 (1979)
69. E.M. Loiola, N.M.M. de Abreu, P.O. Boaventura-Netto, P. Hahn, T. Querido, A survey for the quadratic assignment problem. Eur. J. Oper. Res. **176**, 657-690 (2007)
70. T. Lyche, *Numerical Linear Algebra and Matrix Factorizations* (Springer, 2020)
71. S.T. McCormick, M.R. Rao, G. Rinaldi, Easy and difficult objective functions for max cut. Math. Program. **94**, 459–466 (2003)
72. M. Middendorf, F. Pfeiffer, The max clique problem in classes of string-graphs. Discrete Math. **108**, 365–372 (1992)
73. O.J. Murphy, Computing independent sets in graphs with large girth. Discrete Appl. Math. **35**, 167–170 (1992)
74. P. Nobil, A. Sassano, An $O(n^2 \log n)$ algorithm for the weighted stable set problem in claw-free graphs. Math. Program. **186**, 409–437 (2021)
75. G.I. Orlova, Y.G. Dorfman, Finding maximum cut in a graph. Eng. Cybern. **10**, 502–506 (1972)
76. P. Pandey, Topics in quadratic binary optimization problems, Ph.D. Thesis, Simon Fraser University, 2018
77. P. Pandey, A.P. Punnen, The generalized vertex cover problem. Discrete Optim. **30**, 121–143 (2018)
78. C.H. Papadimitriou, On the compelxity of integer programming. J. Assoc. Comput. Mach. **28**, 765–768 (1981)

79. C.H. Papadimitriou, M. Yannakakis, Optimization, approximation, and complexity classes. J. Comput. Syst. Sci. **43**, 425–440 (1991)
80. P.M. Pardalos, S. Jha, Graph separation techniques for quadratic zero-one programming. Comput. Math. Appl. **21**, 107–113 (1991)
81. P.M. Pardalos, J. Xue, The maximum clique problem. J. Glob. Optim. **4**, 301–328 (1994)
82. M.S. Paterson, Tape bound for time-bounded Turing machine. J. Comput. Syst. Sci. **6**, 116–124 (1972)
83. W.J. Paul, R.E. Tarjan, J.R. Celoni, Space bounds for a game on graphs. Math. Syst. Theory **10**, 239–251 (1976)
84. J.C. Picard, M. Queyranne, Selected applications of minimum cuts in networks. INFOR Inform. Syst. Oper. Res. **20**, 394–422 (1982)
85. J.C. Picard, H.D. Ratliff, Minimum cuts and related problems. Networks **5**, 357–370 (1975)
86. S. Poljak, A note on stable sets and colorings of graphs. Comment. Math. Univ. Carolinae **15**, 307–309 (1974)
87. W.K. Shih, S. Wu, Y.S. Kuo, Unifying maximum cut and minimum cut of a planar graph. IEEE Trans. Comput. **39**, 694–697 (1990)
88. X.L. Sun, C.L. Liu, D. Li, J.J. Gao, On duality gap in binary quadratic programming. J. Glob. Optim. **53**, 255–269 (2012)
89. R.E. Tarjan, Decomposition by clique separators. Discrete Math. **55**, 221–232 (1985)
90. S. Tsukiyama, M. Ide, H. Ariyoshi, I. Shirakawa, A new algorithms for generating all the maximal independent sets. SIAM J. Comput. **6**, 505–517 (1977)
91. M. Yannakakis, Node and edge deletion NP-complete problems, in *Proc. 10th Annual ACM Sympsium on Theory of Computing* (Association for Computing Machinery, New York, 1978), pp. 253–264

Chapter 4
The Boolean Quadric Polytope

Adam N. Letchford

Abstract When developing an exact algorithm for a combinatorial optimisation problem, it often helps to have a good understanding of certain *polyhedra* associated with that problem. In the case of quadratic unconstrained Boolean optimisation, the polyhedron in question is called the *Boolean quadric polytope*. This chapter gives a brief introduction to polyhedral theory, reviews the literature on the Boolean quadric polytope and related polyhedra, and explains the algorithmic implications.

4.1 Introduction

It has been known for some time that *Quadratic unconstrained Boolean optimisation* (QUBO) is equivalent to another well-known combinatorial optimisation problem, known as the *max-cut* problem [7, 14, 53]. The max-cut problem has been proven to be "strongly $\mathcal{NP}$-hard" [26], and therefore the same holds for QUBO. Rather than explaining strong $\mathcal{NP}$-hardness in detail, let us just say that it makes it unlikely that an algorithm can be developed which solves all QUBO instances quickly.

The situation however is far from hopeless. Indeed, for many specific $\mathcal{NP}$-hard problems, algorithms have been developed that can solve many instances of interest to proven optimality (or near-optimality) in reasonable computing times. Many of these algorithms use a method known as *branch-and-cut* (see, e.g., [10, 51, 54]). Branch-and-cut is an enumerative scheme, in which a "tree" of subproblems is explored, and each subproblem is a *linear program* (LP).

One of the keys to designing a successful branch-and-cut algorithm for a given problem is to gain an understanding of certain *polyhedra* associated with that problem (e.g., [1–3, 12]). In the case of QUBO, the polyhedron in question is called the *Boolean quadric polytope* (e.g., [9, 14, 53]).

A. N. Letchford (✉)
Department of Management Science, Lancaster University, Lancaster, UK
e-mail: a.n.letchford@lancaster.ac.uk

A. P. Punnen (ed.), *The Quadratic Unconstrained Binary Optimization Problem*,
https://doi.org/10.1007/978-3-031-04520-2_4

This chapter gives a brief introduction to polyhedral theory, a detailed survey of known results on the Boolean quadric polytope, and a brief discussion of algorithmic implications. The structure of the chapter is as follows. The basics of polyhedral theory are recalled in Sect. 4.2. In Sect. 4.3, we define the Boolean quadric polytope and mention some of its fundamental properties. In Sect. 4.4, we survey some of the known valid inequalities for the Boolean quadric polytope. In Sect. 4.5, we review some connections between the Boolean quadric polytope and some other important polytopes. In Sect. 4.6, we mention some other related convex sets. In Sect. 4.7, we look at the algorithmic implications. Finally, concluding remarks are made in Sect. 4.8.

We use the following conventions and notation throughout the chapter. Given a positive integer n, we sometimes write V_n for $\{1, \ldots, n\}$. We K_n denote the complete graph on the vertex set V_n, and let E_n denote its edge set. Given a vector $\mathbf{v} \in \mathbf{R}^n$, we let $\sigma(\mathbf{v})$ denote $\sum_{i=1}^n v_i$. All matrices are real. Given two matrices $\mathbf{A}, \mathbf{B} \in \mathbf{R}^{m \times n}$, we write $\mathbf{A} \bullet \mathbf{B}$ for the (Frobenius) inner product

$$\sum_{i=1}^{m} \sum_{j=1}^{n} a_{ij} b_{ij} = \mathrm{Tr}(\mathbf{A}^T \mathbf{B}).$$

Given a positive integer k, we let $\mathcal{S}_+^k$ denote the set of positive semidefinite (psd) matrices of order k. We recall that a symmetric matrix $\mathbf{M}$ of order k is psd if and only if all of its eigenvalues are non-negative, or, equivalently, $\mathbf{v}^T \mathbf{M} \mathbf{v} \geq 0$ for all vectors $\mathbf{v} \in \mathbf{R}^k$.

4.2 Elementary Polyhedral Theory

This section draws on material from [31, 52].

Suppose that $\mathbf{x}^1, \ldots, \mathbf{x}^k \in \mathbf{R}^n$ are (column) vectors and $\lambda_1, \ldots, \lambda_k$ are scalars. A vector of the form $\lambda_1 \mathbf{x}^1 + \cdots + \lambda_k \mathbf{x}^k$ is called a *linear combination* of $\mathbf{x}^1, \ldots, \mathbf{x}^k$. It is called a *conical* combination if $\lambda_1, \ldots, \lambda_k$ are non-negative, an *affine* combination if $\sum_{i=1}^k \lambda_i = 1$, and a *convex* combination if it is both conical and affine. Given some non-empty set $S \subset \mathbf{R}^n$, the *convex hull* of S is the set of all convex combinations of the vectors in S. The linear, affine and conical hulls are defined analogously. We will let $\mathrm{conv}(S)$ denote the convex hull of S.

A set $S \subseteq \mathbf{R}^n$ is called *convex* if $\lambda \mathbf{x}^1 + (1 - \lambda)\mathbf{x}^2 \in S$ holds for all $\mathbf{x}^1, \mathbf{x}^2 \in S$ and all $\lambda \in (0, 1)$. A convex set P is called a *polyhedron* if there exists a non-negative integer m, a matrix $\mathbf{A} \in \mathbf{Z}^{mn}$ and a vector $\mathbf{b} \in \mathbf{Z}^m$ such that

$$P = \left\{\mathbf{x} \in \mathbf{R}^n : \mathbf{A}\mathbf{x} \leq \mathbf{b}\right\}.$$

A polyhedron which is bounded (i.e., not of infinite volume) is called a *polytope*. A famous theorem of Weyl [68] states that a set $P \subset \mathbf{R}^n$ is a polytope if and only if it is the convex hull of a finite number of points.

A point $\mathbf{x} \in P$ is called an *extreme point* of P if it is not a convex combination of other points in P. Every polytope is the convex hull of its extreme points.

A set of vectors is called *affinely independent* if no member of the set is an affine combination of the others. The *dimension* of a polyhedron P, denoted by $dim(P)$, is the maximum number of affinely independent vectors in P, minus one. Note that $dim(P) \le n$. If equality holds, P is said to be *full-dimensional*.

A linear inequality $\mathbf{a}^T\mathbf{x} \le a_0$ is *valid* for a polyhedron P if it is satisfied by every point in P. The set

$$F = P \cap \left\{\mathbf{x} \in \mathbf{R}^n : \mathbf{a}^T\mathbf{x} \le a_0\right\}$$

is called the *face* of P *induced* by the given inequality. Note that F is itself a polyhedron. The face F is called a *facet* of P if $dim(F) = dim(P) - 1$.

Example Suppose that S contains the following four points in $\mathbf{R}^3$:

$$\mathbf{x}^1 = \begin{pmatrix}0\\0\\1\end{pmatrix}, \quad \mathbf{x}^2 = \begin{pmatrix}0\\2\\1\end{pmatrix}, \quad \mathbf{x}^3 = \begin{pmatrix}2\\0\\1\end{pmatrix} \quad \mathbf{x}^4 = \begin{pmatrix}1\\1\\1\end{pmatrix}.$$

One can check that:

- the linear hull of S is $\mathbf{R}^3$ itself;
- the conical hull is $\left\{\mathbf{x} \in \mathbf{R}^3 : x_1 \ge 0,\ x_2 \ge 0,\ x_1 + x_2 \le 2x_3\right\}$;
- the affine hull is $\left\{\mathbf{x} \in \mathbf{R}^3 : x_3 = 1\right\}$;
- $\text{conv}(S) = \left\{\mathbf{x} \in \mathbf{R}^3 : x_1 \ge 0,\ x_2 \ge 0,\ x_1 + x_2 \le 2,\ x_3 = 1\right\}$.

Now let $P = \text{conv}(S)$. One can check that (a) P is a polytope, (b) $\dim(P) = 2$, (c) the extreme points of P are $\mathbf{x}^1, \ldots, \mathbf{x}^3$ and (d) P has three facets, induced by the inequalities $x_1 \ge 0$, $x_2 \ge 0$ and $x_1 + x_2 \le 2$. □

We now explain the connection between polyhedra and combinatorial optimisation. Suppose we can formulate our optimisation problem as an *integer linear program* (ILP) of the form

$$\max\left\{\mathbf{cx} : \mathbf{Ax} \le \mathbf{b},\ \mathbf{x} \in \mathbf{Z}^n_+\right\}. \tag{4.1}$$

Replacing the condition $\mathbf{x} \in \mathbf{Z}^n_+$ with the weaker condition $\mathbf{x} \in \mathbf{R}^n_+$, we obtain the so-called *continuous relaxation* of the ILP. The continuous relaxation is an LP, which is likely to be easy to solve. Let $\mathbf{x}^*$ be a (basic) optimal solution to the continuous relaxation. If $\mathbf{x}^*$ is integral, we have solved the ILP. Otherwise we have to do more work, and this is where polyhedra come into play.

The feasible region of the continuous relaxation is the polyhedron

$$P = \{\mathbf{x} \in \mathbf{R}^n_+ : \mathbf{A}\mathbf{x} \le \mathbf{b}\},$$

and the set of feasible solutions to the ILP is $S = P \cap \mathbf{Z}^n_+$. The convex hull of S is also a polyhedron, called the *integral hull* of P. We will denote it by P_I. By definition, we have $P_I \subseteq P$. Also, if $\mathbf{x}^*$ is not integral, then P_I is strictly contained in P, and there must exist a linear inequality that is valid for P_I but violated by $\mathbf{x}^*$. Such an inequality is called a *cutting plane*.

Example Consider the ILP

$$\begin{aligned} \max \quad & x_1 + x_2 \\ \text{s.t.} \quad & 4x_1 + 2x_2 \le 15 \\ & 2x_1 - 2x_2 \le 5 \\ & -2x_1 + 2x_2 \le 3 \\ & -6x_1 - 10x_2 \le -15 \\ & 2x_2 \le 5 \\ & \mathbf{x} \in \mathbf{R}^2_+. \end{aligned}$$

On the left of Fig. 4.1, we show the polyhedron P. Points with integer coordinates are represented by small circles. On the right of Fig. 4.1, the points in S are represented as larger circles. One can check that there are two optimal solutions to the ILP, $\binom{2}{2}$ and $\binom{3}{1}$, each with profit 4. The solution to the continuous relaxation, on the other hand, is $\mathbf{x}^* = \binom{2.5}{2.5}$, giving an upper bound of 5. On the left of Fig. 4.2, we show the integral hull P_I. Finally, on the right of Fig. 4.2, we show both P and P_I, together with a possible cutting plane, represented by a dashed line. □

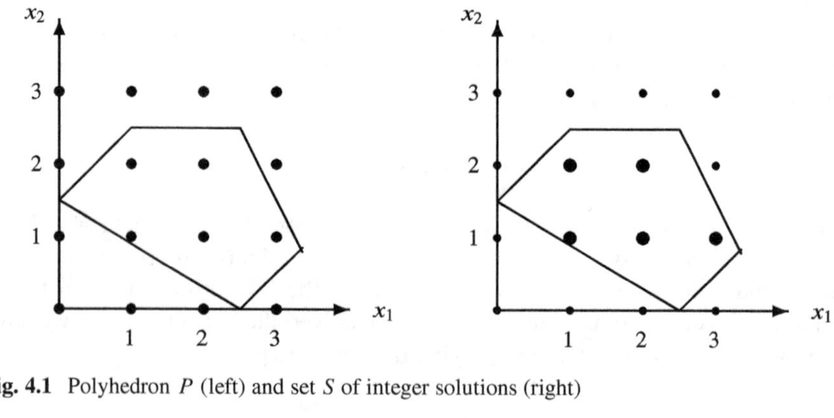

Fig. 4.1 Polyhedron P (left) and set S of integer solutions (right)

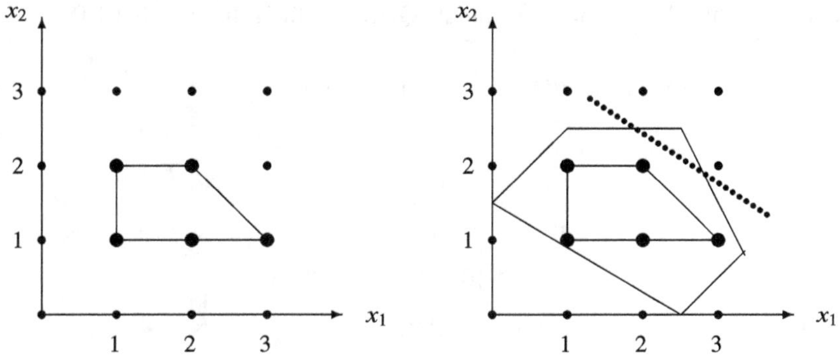

Fig. 4.2 Polyhedron P_I (left) and a possible cutting plane (right)

For simplicity and brevity, we assume from now on that P_I (and therefore also P) is a full-dimensional polytope. Under this assumption, the strongest possible cutting planes for a given ILP are those that induce facets of P_I.

At this point we should mention some negative results from Karp and Papadimitriou [39]. They showed that, if a combinatorial optimisation problem is $\mathcal{NP}$-hard, then, regardless of how it is formulated as an ILP, it is $\mathcal{NP}$-hard to check if a given linear inequality is valid for the associated polytope P_I. They also show that it is $\mathcal{NP}$-hard to check if a given inequality induces a facet of P_I.

Although the above-mentioned results may appear discouraging, there is also good news: for many important combinatorial optimisation problems (such as the knapsack problem, the travelling salesman problem, the stable set problem, and QUBO itself), researchers have discovered several large families of facet-inducing inequalities (see, e.g., [1–3, 12, 31, 52]). These inequalities can be used as cutting planes in branch-and-cut algorithms.

4.3 The Boolean Quadric Polytope

Now consider a QUBO instance of the form:

$$\begin{aligned} \max \quad & \mathbf{x}^T \mathbf{Q} \mathbf{x} \\ \text{s.t. } & \mathbf{x} \in \{0, 1\}^n, \end{aligned}$$

where, without loss of generality, we assume that $\mathbf{Q}$ is symmetric. Glover and Woolsey [28] proposed to replace each quadratic term $x_i x_j$ with a new binary variable y_{ij}. This allows one to formulate QUBO as the following 0-1 LP:

$$\max \sum_{i \in V_n} q_{ii} x_i + 2 \sum_{\{i,j\} \in E_n} q_{ij} y_{ij} \tag{4.2}$$

$$\text{s.t.} \quad y_{ij} \le x_i \quad (\{i,j\} \in E_n) \tag{4.3}$$

$$y_{ij} \le x_j \quad (\{i,j\} \in E_n) \tag{4.4}$$

$$x_i + x_j \le y_{ij} + 1 \quad (\{i,j\} \in E_n) \tag{4.5}$$

$$\mathbf{x} \in \{0,1\}^n \tag{4.6}$$

$$\mathbf{y} \in \{0,1\}^{\binom{n}{2}}. \tag{4.7}$$

For a given $n \ge 2$, the convex hull of pairs $(\mathbf{x}, \mathbf{y})$ satisfying (4.3)–(4.7) is called the *Boolean quadric polytope of order n* and denoted by BQP_n [14, 53]. (Some authors call it the *correlation polytope* instead; see, e.g., [21, 55].)

To make this clear, consider the case $n = 2$. To obtain a feasible solution to the 0-1 LP, we require:

$$\begin{pmatrix} x_1 \\ x_2 \\ y_{12} \end{pmatrix} \in \left\{ \begin{pmatrix} 0 \\ 0 \\ 0 \end{pmatrix}, \begin{pmatrix} 1 \\ 0 \\ 0 \end{pmatrix}, \begin{pmatrix} 0 \\ 1 \\ 0 \end{pmatrix}, \begin{pmatrix} 1 \\ 1 \\ 1 \end{pmatrix} \right\}.$$

One can check that the four points in question are affinely independent. Thus, BQP_2 is a tetrahedron, as shown in Fig. 4.3. Its facets are induced by the inequalities $y_{12} \le x_1$, $y_{12} \le x_2$, $y_{12} \ge x_1 + x_2 - 1$ and $y_{12} \ge 0$.

Padberg [53] proved that BQP_n is full-dimensional and that the inequalities (4.3)–(4.5), along with the non-negativity inequalities $y_{ij} \ge 0$, always induce facets. He also derived some additional inequalities, which we review in Sect. 4.4.

The Boolean quadric polytope has some remarkable properties. For one thing, every extreme point of BQP_n is adjacent to every other one [60]. Moreover, BQP_n has a high degree of *symmetry*. In particular, BQP_n is invariant under two

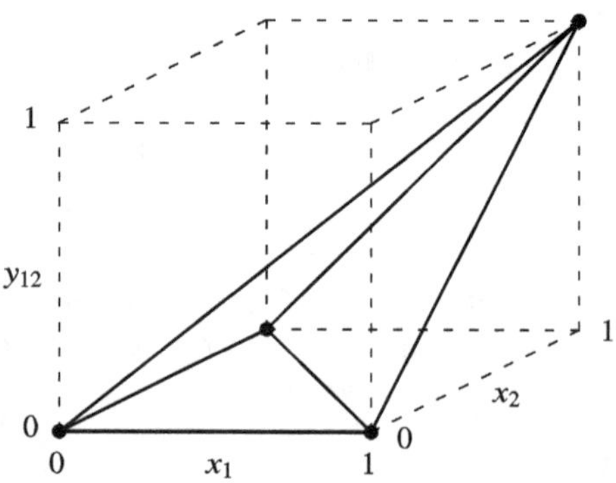

Fig. 4.3 The Boolean quadric polytope of order 2

transformations, called *permutation* and *switching* [21, 53, 55]. These are defined as follows.

Definition 4.1 (Permutation) Let $\pi : V_n \mapsto V_n$ be an arbitrary permutation. Consider the linear transformation $\phi^\pi : \mathbf{R}^{n+\binom{n}{2}} \mapsto \mathbf{R}^{n+\binom{n}{2}}$ that:

- replaces x_i with $x_{\pi(i)}$ for all $i \in V_n$,
- replaces y_{ij} with $y_{\pi(i),\pi(j)}$ for all $\{i, j\} \in E_n$.

By abuse of terminology, we call this transformation itself a "permutation".

Definition 4.2 (Switching) For an arbitrary set $S \subset V_n$, let $\psi^S : \mathbf{R}^{n+\binom{n}{2}} \mapsto \mathbf{R}^{n+\binom{n}{2}}$ be the affine transformation that:

- replaces x_i with $1 - x_i$ for all $i \in S$,
- replaces y_{ij} with $x_i - y_{ij}$ for all $i \in \{1, \ldots, n\} \setminus S$ and all $j \in S$,
- replaces y_{ij} with $1 - x_i - x_j + y_{ij}$ for all $\{i, j\} \subset S$,
- leaves all other x_i and y_{ij} variables unchanged.

Applying the transformation ψ^S is called "switching" (on S).

It is fairly easy to show that BQP_n is invariant under permutation. (That is, for any n and any permutation π of $\{1, \ldots, n\}$, we have $\phi^\pi(\text{BQP}_n) = \text{BQP}_n$.) To make this chapter self-contained, we now show that the same holds for switching:

Proposition 4.1 BQP_n *is invariant under switching. That is, for any n and any* $S \subset V_n$, $\psi^S(\text{BQP}_n) = \text{BQP}_n$.

Proof Let $(\bar{\mathbf{x}}, \bar{\mathbf{y}})$ be an extreme point of BQP_n. By definition, we have $\bar{x}_i \in \{0, 1\}$ for $i \in V_n$ and $\bar{y}_{ij} = \bar{x}_i \bar{x}_j$ for $\{i, j\} \in E_n$. Now let $(\tilde{\mathbf{x}}, \tilde{\mathbf{y}}) = \psi^S(\bar{\mathbf{x}}, \bar{\mathbf{y}})$. From the definition of switching, we have $\tilde{x}_i \in \{0, 1\}$ for $i \in V_n$ and $\tilde{y}_{ij} = \tilde{x}_i \tilde{x}_j$ for $\{i, j\} \in E_n$. Thus, $(\tilde{\mathbf{x}}, \tilde{\mathbf{y}})$ is also an extreme point of BQP_n. This shows that every extreme point of $\psi^S(\text{BQP}_n)$ is an extreme point of BQP_n. A similar argument shows that every extreme point of BQP_n is an extreme point of $\psi^S(\text{BQP}_n)$. Now, recall that BQP_n is a polytope. Given that switching is an affine transformation, $\psi^S(\text{BQP}_n)$ must be a polytope as well. Thus, BQP_n and $\psi^S(\text{BQP}_n)$ are polytopes with the same extreme points, and are therefore equal. □

The permutation and switching transformations are very useful, because they enable one to convert valid linear inequalities for BQP_n into other valid linear inequalities that induce faces of the same dimension. For example, if we take the inequality $y_{ij} \geq 0$ and switch on $\{i\}$ or $\{j\}$, we obtain the inequalities $y_{ij} \leq x_j$ and $y_{ij} \leq x_i$, respectively. If we switch on $\{i, j\}$ instead, we obtain the inequality $y_{ij} \geq x_i + x_j - 1$.

We remark that switching on S and then switching on T is equivalent to switching on the set $(S \cup T) \setminus (S \cap T)$. Thus, given any valid (or facet-inducing) inequality for BQP_n, we can obtain up to $2^n - 1$ other valid (or facet-inducing) inequalities by switching.

4.4 Some More Valid Inequalities

In this section, we review some additional valid inequalities for BQP_n.

Padberg [53] derived three additional families of inequalities. The first are the following *triangle* inequalities:

$$x_i + x_j + x_k \le y_{ij} + y_{ik} + y_{jk} + 1 \quad \big(\{i, j, k\} \subseteq V_n\big) \tag{4.8}$$

$$y_{ij} + y_{ik} \le x_i + y_{jk} \quad \big(i \in V_n,\ \{j, k\} \subseteq V_n \setminus \{i\}\big). \tag{4.9}$$

To see how these might be useful as cutting planes, observe that fractional points with $x_i = x_j = x_k = 1/2$ and $y_{ij} = y_{ik} = y_{jk} = 0$ satisfy (4.3)–(4.5), but violate (4.8). Similarly, fractional points with $x_i = x_j = x_k = y_{ij} = y_{ik} = 1/2$ and $y_{jk} = 0$ satisfy (4.3)–(4.5), but violate (4.9). Note that (4.9) can be obtained from (4.8) by switching on $\{i\}$.

Padberg's second family are called *clique* inequalities. The easiest way to derive them is to note that, given any integer s, we have $s(s+1) \ge 0$. Thus, for any $S \subseteq V_n$ and any integer s, we have

$$\left(\sum_{i \in S} x_i - s\right)\left(\sum_{i \in S} x_i - s - 1\right) \ge 0.$$

Expanding this and re-arranging yields

$$(2s+1)\sum_{i \in S} x_i - \sum_{i \in S} x_i^2 \le 2 \sum_{\{i,j\} \subseteq S} x_i x_j + s(s+1).$$

Linearising and dividing by two yields the clique inequalities:

$$s \sum_{i \in S} x_i \le \sum_{\{i,j\} \subseteq S} y_{ij} + \binom{s+1}{2} \quad \big(S \subseteq V_n,\ s = 0, \ldots, |S| - 1\big). \tag{4.10}$$

Padberg showed that these induce facets when $|S| \ge 3$ and $1 \le s \le |S| - 2$.

Note that the clique inequalities (4.10) reduce to the triangle inequalities (4.8) when $|S| = 3$ and $s = 1$. Moreover, the inequalities (4.5) can be regarded as "degenerate" clique inequalities with $|S| = 2$ and $s = 1$. In a similar way, the non-negativity inequalities $y_{ij} \ge 0$ can be regarded as "degenerate" clique inequalities with $|S| = 2$ and $s = 0$.

Padberg's last family are called *cut* inequalities. They can be derived from the fact that, for any disjoint sets $S, T \subset V_n$, we have

$$\left(\sum_{i \in S} x_i - \sum_{i \in T} x_i\right)\left(\sum_{i \in S} x_i - \sum_{i \in T} x_i - 1\right) \ge 0.$$

They take the form:

$$\sum_{i\in S, j\in T} y_{ij} \leq \sum_{i\in T} x_i + \sum_{\{i,j\}\subseteq S} y_{ij} + \sum_{\{i,j\}\subseteq T} y_{ij} \qquad (S, T \subseteq V_n, S\cap T = \emptyset). \tag{4.11}$$

They induce facets when $|S| \geq 1$ and $|T| \geq 2$.

Note that the cut inequalities (4.11) reduce to the triangle inequalities (4.9) when $|S| = 2$ and $|T| = 1$. Moreover, the inequalities (4.3) and (4.4) can be regarded as "degenerate" cut inequalities with $|S| = |T| = 1$.

Next, we observe that the arguments for proving the validity of the clique and cut inequalities can be easily generalised. Indeed, for any disjoint sets $S, T \subset V_n$ and any $s \in \mathbf{Z}$, we have

$$\left(\sum_{i\in S} x_i - \sum_{i\in T} x_i - s\right)\left(\sum_{i\in S} x_i - \sum_{i\in T} x_i - s - 1\right) \geq 0.$$

Expanding and linearising yields

$$s\sum_{i\in S} x_i + \sum_{i\in S, j\in T} y_{ij} \leq (s+1)\sum_{i\in T} x_i + \sum_{\{i,j\}\subseteq S} y_{ij} + \sum_{\{i,j\}\subseteq T} y_{ij} + \binom{s+1}{2}. \tag{4.12}$$

These inequalities, which include all those mentioned so far, have been rediscovered many times (e.g., [9, 15, 21, 69]). They define facets when $|S| + |T| \geq 3$ and $1 - |T| \leq s \leq |S| - 2$. We remark that they can also be derived by taking the clique inequality (4.10), and switching on T.

An even larger family of valid inequalities was found by Boros and Hammer [9]. Take an arbitrary vector $\mathbf{v} \in \mathbf{Z}^n$ and integer s, and consider the quadratic inequality $(\mathbf{v}^T\mathbf{x} - s)(\mathbf{v}^T\mathbf{x} - s - 1) \geq 0$. Expanding and linearising yields:

$$\sum_{i\in V_n} v_i(2s + 1 - v_i)x_i \leq 2 \sum_{1\leq i<j\leq n} v_i v_j y_{ij} + s(s+1). \tag{4.13}$$

Although the Boros-Hammer inequalities are infinite in number, it is known that they define a polytope [45]. That is, a finite number of them dominate all the others. At the time of writing, however, a necessary and sufficient condition for a Boros-Hammer inequality to define a facet of BQP_n is not known.

We remark that switching a Boros-Hammer inequality is remarkably easy. Indeed, to switch on a set $S \subset V_n$, it suffices to change the sign of v_i for all $i \in S$.

Still more valid inequalities for BQP_n can be derived from a connection between BQP_n and the *cut polytope*. This is explained in the next section.

4.5 Some Related Polytopes

We now review some polytopes that are closely related to the Boolean quadric polytope. Section 4.5.1 deals with the cut polytope, and Sect. 4.5.2 deals with polytopes that exploit sparsity in the objective function.

4.5.1 *The Cut Polytope*

As before, let $K_n = (V_n, E_n)$ denote the complete graph on n nodes. Given any set $S \subseteq V_n$, we let $\delta(S)$ denote the set of edges in E_n that have exactly one end-node in S. The set $\delta(S)$ is called an *edge-cutset* or simply *cut*. Given an integer $n \geq 3$ and a *weight* $w_e \in \mathbf{Q}$ for all $e \in E_n$, the *max-cut* problem calls for a cut of maximum total weight.

It is well-known (e.g., [7, 14]) that any QUBO instance with n variables can be converted into a max-cut instance with $n + 1$ variables, and vice-versa. This result turns out to have a polyhedral counterpart. Before explaining this, we first present the standard 0-1 LP formulation of the max-cut problem.

For all $e \in E_n$, let z_e be a binary variable, taking the value 1 if and only if e belongs to the cut. The max-cut problem can be formulated as:

$$\max \quad \sum_{e \in E_n} w_e z_e \tag{4.14}$$

$$\text{s.t. } z_{ij} + z_{ik} + z_{jk} \leq 2 \ \left(\{i, j, k\} \subseteq V_n\right) \tag{4.15}$$

$$z_{ij} - z_{ik} - z_{jk} \leq 0 \ \left(\{i, j\} \in E_n, \ k \in V_n \setminus \{i, j\}\right) \tag{4.16}$$

$$\mathbf{z} \in \{0, 1\}^{\binom{n}{2}}. \tag{4.17}$$

The constraints (4.15), (4.16) are (somewhat confusingly) also called *triangle inequalities*.

For a given $n \geq 3$, the convex hull of vectors $\mathbf{z}$ satisfying (4.15)–(4.17) is called the *cut polytope* and denoted by CUT_n [6]. To make this clear, consider the case $n = 3$. There are four cut vectors:

$$\begin{pmatrix} z_{12} \\ z_{13} \\ z_{23} \end{pmatrix} \in \left\{ \begin{pmatrix} 0 \\ 0 \\ 0 \end{pmatrix}, \begin{pmatrix} 0 \\ 1 \\ 1 \end{pmatrix}, \begin{pmatrix} 1 \\ 0 \\ 1 \end{pmatrix}, \begin{pmatrix} 1 \\ 1 \\ 0 \end{pmatrix} \right\}.$$

One can check that these vectors are affinely independent. Thus, CUT_3 is a tetrahedron, as shown in Fig. 4.4.

One can check that the tetrahedron in question is defined by the triangle inequalities $z_{12} + z_{13} + z_{23} \leq 2$, $z_{12} - z_{13} - z_{23} \leq 0$, $z_{13} - z_{13} - z_{23} \leq 0$

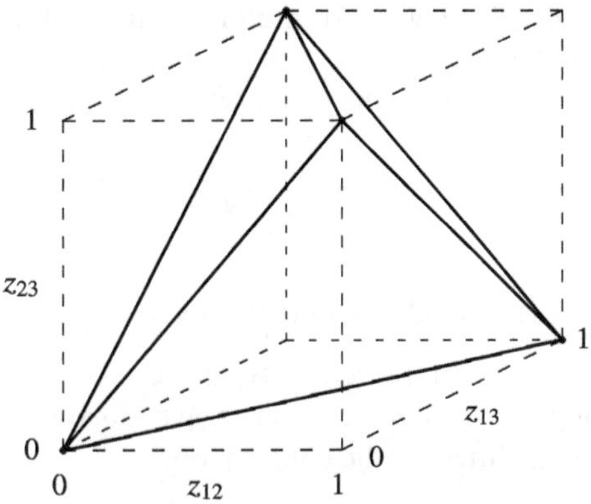

Fig. 4.4 The cut polytope of order 3

and $z_{23} - z_{12} - z_{13} \leq 0$. In other words, for $n = 3$, the triangle inequalities give a complete linear description of CUT_n.

Now recall that BQP_2 was also a tetrahedron. It turns out that BQP_n and CUT_{n+1} are congruent to each other, under a simple (invertible) linear transformation [14, 19, 53]:

Theorem 4.1 *Let* $\mathbf{x}^* \in \mathbf{R}^n$ *and* $\mathbf{y}^* \in \mathbf{R}^{\binom{n}{2}}$ *be given. Construct a vector* $\mathbf{z}^* \in \mathbf{R}^{\binom{n+1}{2}}$ *as follows:*

$$\begin{aligned} z^*_{i,n+1} &= \qquad x^*_i \qquad\quad \big(i \in V_n\big) \\ z^*_{ij} &= x^*_i + x^*_j - 2y^*_{ij} \;\; \big(\{i,j\} \in E_n\big). \end{aligned}$$

Then $(\mathbf{x}^*, \mathbf{y}^*) \in \text{BQP}_n$ *if and only if* $\mathbf{z} \in \text{CUT}_{n+1}$.

The linear transformation in Theorem 4.1 has come to be known as the *covariance map* [21]. A consequence of Theorem 4.1 is that the inequality $\alpha^T z \leq \beta$ is valid for CUT_{n+1} if and only if the inequality

$$\sum_{i \in V_n} \left(\sum_{j \in V_{n+1} \setminus \{i\}} \alpha_{ij} \right) x_i - 2 \sum_{e \in E_n} \alpha_e y_e \leq \beta$$

is valid for BQP_n. This enables one to easily convert valid (or facet-defining) inequalities for the cut polytope into valid (or facet-defining) inequalities for the Boolean quadric polytope, and vice-versa.

Example If we take the inequalities (4.15) and apply the covariance map, we can obtain the inequalities (4.5) (if $k = n + 1$) or (4.8) (if $n + 1 \notin \{i, j, k\}$). Similarly, if we take the inequalities (4.16) and apply the covariance map, we can obtain the inequalities (4.3) and (4.4) (if we set i or j to $n + 1$), the non-negativity inequality $y_{ij} \geq 0$ (if we set k to $n + 1$), or the inequality (4.9) (if $n + 1 \notin \{i, j, k\}$). □

Example If we take the clique inequalities (4.10) with $|S|$ odd and $s = (|S| - 1)/2$, and apply the covariance map, we obtain (with a little work) the following inequalities for the cut polytope:

$$\sum_{\{i,j\} \subset S} z_{ij} \le \left\lfloor |S|^2/4 \right\rfloor \qquad \big(S \subseteq V_n : |S| \text{ odd}\big). \tag{4.18}$$

These inequalities were discovered by Barahona and Mahjoub [6]. □

Example More generally, if we take the Boros-Hammer inequalities (4.13), and apply the covariance map, we obtain (again with a little work) the following inequalities for the cut polytope:

$$\sum_{\{i,j\} \in E_n} v_i v_j z_{ij} \le \left\lfloor \frac{\sigma(\mathbf{v})^2}{4} \right\rfloor \qquad \big(\mathbf{v} \in \mathbf{Z}^n : \sigma(\mathbf{v}) \text{ odd}\big). \tag{4.19}$$

These inequalities were discovered by Deza (see [21]). □

We remark that Laurent and Poljak [44] derived a family of inequalities for CUT_n, called *gap* inequalities, that are even more general than (4.19). In [24], the gap inequalities are adapted to BQP_n, and then generalised to the case of general mixed-integer quadratic programs.

One can also define a switching operation for the cut polytope [6, 21].

Proposition 4.2 (Switching for the Cut Polytope) *For an arbitrary set $S \subseteq V_n$, let $\pi^S : \mathbf{R}^{\binom{n}{2}} \mapsto \mathbf{R}^{\binom{n}{2}}$ be the affine transformation that:*

- *replaces z_e with $1 - z_e$ for all $e \in \delta(S)$,*
- *leaves z_e unchanged for all $e \in E_n \setminus \delta(S)$.*

CUT_n *is invariant under this operation.*

This switching operation enables one to take any valid inequality for CUT_n and generate other valid inequalities. For example, if we take the triangle inequality (4.15) and switch on $\{k\}$, we obtain the triangle inequality (4.16).

Many other valid inequalities have been discovered for the cut polytope (see [21] for a survey). Among them, we mention only the *odd bicycle wheel* inequalities [6] and the *2-circulant* inequalities [57]. We will see in Sect. 4.7 that those particular inequalities are "well-behaved" from an algorithmic viewpoint.

For some other polytopes related to BQP_n see, e.g., [36, 45, 49, 61, 63]. We close this subsection with a remark about the strength of the LP relaxation of the 0-1 LP (4.14)–(4.17). Poljak and Tuza [58] showed that, even if all edge-weights are non-negative, the upper bound from the relaxation can be as large as twice the optimum. In other words, the integrality gap can be as large as 100%. For a generalisation of this result, see [5].

4.5.2 Polytopes Which Exploit Sparsity

A matrix is said to *sparse* if the majority of its elements are zero. Consider a QUBO instance whose quadratic cost matrix $\mathbf{Q}$ is sparse, and assume w.l.o.g. that $\mathbf{Q}$ is symmetric. For all $\{i, j\} \in E_n$ such that $q_{ij} = 0$, we can delete the variable y_{ij} from the formulation (4.2)–(4.7), along with the associated constraints. This makes the formulation much smaller and, if we are lucky, much easier to solve. On the other hand, we must take care when deriving valid inequalities: we can no longer use inequalities that involve the variables that have been deleted.

To deal with this from a polyhedral point of view, we need a bit of notation. Let $E = \big\{\{i, j\} \in E_n : q_{ij} \neq 0\big\}$, let $m = |E|$, and let $G = \big(V_n, E\big)$. We define the polytope:

$$\text{BQP}(G) = \text{conv}\Big\{(\mathbf{x}, \mathbf{y}) \in \{0, 1\}^{n+m} : y_{ij} = x_i x_j \,\big(\{i, j\} \in E\big)\Big\}.$$

Geometrically speaking, BQP(G) is the *projection* of BQP_n into $\mathbf{R}^{n+m}$.

Unfortunately, projecting a polytope into a subspace is difficult computationally. This makes it harder to derive valid inequalities for BQP(G) than for BQP_n. Nevertheless, some useful inequalities are known. In particular, Padberg [53] derived some inequalities called *odd cycle* inequalities, and proved that they define facets of BQP(G). We do not go into details, however, since the notation is rather burdensome.

We can exploit sparsity in the case of the max-cut problem as well. Consider a max-cut instance defined on a graph $G = \big(V_n, E\big)$. For a given $S \subseteq V_n$, we let $\delta_G(S)$ denote the set of edges in E that have exactly one end-node in S. We then define the polytope:

$$\text{CUT}(G) = \text{conv}\Big\{\mathbf{z} \in \{0, 1\}^m : \exists S \subseteq V_n : z_e = 1 \iff e \in \delta_G(S)\Big\}.$$

As one might expect, CUT(G) is the projection of CUT_n into $\mathbf{R}^m$.

Barahona and Mahjoub [6] proved the following. Let C be the set of chordless simple cycles in G. Then a vector $\mathbf{z} \in \{0, 1\}^m$ belongs to CUT(G) if and only if it satisfies the following inequalities:

$$\sum_{e \in C \setminus D} z_e \geq \sum_{e \in D} z_e - |D| + 1 \qquad \big(C \in \mathcal{C},\ D \subseteq C : |D| \text{ odd}\big).$$

These inequalities are called *co-circuit* inequalities. Their validity follows from the fact that every cut intersects every cycle an even number of times. Note that the number of co-circuit inequalities can grow exponentially with n. Note also that, when $G = K_n$, every chordless simple cycle is a triangle, and the co-circuit inequalities reduce to the triangle inequalities (4.15), (4.16).

It turns out that Padberg's odd cycle inequalities for BQP(G) are precisely the inequalities that can be obtained from the co-circuit inequalities via the covariance map. We omit the proof, for brevity.

4.6 Some Other Related Convex Sets

In this section, we mention some other important convex sets related to BQP_n. Section 4.6.1 presents a non-polyhedral convex set that contains BQP_n, and Sect. 4.6.2 presents the analogous set for CUT_n. Then, Sect. 4.6.3 deals with certain convex cones.

4.6.1 A Non-polyhedral Convex Set

Our first convex set arises from a certain *semidefinite programming* (SDP) relaxation of QUBO. This set turns out to be non-polyhedral, because an infinite number of linear inequalities are needed to define it. (Informally speaking, it has a "curved" surface.)

The idea of applying SDP to 0-1 quadratic programs is due to Shor [66], and was developed in, e.g., [35, 40, 59]. The basic idea is as follows. We define the $n \times n$ symmetric matrix $\hat{\mathbf{X}} = \mathbf{x}\mathbf{x}^T$, along with the augmented matrix

$$\hat{\mathbf{X}}^+ := \begin{pmatrix} 1 \\ \mathbf{x} \end{pmatrix} \begin{pmatrix} 1 \\ \mathbf{x} \end{pmatrix}^T = \begin{pmatrix} 1 & \mathbf{x}^T \\ \mathbf{x} & \hat{\mathbf{X}} \end{pmatrix}.$$

Since $\hat{\mathbf{X}}^+$ is defined as the product of a vector and its transpose, it should be psd. Moreover, given that $x_i = x_i^2$ for all i, the main diagonal of $\hat{\mathbf{X}}$ should be equal to $\mathbf{x}$. This leads immediately to the following SDP relaxation of QUBO:

$$\max \left\{ \mathbf{Q} \bullet \hat{\mathbf{X}} : \; \text{diag}(\hat{\mathbf{X}}) = \mathbf{x}, \; \hat{\mathbf{X}}^+ \in \mathcal{S}_+^{n+1} \right\}.$$

To someone who is unfamiliar with nonlinear optimisation, this SDP relaxation may look somewhat mysterious. Fortunately, it can be interpreted in the space of the x and y variables. Indeed, $\hat{\mathbf{X}}^+$ is psd if and only if

$$\begin{pmatrix} s \\ \mathbf{v} \end{pmatrix}^T \begin{pmatrix} 1 & \mathbf{x}^T \\ \mathbf{x} & \hat{\mathbf{X}} \end{pmatrix} \begin{pmatrix} s \\ \mathbf{v} \end{pmatrix} \geq 0 \qquad \left(s \in \mathbf{R}, \; \mathbf{v} \in \mathbf{R}^n \right). \tag{4.20}$$

Moreover, we have $\hat{x}_{ii} = x_i$ for all $i \in V_n$, and $\hat{x}_{ij} = \hat{x}_{ji} = y_{ij}$ for all $\{i, j\} \in E_n$. Thus, we can write the inequalities (4.20) in the following form:

$$\sum_{i \in V_n} v_i \big(2s + v_i\big) x_i + 2 \sum_{\{i,j\} \in E_n} v_i v_j y_{ij} + s^2 \geq 0 \qquad \big(s \in \mathbf{R},\ \mathbf{v} \in \mathbf{R}^n\big). \tag{4.21}$$

We will call these *psd* inequalities. Note that the psd inequalities are infinite in number.

The psd inequalities include some important inequalities as special cases. For example, if we set v_i to 1, s to 0, and all other components of $\mathbf{v}$ to 0 in (4.21), we obtain the non-negativity inequality $x_i \geq 0$. If we change s to -1, we obtain $-x_i + 1 \geq 0$, which is equivalent to the upper bound $x_i \leq 1$.

Now consider the following principle submatrix of $\hat{\mathbf{X}}$:

$$\begin{pmatrix} \hat{x}_{ii} & \hat{x}_{ij} \\ \hat{x}_{ij} & \hat{x}_{jj} \end{pmatrix}.$$

Given that the SDP relaxation includes the constraints $x_i = \hat{x}_{ii}$ and $x_j = \hat{x}_{jj}$, this submatrix must be equal to

$$\begin{pmatrix} x_i & \hat{x}_{ij} \\ \hat{x}_{ij} & x_j \end{pmatrix}.$$

Moreover, given that $\hat{\mathbf{X}}$ is psd, this submatrix must have non-negative determinant. Thus, all feasible solutions to the SDP relaxation satisfy $\hat{x}_{ij}^2 \leq x_i x_j$. In other words, the projection into $(\mathbf{x}, \mathbf{y})$-space satisfies $y_{ij}^2 \leq x_i x_j$. This implies in particular that $y_{ij} \leq 1$.

On the other hand, the projection does not satisfy the non-negativity inequalities of the form $y_{ij} \geq 0$. Indeed, one can check that, when $n = 2$, we obtain a feasible solution to the SDP by setting x_1, x_2, $\hat{x}_{11}$ and $\hat{x}_{22}$ to $1/4$, and setting $\hat{x}_{12}$ to $-1/8$. In other words, the SDP relaxation can be strengthened by adding the inequalities $\hat{x}_{ij} \geq 0$ for $1 \leq i < j \leq n$.

4.6.2 A Convex Set Related to Max-Cut

As one might expect, the results in the previous subsection have an analogue for the max-cut problem. The SDP relaxation of max-cut was suggested by Schrijver (unpublished) and analysed in, e.g. [29, 43, 56].

We now show how to derive the SDP relaxation. Recall the definitions of $\delta(S)$ and w_e from Sect. 4.5.1. For each $i \in V_n$, let μ_i be a variable that takes the value 1 if $i \in S$, and -1 otherwise. One can formulate max-cut as the following bivalent quadratic program:

$$\max \left\{ \frac{1}{2} \sum_{\{i,j\} \in E_n} w_{ij} \left(1 - \mu_i \mu_j\right) : \boldsymbol{\mu} \in \{-1, +1\}^n \right\}.$$

To see that this formulation is valid, note that the quantity $\frac{1}{2}(1 - \mu_i \mu_j)$ equals 1 if nodes i and j are on opposite shores of the cut, and 0 if they are on the same shore.

Next, we define the matrix $\mathbf{M} = \boldsymbol{\mu}\boldsymbol{\mu}^T$. Note that $\mathbf{M}$ is psd and has 1s on the main diagonal (since $\boldsymbol{\mu}_i^2 = 1$ for all $i \in V_n$). This leads immediately to the SDP relaxation

$$\max \left\{ \frac{1}{2} \sum_{\{i,j\} \in E_n} w_{ij} \left(1 - m_{ij}\right) : m_{ii} = 1 \left(i \in V_n\right), \mathbf{M} \in \mathcal{S}^n_+ \right\}.$$

The feasible region of this SDP is called the *elliptope* [43].

Note that the matrix $\mathbf{M}$ is related to the traditional z variables via the identities $m_{ij} = 1 - 2z_{ij}$ for $\{i, j\} \in E_n$. Using this fact, Laurent and Poljak [43] projected the elliptope into $\mathbf{z}$-space. The resulting convex set is defined by the following inequalities:

$$\sum_{\{i,j\} \in E_n} v_i v_j z_{ij} \leq \sigma(\mathbf{v})^2/4 \qquad (\mathbf{v} \in \mathbf{R}^n). \tag{4.22}$$

One can check that these inequalities are equivalent to the psd inequalities (4.21), via the covariance map.

It is not hard to see that the inequalities (4.19) dominate the inequalities (4.22). This implies in turn that the Boros-Hammer inequalities (4.13) dominate the psd inequalities (4.21). See [21] for detailed proofs.

4.6.3 Cones

There are also several important *convex cones* that are related to BQP_n. To explain them properly, we need to define two more families of inequalities:

- When $\sigma(\mathbf{v}) = 1$, the inequalities (4.19) reduce to

$$\sum_{\{i,j\} \in E_n} v_i v_j z_{ij} \leq 0 \qquad \left(\mathbf{v} \in \mathbf{Z}^n : \sigma(\mathbf{v}) = 1\right). \tag{4.23}$$

  These are called *hypermetric* inequalities [17, 18].

- When $\sigma(\mathbf{v}) = 0$, the inequalities (4.22) reduce to

$$\sum_{\{i,j\} \in E_n} v_i v_j z_{ij} \le 0 \qquad (\mathbf{v} \in \mathbf{R}^n : \sigma(\mathbf{v}) = 0). \tag{4.24}$$

  These are called *negative-type* inequalities [19, 64].

Note that the triangle inequalities (4.16) are hypermetric inequalities. It is also known that the hypermetric inequalities dominate the negative-type inequalities [19].

We now define four convex cones:

- The *cut* cone of order n, which we will call CC_n, is the conic hull of the vectors $\mathbf{z}$ that lie in CUT_n.
- The *metric* cone of order n, denoted by MET_n, is the set of points $\mathbf{z}$ satisfying the triangle inequalities (4.16).
- The *hypermetric* cone, HYP_n, is the cone defined by the hypermetric inequalities (4.23).
- The *negative-type* cone, NEG_n, is the cone defined by the negative-type inequalities (4.24).

From the above considerations, we have $\text{CC}_n \subset \text{HYP}_n \subset \text{MET}_n \cap \text{NEG}_n$. By definition, the cut and metric cones are polyhedral. It has also been shown that the hypermetric cone is polyhedral [22]. The negative-type cone, however, is not. All four cones have interesting applications to the theory of metric spaces and the geometry of numbers; see again [21] for details.

Note that, if we are given a complete linear description of CC_n, we can use switching to get a complete linear description of CUT_n. To see this, let $\bar{\mathbf{z}}$ be an extreme point of CUT_n that is not the origin, and let $\delta(S)$ be the corresponding cut in K_n. If we switch on S, $\bar{\mathbf{z}}$ is mapped to the origin, and any facet containing $\bar{\mathbf{z}}$ is mapped to a facet containing the origin. Reversing this argument, we can obtain any inequality that defines a facet of CUT_n by switching an inequality that defines a facet of CC_n.

Using the covariance map, one can derive analogous inequalities and cones in $(\mathbf{x}, \mathbf{y})$-space. For the sake of brevity, we do not go into details. We just mention that the hypermetric inequalities (4.23) map to the following inequalities for BQP_n:

$$\sum_{i \in V_n} v_i \big(1 - v_i\big) x_i \le 2 \sum_{1 \le i < j \le n} v_i v_j y_{ij} \qquad \big(\mathbf{v} \in \mathbf{Z}^n\big). \tag{4.25}$$

These are called *hypermetric correlation* inequalities [20]. We remark that they can be viewed as the special case of the Boros-Hammer inequalities (4.13) in which $s = 0$.

4.7 Algorithms

Now we turn to the algorithmic implications of the above results. Section 4.7.1 describes a (fairly) simple exact algorithm, called *cut-and-branch*. Section 4.7.2 concerns subroutines that search for useful cutting planes. Finally, Sect. 4.7.3 describes a more sophisticated algorithmic framework, called *branch-and-cut*.

4.7.1 *Cut-and-Branch*

Suppose we wish to solve an ILP of the form (4.1). The continuous relaxation of the ILP is an LP, which can be solved with, e.g., the simplex method. This yields a solution, say $\mathbf{x}^*$. If $\mathbf{x}^*$ is integral, we have solved the ILP. If not, the quantity $\mathbf{c}^T\mathbf{x}^*$ gives an *upper bound* on the optimal profit.

At this point, we can resort to an old-fashioned solution method, such as Gomory's cutting-plane method [30] or branch-and-bound [42]. A more effective approach, first suggested by Crowder et al. [13], is to add some *strong* (preferably facet-defining) valid inequalities to the formulation, and then invoke branch-and-bound. The resulting algorithm, which has come to be known as "cut-and-branch", is outlined in Algorithm 1.

Algorithm 1: Cut-and-branch algorithm

input : positive integers n and m; matrix $\mathbf{A}$; vectors $\mathbf{b}$, $\mathbf{c}$
Solve the LP relaxation and let $\mathbf{x}^*$ be the solution;
repeat
    if $\mathbf{x}^*$ *is integer* **then**
        Output $\mathbf{x}^*$ and quit;
    end
    Search for strong valid inequalities that are violated by $\mathbf{x}^*$;
    if *at least one inequality has been found* **then**
        Add one or more inequalities to the LP as cutting planes;
        Re-optimise the LP and update $\mathbf{x}^*$;
    end
until *no more violated inequalities are found*;
Optional: Delete all cutting planes that have a positive slack;
Declare all variables integer;
Feed the resulting ILP into a branch-and-bound solver;
Let $\mathbf{x}^*$ be the solution;
output: Optimal solution $\mathbf{x}^*$

The idea behind cut-and-branch is that the cutting planes typically yield a significant decrease in the upper bound. This in turn leads to a reduction in the size of the branch-and-bound tree. The results in [13] indicate that, for many ILPs arising in practice, the reduction in the size of the tree can be dramatic, and enable one to solve ILPs that are unsolvable with traditional branch-and-bound.

Cut-and-branch algorithms for QUBO and related problems can be found in, e.g., [7, 23, 35, 49].

4.7.2 Separation Algorithms

In Algorithm 1, there is a line that says "Search for strong valid inequalities that are violated by $\mathbf{x}^*$". Geometrically speaking, finding such inequalities (cutting planes) amounts to finding a hyperplane that "separates" the current fractional LP solution from the feasible integer solutions. For this reason, algorithms that search for cutting planes are called "separation algorithms" [33]. More precisely:

- An *exact separation algorithm* for a given family of valid inequalities is an algorithm that takes a fractional LP solution as input, and outputs one or more violated inequalities in the given family, if any exist.
- A *heuristic separation algorithm* for a given family of valid inequalities is an algorithm that takes a fractional LP solution as input, and outputs either one or more violated inequalities in the given family, or a failure message.

In the context of QUBO, we can assume that the fractional solution is a pair $(\mathbf{x}^*, \mathbf{y}^*) \in [0, 1]^{n+\binom{n}{2}}$.

The separation problem for the triangle inequalities (4.8), (4.9) can be solved in $O(n^3)$ time by brute-force enumeration. The complexity of the separation problems for the inequalities (4.10)–(4.13) is unknown, but we suspect that they are all $\mathcal{NP}$-hard. Greedy separation heuristics for the inequalities (4.10)–(4.12) can be found in, e.g., [47, 67, 69].

The separation problem for the psd inequalities (4.21) can be solved in polynomial time [33]. The following method works well in practice (e.g., [34, 65]). Given $(\mathbf{x}^*, \mathbf{y}^*)$, construct the matrix $\hat{\mathbf{X}}^+$. Find the minimum eigenvalue of $\hat{\mathbf{X}}^+$, to some desired precision. If the eigenvalue is non-negative, stop. Otherwise, find the associated eigenvector, again to the desired precision. Write the eigenvector as $\binom{s^*}{\mathbf{v}^*}$. This eigenvector yields a violated psd inequality. To see why, let $\lambda < 0$ be the eigenvalue, and note that

$$\begin{pmatrix} s^* \\ \mathbf{v}^* \end{pmatrix}^T \begin{pmatrix} 1 & \mathbf{x}^T \\ \mathbf{x} & \hat{\mathbf{X}} \end{pmatrix} \begin{pmatrix} s^* \\ \mathbf{v}^* \end{pmatrix} = \begin{pmatrix} s^* \\ \mathbf{v}^* \end{pmatrix}^T \left(\lambda \begin{pmatrix} s^* \\ \mathbf{v}^* \end{pmatrix} \right) = \lambda \left\| \begin{pmatrix} s^* \\ \mathbf{v}^* \end{pmatrix} \right\|_2^2 < 0.$$

Several polynomial-time separation algorithms are known for the cut polytope. The separation problem for the triangle inequalities (4.15), (4.16) can be solved in $O(n^3)$ time by enumeration. Gerards [27] presented an $O(n^5)$ separation algorithm for the odd bicycle wheel inequalities. There are also $O(n^5)$ separation algorithms for various generalisations of the 2-circulant inequalities [37, 38, 45]. The separation problems for the inequalities (4.22) and (4.24) can be solved in a similar way to the psd inequalities (see [21]).

At the time of writing, the complexity of separation is unknown for the remaining inequalities for the cut polytope. In [35], a greedy separation heuristic is presented for the inequalities (4.18) and their switchings. Separation heuristics for the inequalities (4.19) can be found in [16, 25, 34, 35]. A separation heuristic for the gap inequalities is given in [25].

Letchford and Sørensen [46] showed that the separation problems for the inequalities (4.13), (4.19), (4.23) and (4.25) are equivalent. That is, either all of them can be solved in polynomial time, or none of them can. See also Avis [4].

Finally, we mention that there are several exact and heuristic separation algorithms designed for *sparse* QUBO and max-cut instances (e.g., [6–8, 11, 48]). We omit details, for brevity.

4.7.3 Branch-and-Cut

Now that we have explained the concept of separation, we return to solution algorithms. Recall that, in cut-and-branch, we are permitted to add cutting planes only *before* running branch-and-bound. A natural extension is to permit the addition of cutting planes *while* running branch-and-bound.

To make this more precise, we recall that branch-and-bound is a recursive algorithm, which solves a series of LP subproblems, arranged in a tree structure. Suppose that we have just solved the LP that corresponds to one particular branch of the tree. If the solution is fractional, we can run one or more separation algorithms, in an attempt to cut it off. If any cutting planes are found, we can add them to the LP, re-optimise, and repeat. This causes the upper bound at that branch to decrease, which may allow one to eliminate the branch from consideration.

This approach was discovered by several authors, apparently independently (e.g., [32, 50, 54]). It was given the name *branch-and-cut* by Padberg and Rinaldi [54]. It works remarkably well, and has been applied to a wide range of problems in integer programming and combinatorial optimisation [10, 51].

Although branch-and-cut is conceptually simple, it requires considerably more programming effort than cut-and-branch. The main reason is that the branch-and-bound solver can no longer be treated as a "black box". Moreover, some additional implementation "tricks" are needed to make the approach work efficiently. Several such tricks are given in [54], such as (a) starting with a subset of the variables and generating the others dynamically, (b) storing cutting planes in a "cut pool", (c) scanning the cut pool before calling the more time-consuming separation routines, (d) deleting non-binding cutting planes to save memory, and (e) producing heuristic integer solutions by rounding fractional values to integers.

Oddly, no-one has yet designed and implemented a full branch-and-cut algorithm for QUBO. Indeed, at present, the most effective exact algorithms for QUBO use

SDP relaxations, with triangle inequalities incorporated via Lagrangian relaxation (see, e.g., [41, 62]).

4.8 Concluding Remarks

The Boolean quadric and cut polytopes have been studied in depth, and many families of strong valid linear inequalities are now known. For some of the families, we also have efficient exact or heuristic separation algorithms.

There remain several interesting directions for possible future research. Among them, we mention the following:

- Determine whether or not the separation problem for the hypermetric inequalities (4.23) can be solved in polynomial time.
- Design, implement and test a full branch-and-cut algorithm for QUBO and related problems.
- Understand better the relative advantages and disadvantages of LP-based and SDP-based approaches to QUBO.
- Provide an open source library of separation algorithms for QUBO and related problems.

References

1. K. Aardal, C.P.M. Van Hoesel, Polyhedral techniques in combinatorial optimization I: theory. Stat. Neerlandica **50**, 3–26 (1996)
2. K. Aardal, C.P.M. Van Hoesel, Polyhedral techniques in combinatorial optimization II: applications and computations. Stat. Neerlandica **53**, 131–177 (1999)
3. K. Aardal, R. Weismantel, Polyhedral combinatorics, in *Annotated Bibliographies in Combinatorial Optimization*, ed. by M. Dell'Amico, F. Maffioli, S. Martello (Wiley, New York, 1997), pp. 31–44
4. D. Avis, On the complexity of testing hypermetric, negative type, k-gonal and gap inequalities, in *Discrete and Computational Geometry*, ed. by J. Akiyama, M. Kanö (Springer, Berlin, 2003), pp. 51–59
5. D. Avis, J. Umemoto, Stronger linear programming relaxations of max-cut. Math. Program. **97**, 451–469 (2003)
6. F. Barahona, A. Mahjoub, On the cut polytope. Math. Program. **36**, 157–173 (1986)
7. F. Barahona, M. Jünger, G. Reinelt, Experiments in quadratic 0-1 programming. Math. Program. **44**, 127–137 (1989)
8. T. Bonato, M. Jünger, G. Reinelt, G. Rinaldi, Lifting and separation procedures for the cut polytope. Math. Program. **146**, 351–378 (2014)
9. E. Boros, P.L. Hammer, Cut-polytopes, Boolean quadric polytopes and nonnegative quadratic pseudo-Boolean functions. Math. Oper. Res. **18**, 245–253 (1993)
10. A. Caprara, M. Fischetti, Branch-and-cut algorithms, in *Annotated Bibliographies in Combinatorial Optimization*, ed. by M. Dell'Amico, F. Maffioli, S. Martello (Wiley, New York, 1997), pp. 45–64

11. E. Cheng, Separating subdivision of bicycle wheel inequalities over cut polytopes. Oper. Res. Lett. **23**, 13–19 (1998)
12. W. Cook, Fifty-plus years of combinatorial integer programming, in *50 Years of Integer Programming*, eds. by M. Juenger et al. (Heidelberg, Springer, 2010), pp. 387–430
13. H. Crowder, E.L. Johnson, M. Padberg, Solving large-scale zero-one linear programming problems. Oper. Res. **31**, 803–834 (1983)
14. C. De Simone, The cut polytope and the Boolean quadric polytope. Discrete Math. **79**, 71–75 (1989)
15. C. De Simone, A note on the Boolean quadric polytope. Oper. Res. Lett. **19**, 115–116 (1996)
16. C. De Simone, G. Rinaldi, A cutting plane algorithm for the max-cut problem. Optim. Methods Softw. **3**, 195–214 (1994)
17. M. Deza, On the Hamming geometry of unitary cubes. Sov. Phys. Dokl. **5**, 940–943 (1961)
18. M. Deza, Realizability of distance matrices in unit cubes (in Russian). Problemy Kibernitiki **7**, 31–42 (1962)
19. M. Deza, Matrices de formes quadratiques non négatives pour des arguments binaires. Comptes Rendus Acad. Sci. Paris **277**, 873–875 (1973)
20. M. Deza, V.P. Grishukhin, Voronoi L-decomposition of PSD_n and the hypermetric correlation cone, in *Voronoi's Impact on Modern Science*, ed. by P. Engel, H. Syta (Institute of Mathematics of the National Academy of Science, Kiev, 1998)
21. M. Deza, M. Laurent, *Geometry of Cuts and Metrics* (Springer, Berlin, 1997)
22. M. Deza, V.P. Grishukhin, M. Laurent, The hypermetric cone is polyhedral. Combinatorica **13**, 1–15 (1993)
23. F.D. Fomeni, K. Kaparis, A.N. Letchford, A cut-and-branch algorithm for the quadratic knapsack problem. Discrete Optim. (2020). https://doi.org/10.1016/j.disopt.2020.100579
24. L. Galli, K. Kaparis, A.N. Letchford, Gap inequalities for non-convex mixed-integer quadratic programs. Oper. Res. Lett. **39**, 297–300 (2011)
25. L. Galli, K. Kaparis, A.N. Letchford, Gap inequalities for the max-cut problem: a cutting-plane algorithm, in *Combinatorial Optimization: 2nd International Symposium*, ed. by A.R. Mahjoub et al. (Springer, Berlin, 2012), pp. 178–188
26. M.R. Garey, D.S. Johnson, L. Stockmeyer. Some simplified NP-complete graph problems. Theor. Comput. Science **1**, 237–267 (1976)
27. A.M.H. Gerards, Testing the odd bicycle wheel inequalities for the bipartite subgraph polytope. Math. Oper. Res. **10**, 359–360 (1985)
28. F. Glover, E. Woolsey, Converting the 0–1 polynomial programming problem to a 0–1 linear program. Oper. Res. **22**, 180–182 (1974)
29. M.X. Goemans, D.P. Williamson, Improved approximation algorithms for maximum cut and satisfiability problems using semidefinite programming. J. ACM **42**(6), 1115–1145 (1995)
30. R.E. Gomory, Outline of an algorithm for integer solutions to linear programs. Bull. Am. Math. Soc. **64**, 275–278 (1958)
31. M. Grötschel, M.W. Padberg, Polyhedral theory, in *The Travelling Salesman Problem: A Guided Tour of Combinatorial Optimization*, ed. by E.L. Lawler et al. (Wiley, New York, 1985)
32. M. Grötschel, M. Jünger, G. Reinelt, A cutting plane algorithm for the linear ordering problem. Oper. Res. **32**, 1195–1220 (1984)
33. M. Grötschel, L. Lovász, A.J. Schrijver, *Geometric Algorithms and Combinatorial Optimization* (Wiley, New York, 1988)
34. G. Gruber, On semidefinite programming and applications in combinatorial optimization. PhD thesis, Department of Mathematics, University of Klagenfurt, 2000
35. C. Helmberg, F. Rendl, Solving quadratic (0,1)-problems by semidefinite programs and cutting planes. Math. Program. **82**, 291–315 (1998)
36. M. Jünger, V. Kaibel, Box-inequalities for quadratic assignment polytopes. Math. Program. **91**, 175–197 (2001)
37. K. Kaparis, A.N. Letchford, A note on the 2-circulant inequalities for the max-cut problem. Oper. Res. Lett. **46**, 443–447 (2018)

38. K. Kaparis, A.N. Letchford, Y. Mourtos, Generalised 2-circulant inequalities for the max-cut problem, in *Working Paper*, Department of Management Science, Lancaster University, UK (2021)
39. R.M. Karp, C.H. Papadimitriou, On linear characterizations of combinatorial optimization problems. SIAM J. Comput. **11**, 620–632 (1982)
40. F. Körner, A tight bound for the Boolean quadratic optimization problem and its use in a branch and bound algorithm. Optimization **19**, 711–721 (1988)
41. N. Krislock, J. Malick, F. Roupin, Improved semidefinite bounding procedure for solving max-cut problems to optimality. Math. Program. **143**, 61–86 (2014)
42. A.H. Land, A.G. Doig, An automatic method of solving discrete programming problems. Econometrica **28**, 497–520 (1960)
43. M. Laurent, S. Poljak, On a positive semidefinite relaxation of the cut polytope. Linear Algebra Appl. **223/224**, 439–461 (1995)
44. M. Laurent, S. Poljak, Gap inequalities for the cut polytope. Eur. J. Combin. **17**, 233–254 (1996)
45. A.N. Letchford, M.M. Sørensen, Binary positive semidefinite matrices and associated integer polytopes. Math. Program. **131**, 253–271 (2012)
46. A.N. Letchford, M.M. Sørensen, A new separation algorithm for the Boolean quadric and cut polytopes. Discrete Optim. **14**, 61–71 (2014)
47. E.M. Macambira, C.C. de Souza, The edge-weighted clique problem: Valid inequalities, facets and polyhedral computations. Eur. J. Oper. Res. **123**, 346–371 (2000)
48. G.P. McCormick, Converting general nonlinear programming problems to separable nonlinear programming problems. Report T—267, The George Washington University (1972)
49. A. Mehrotra, Cardinality constrained Boolean quadratic polytope. Discrete Appl. Math. **79**, 137–154 (1997)
50. P. Miliotis, Integer programming approaches to the travelling salesman problem. Math. Program. **10**, 367–378 (1976)
51. J.E. Mitchell, Branch and cut, in *Encyclopedia of Operations Research and Management Science*, ed. by J.J. Cochran et al. (Wiley, New York, 2010)
52. G.L. Nemhauser, L.A. Wolsey, *Integer and Combinatorial Optimization* (Wiley, New York, 1988)
53. M. Padberg, The Boolean quadric polytope: Some characteristics, facets and relatives. Math. Program. **45**, 134–172 (1989)
54. M.W. Padberg, G. Rinaldi, A branch-and-cut algorithm for the resolution of large-scale symmetric travelling salesman problems. SIAM Rev. **33**, 60–100 (1991)
55. I. Pitowsky, Correlation polytopes: their geometry and complexity. Math. Program. **50**, 395–414 (1991)
56. S. Poljak, F. Rendl, Nonpolyhedral relaxations of graph-bisection problems. SIAM J. Optim. **5**, 467–487 (1995)
57. S. Poljak, D. Turzik, Max-cut in circulant graphs. Discrete Math. **108**, 379–392 (1992)
58. S. Poljak, Z. Tuza, The expected relative error of the polyhedral approximation of the max-cut problem. Oper. Res. Lett. **16**, 191–198 (1994)
59. S. Poljak, F. Rendl, H. Wolkowicz, A recipe for semidefinite relaxation for (0,1)-quadratic programming. J. Glob. Optim. **7**, 51–73 (1995)
60. A.K. Pujari, A.K. Mittal, S.K. Gupta, A convex polytope of diameter one. Discrete Appl. Math. **5**, 241–242 (1983)
61. D.J. Rader, Valid inequalities and facets of the quadratic 0-1 knapsack polytope. RUTCOR Research Report 16-97, Rutgers University (1997)
62. F. Rendl, G. Rinaldi, A. Wiegele, Solving max-cut to optimality by intersecting semidefinite and polyhedral relaxations. Math. Program. **121**, 307–335 (2010)
63. H. Saito, T. Fujie, T. Matsui, S. Matuura, A study of the quadratic semi-assignment polytope. Discrete Optim. **6**, 37–50 (2009)
64. I.J. Schoenberg, Metric spaces and positive definite functions. Trans. Am. Math. Soc. **44**, 522–536 (1938)

65. H.D. Sherali, B.M.P. Fraticelli, Enhancing RLT relaxations via a new class of semidefinite cuts. J. Glob. Optim. **22**, 233–261 (2002)
66. N.Z. Shor, Quadratic optimization problems. Tekhnich. Kibernetika **1**, 128–139 (1987)
67. M.M. Sørensen, New facets and a branch-and-cut algorithm for the weighted clique problem. Eur. J. Oper. Res. **154**, 57–70 (2004)
68. H. Weyl, Elementare Theorie der konvexen Polyeder. Comment. Math. Helvet. **7**, 290–306 (1935)
69. Y. Yajima and T. Fujie, A polyhedral approach for nonconvex quadratic programming problems with box constraints. J. Glob. Optim. **13**, 151–170 (1998)

Chapter 5
Autarkies and Persistencies for QUBO

Endre Boros

Abstract In many classes of discrete optimization problems, we do not know (yet) an efficient algorithm that would guarantee an optimal solution for all instances. However, for some instances we may be able to identify provably optimal values for some of its variables. Persistency and autarky are properties of partial assignments that allow us to argue about their optimality. This chapter investigates these and some related notions from the literature and recalls algorithms for their recognition. These techniques provide efficient pre-processing for QUBO problems and in some applications these result in substantial reduction in problem sizes.

5.1 Introduction

In this section we focus on the *persistency* property of optimization problems, which in fact can be defined in several ways. Loosely speaking, a continuous relaxation of a binary optimization problem has the (weak) persistency property if for an arbitrary optimal solution $x \in [0, 1]^n$ of it there exists an optimal solution $x^* \in \{0, 1\}^n$ of the binary problem such that $x_i^* = x_i$ whenever $x_i \in \{0, 1\}$. If the same hold for all optimal solutions of the binary problem, then we talk about strong persistency. This is of course, if we consider optimization problems in their polyhedral form, when a continuous relaxation maybe much simpler to solve, e.g., by linear programming. As previous chapters show, many problems can be framed as unconstrained binary optimization, e.g., QUBO. In that setting persistency is a partial assignment that can be extended to an optimal solution. In this chapter we consider several notions of this type, and show their relations. In particular we recall and provide efficient algorithmic result for QUBO problems.

Let us note that for many optimization problems there may not exists persistency that could be derived in an efficient way. It turns out however that for particular

E. Boros (✉)
RUTCOR and MSIS Department, Rutgers Business School, Rutgers University, New Brunswick, NJ, USA
e-mail: Endre.Boros@rutgers.edu

A. P. Punnen (ed.), *The Quadratic Unconstrained Binary Optimization Problem*,
https://doi.org/10.1007/978-3-031-04520-2_5

types of problems arising in certain applications, these techniques could lead to substantial reduction in the size of the problem. For instance, it turns out the QUBO formulations are quite natural in computer visions problems, and that family of problems tend to have persistencies. Efficient algorithms to derive persistencies for QUBO problems gained prominent application in computer vision, see e.g., [9–11, 15, 17, 23, 24]. Such important examples motivate our chapter on these notions and techniques.

5.2 Basic Definitions

To arrive to precise definitions and results, we need to recall some terminology and corresponding notation. We denote by $V = \{1, 2, \dots, n\}$ the set of indices, by $\mathbf{x} = (x_j \mid j \in V)$ the vector of our variables, and call $\bar{x}_j = 1 - x_j$ for $j \in V$ *complemented variables*. The $2n$ element set $L = \{x_j, \bar{x}_j \mid j \in V\}$ is called the set of *literals*.

When considering optimization, in this chapter we always consider minimization. While generally minimization and maximization are equivalent problems for several definitions and results, considering minimization will simplify our presentation.

Let us first note that any real valued mapping $f : \{0, 1\}^n \mapsto \mathbb{R}$, called a pseudo-Boolean function (or PBF, in short), has a unique multilinar polynomial form

$$f(\mathbf{x}) = \sum_{S \subseteq V} \alpha_S \prod_{j \in S} x_j, \tag{5.1}$$

where $\alpha_S \in \mathbb{R}$ for all $S \subseteq V$ (see [12]). The degree of f, denoted by $deg(f)$, is the largest cardinality $|S|$ with $\alpha_S \neq 0$. The unconstrained binary optimization problem (UBO) is the problem

$$\min_{\mathbf{x} \in \{0,1\}^n} f(\mathbf{x}) \tag{5.2}$$

where we assume that f is given by its unique multilinear polynomial. QUBO is the special case when $deg(f) \leq 2$.

A *posiform* is a multilinear polynomial in terms of the $2n$ literals

$$\phi(\mathbf{x}) = \sum_{\substack{P,N \subseteq V \\ P \cap N = \emptyset}} \beta_{P,N} \prod_{j \in P} x_j \prod_{j \in N} \bar{x}_j \tag{5.3}$$

satisfying $\beta_{P,N} \geq 0$ whenever $P \cup N \neq \emptyset$. Note that $\beta_{\emptyset,\emptyset} < 0$ is possible. The degree of such a posiform, denoted by $deg(\phi)$, is the largest cardinality $|P \cup N|$ such that $\beta_{P,N} > 0$. We say that a posiform ϕ represents a pseudo-Boolean function f, if $f(\mathbf{x}) = \phi(\mathbf{x})$ for all $\mathbf{x}\{0, 1\}^n$. Note that in this case $\beta_{\emptyset,\emptyset}$ is a lower bound on the minimum value of (5.2). It is well-known that every pseudo-Boolean function f can be represented by posiforms, typically in a non unique way, and $deg(f) \leq deg(\phi)$ for all such posiforms.

Example 5.1 For instance, let us consider the following example in $n = 3$ variables:

$$\begin{aligned} f(\mathbf{x}) &= -2 - x_1 - x_2 - x_3 + x_1x_2 + x_1x_3 + x_2x_3 && \text{quadratic PBF} \\ &= -5 + \overline{x}_1 + \overline{x}_2 + \overline{x}_3 + x_1x_2 + x_1x_3 + x_2x_3 && \text{a quadratic posiform} \\ &= -4 + \overline{x}_3 + \overline{x}_1\overline{x}_2 + x_1x_3 + x_2x_3 && \text{another quadratic posiform} \\ &= -3 + x_1x_2x_3 + \overline{x}_1\overline{x}_2\overline{x}_3 && \text{a cubic posiform} \end{aligned}$$

As we can see, posiform representations are not unique, and maybe of higher degree than f. Note that in the above small example, the largest constant term, -3 is in fact the minimum value of this quadratic PBF. It is known and easy to see that for every PBF $f(\mathbf{x})$ has a posiform representation $\phi(\mathbf{x})$ of the form (5.3) such that its constant term $\beta_{\emptyset,\emptyset}$ is the minimum value of $f(\mathbf{x})$. In fact problem (5.2) can equivalently be viewed as the problem of finding a posiform representation with the largest constant term. For such a posiform there exists a binary assignment to the variables that make all nontrivial terms vanish. Even when the constant term of a posiform representation of a PBF $f(\mathbf{x})$ is not equal to the minimum value of $f(\mathbf{x})$ we still maybe able to find (partial) assignments that make many of the nontrivial terms vanish. This dual (equivalent) view of UBO problems and the above observation about (partial) assignments lead us to another interpretations of what a persistency is (or could be). We shall see that we can in fact arrive to equivalent notions with increased algorithmic opportunities.

Remark 5.1 When talking about algorithmic efficiency and computational complexity, one has to be careful switching between these different representations. Even quadratic pseudo-Boolean functions have posiform representations that are exponentially larger then their unique multilinear polynomial one, and vice versa. For instance, the posiform $\bar{x}_1\bar{x}_2\ldots\bar{x}_n$ defines a PBF that has exponentially many terms in its unique multilinear polynomial form.

The so called partial assignments play an important role in our claims and analysis. For a subset $S \subseteq V$ a binary vector $\mathbf{y} \in \{0,1\}^S$ is called a *partial assignment*. For a binary vector $\mathbf{x} \in \{0,1\}^V$ and partial assignment $\mathbf{y} \in \{0,1\}^S$ we denote by $\mathbf{z} = \mathbf{x}^{\mathbf{y}}$ the *switch* of $\mathbf{x}$ by $\mathbf{y}$, defined by

$$z_j = \begin{cases} y_j & \text{for } j \in S, \\ x_j & \text{for } j \in V \setminus S. \end{cases}$$

5.3 Functional Pesistency

First we consider possible conditions guaranteeing that a partial assignment in a UBO problem can be extended to an optimal one. The following definition, from [7] is one of the simplest ways one can achieve that.

Definition 5.1 A partial assignment $\mathbf{y} \in \{0, 1\}^S$, $S \subseteq V$ is called a *persistency* for a pseudo-Boolean function f if $f(\mathbf{x}^{\mathbf{y}}) \leq f(\mathbf{x})$ for all $\mathbf{x} \in \{0, 1\}^V$.

Proposition 5.1 *If a partial assignment* $\mathbf{y} \in \{0, 1\}^S$, $S \subseteq V$ *is a* persistency *for a pseudo-Boolean function* f*, then problem* (5.2) *has an optimal solution that agrees with* $\mathbf{y}$ *in the components belonging to* S.

Proof Consider an arbitrary optimal solution $\mathbf{x} \in \{0, 1\}^V$ of problem (5.2). Then, by the above persistency property of $\mathbf{y}$, the vector $\mathbf{x}^{\mathbf{y}}$ is also an optimal solution of problem (5.2). □

Next we argue that the family of subsets for which there exists a persistent binary (partial) assignment is closed under union.

Proposition 5.2 *If* $S_1, S_2 \subseteq V$ *are two subsets such that we have binary assignments* $\mathbf{y}^1 \in \{0, 1\}^{S_1}$ *and* $\mathbf{y}^2 \in \{0, 1\}^{S_2}$ *that are both persistencies for a PBF* f*, then the (partial) assignment* $\mathbf{z} \in \{0, 1\}^{S_1 \cup S_2}$ *defined by*

$$\mathbf{z}_j = \begin{cases} y_j^1 & \text{for } j \in S_1, \\ y_j^2 & \text{for } j \in S_2 \setminus S_1 \end{cases}$$

is also a persistency for f.

Proof Since both $\mathbf{y}^1$ and $\mathbf{y}^2$ ar persistencies for f, we have the following chain of inequalities for an arbitrary $\mathbf{x} \in \{0, 1\}^V$:

$$f(\mathbf{x}^{\mathbf{z}}) = f\left(\left(\mathbf{x}^{\mathbf{y}^2}\right)^{\mathbf{y}^1}\right) \leq f\left(\mathbf{x}^{\mathbf{y}^2}\right) \leq f(\mathbf{x}).$$

□

Remark 5.2 Every pseudo-Boolean function f has some persistencies. For instance, for $S = V$ and the optimal solution $\mathbf{x}^*$ of (5.2), the vector $\mathbf{x}^*$ itself is a persistency for f.

Remark 5.3 While this notion is simple and looks useful, it is computationally not easy to apply. For instance, given a PBF f (by either its unique multilinear polynomial, or by one its posiform representations) and given a partial assignment $\mathbf{y} \in \{0, 1\}^S$ it is NP-complete to decide if $\mathbf{y}$ is a persistency for f, or not.

5.4 Autarkies

The next notion is motivated by the satisfiability literature (see e.g., [18]), and the form of UBO, in which we minimize a given posiform. In fact we try to characterize the case when we can achieve the best possible improvement with a subset of the variables toward the goal of making the terms of the posiform vanish.

Definition 5.2 A partial assignment $\mathbf{y} \in \{0, 1\}^S$, $S \subseteq V$ is called an *autarky* for a posiform ϕ given by (5.3) if for every pair of subsets $P, N \subseteq V$ with $P \cap N = \emptyset$, $S \cap (P \cup N) \neq \emptyset$ and $\beta_{P,N} > 0$ we have

$$\prod_{j \in P \cap S} y_j \prod_{j \in N \cap S} \bar{y}_j = 0.$$

Simply saying, $\mathbf{y}$ is an autarky for posiform ϕ, if fixing the variables with indices in S at the values y_j, $j \in S$ makes all terms of ϕ vanish that involves any of these variables. Clearly, one cannot change the value of a term with a variable that does not appear in that term, thus this notion of autarky corresponds to the "lucky" assignments that makes the value of ϕ as small as it is at all possible.

Proposition 5.3 *If a partial assignment* $\mathbf{y} \in \{0, 1\}^S$ *for some* $S \subseteq V$ *is an autarky for a posiform* ϕ *representing the PBF* f*, then it is also a persistency for* f*.*

Proof To see this claim, let us group the terms of ϕ given as in (5.3)

$$\phi(\mathbf{x}) = \beta_{\emptyset,\emptyset} + A(\mathbf{x}) + B(\mathbf{x}),$$

where $A(\mathbf{x})$ is the sum of all those terms that involve a variable (either positively, or negated) with index in S, and $B(\mathbf{x})$ is the sum of all other terms. Then for an arbitrary $\mathbf{x} \in \{0, 1\}^V$ we have both $A(\mathbf{x}) \geq 0$ and $B(\mathbf{x}) \geq 0$, since in a posiform only the constant term can be negative. Furthermore by our grouping, we also have $B(\mathbf{x}^\mathbf{y}) = B(\mathbf{x})$, since terms in B do not involve any of the variables the value of which may be switched in $\mathbf{x}^\mathbf{y}$. Finally, by the autarky property of $\mathbf{y}$ we must have $A(\mathbf{x}^\mathbf{y}) = 0$. Thus

$$f(\mathbf{x}^\mathbf{y}) = \beta_{\emptyset,\emptyset} + A(\mathbf{x}^\mathbf{y}) + B(\mathbf{x}^\mathbf{y}) = \beta_{\emptyset,\emptyset} + 0 + B(\mathbf{x}) \leq \phi(\mathbf{x}) = f(\mathbf{x}).$$

□

Remark 5.4 In contrast to the computational difficulty of recognizing the persistency property, one can check easily if a given partial assignment is an autarky for a given posiform, in time, linear in the size of the posiform.

Thus, in our search for finding some nontrivial persistency for a given PBF, it is very useful to consider its posiform representations. Unfortunately, not all posiforms have autarkies. For instance in Example 5.1 the first two quadratic posiforms do not have any autarkies, while $\mathbf{y} = (0, 1) \in \{0, 1\}^{\{1,2\}}$ is an autarky for the third posiform representation. Thus, we need special algorithmic results to transform a posiform into an equivalent posiform that may have more autarkies. In fact we can develop such techniques for quadratic pseudo-Boolean functions.

Let us also note that "persistency" was introduced for quadratic pseudo-Boolean functions by Hammer et al. [13] via defining a particular type of autarky (without actually using this term) and using implicitly Proposition 5.3.

Before we describe the application of these notions to QUBO problems, let us study autarkies in a general setting.

Proposition 5.4 *For every posiform ϕ there exists a unique maximal subset $S \subseteq V$ such that some (partial) assignment $\mathbf{y} \in \{0, 1\}^S$ is an autarky for ϕ.*

Proof To prove this statement, assume that $S_1, S_2 \subseteq V$ are two incomparable subsets such that we have binary assignments $\mathbf{y}^1 \in \{0, 1\}^{S_1}$ and $\mathbf{y}^2 \in \{0, 1\}^{S_2}$ that are both autarkies for ϕ. We claim that in this case $\mathbf{z} \in \{0, 1\}^{S_1 \cup S_2}$ defined by

$$\mathbf{z}_j = \begin{cases} y_j^1 & \text{for } j \in S_1, \\ y_j^2 & \text{for } j \in S_2 \setminus S_1 \end{cases}$$

is also an autarky for ϕ. To see this claim, partition the nontrivial terms of ϕ into three groups:

$$\phi(\mathbf{x}) = \beta_{\emptyset,\emptyset} + A(\mathbf{x}) + B(\mathbf{x}) + C(\mathbf{x}),$$

where A consist of the sum of all terms of ϕ that have a variable x_j with $j \in S_1$ involved (either positively, or negated), B consist of sum of the all terms of ϕ that have a variable x_j with $j \in S_2$ involved but no variable with $j \in S_1$ (either positively, or negated), and C consists the sum of those terms of ϕ that do not involve any variables with indices $j \in S_1 \cup S_2$. Let us consider now an arbitrary assignment $\mathbf{x} \in \{0, 1\}^V$. Then, $A(\mathbf{x}^\mathbf{z}) = A(\mathbf{x}^{\mathbf{y}^1}) = 0$ by the autarctic property of $\mathbf{y}^1$ and $B(\mathbf{x}^\mathbf{z}) = B(\mathbf{x}^{\mathbf{y}^2}) = 0$ since the terms in B do not involve any variables with indices $j \in S_1 \cap S_2$ and $\mathbf{y}^2$ is an autarky for ϕ. Thus we get

$$\begin{aligned} \phi(\mathbf{x}^\mathbf{z}) &= \beta_{\emptyset,\emptyset} + A(\mathbf{x}^\mathbf{z}) + B(\mathbf{x}^\mathbf{z}) + C(\mathbf{x}^\mathbf{z}) \\ &= \beta_{\emptyset,\emptyset} + 0 + 0 + C(\mathbf{x}^\mathbf{z}) \\ &\leq \beta_{\emptyset,\emptyset} + A(\mathbf{x}) + B(\mathbf{x}) + C(\mathbf{x}) \\ &= \phi(\mathbf{x}), \end{aligned}$$

since $A(\mathbf{x}) + B(\mathbf{x}) \geq 0$ and $C(\mathbf{x}^\mathbf{z}) = C(\mathbf{x})$ because C does not include any terms that involves variables x_j with $j \in S_1 \cup S_2$. □

Let us denote by $S(\phi) \subseteq V$ this unique maximal set for which there exists a partial assignment $\mathbf{y} \in \{0, 1\}^{S(\phi)}$ that is an autarky for ϕ. Note that $S(\phi)$ may be the empty set, as our Example 5.1 shows.

Let us close this section about autarkies with pointing out a few more important properties of them.

Proposition 5.5 *Every persistency of a pseudo-Boolean function f is an autarky of some its posiform representations.*

Proof Assume that $\mathbf{y} \in \{0,1\}^S$ for some subset $S \subseteq V$ is a persistency for function f. To simplify our notation, for this proof, we assume that we view $V \setminus S$ as the initial segment of the binary vectors in $\{0,1\}^V$, thus in particular we can view every binary vector as a concatenation $\mathbf{x} = (\alpha, \beta)$, where $\alpha \in \{0,1\}^{V\setminus S}$ and $\beta \in \{0,1\}^S$. Note that with this convention in mind we have $(\alpha, \beta)^{\mathbf{y}} = (\alpha, \mathbf{y})$ for all vectors. Let us further introduce

$$Z = \min_{\mathbf{x} \in \{0,1\}^V} f(\mathbf{x}),$$

and consider the following expression:

$$\begin{aligned}\phi(\mathbf{x}) = Z + &\sum_{\mathbf{z} \in \{0,1\}^{V\setminus S}} (f(\mathbf{z},\mathbf{y}) - Z) \prod_{\substack{j \in V\setminus S \\ z_j = 1}} x_j \prod_{\substack{j \in V\setminus S \\ z_j = 0}} \bar{x}_j \\ + &\sum_{\mathbf{z} \in \{0,1\}^V} (f(\mathbf{z}) - f(\mathbf{z}^{\mathbf{y}})) \prod_{\substack{j \in V \\ z_j = 1}} x_j \prod_{\substack{j \in V \\ z_j = 0}} \bar{x}_j\end{aligned}$$

Let us note first that ϕ defined above is a posiform. Indeed, $f(\mathbf{x},\mathbf{y}) \geq Z$ for all $\mathbf{z} \in \{0,1\}^{V\setminus S}$ by our choice of Z, and $f(\mathbf{z}) \geq f(\mathbf{z}^{\mathbf{y}})$ for all $\mathbf{z} \in \{0,1\}^V$ since $\mathbf{y}$ is a persistency for f.

Let us note next that ϕ is a representation of f, that is $\phi(\mathbf{x}) = f(\mathbf{x})$ for all $\mathbf{x} \in \{0,1\}^V$. This is because for a fixed binary vector $\mathbf{x} \in \{0,1\}^V$ there are unique non-vanishing terms in both summation: in the first one it corresponds to the vector $\mathbf{z} \in \{0,1\}^{V\setminus S}$ for which $\mathbf{x}^{\mathbf{y}} = (\mathbf{z},\mathbf{y})$; in the second summation it corresponds to $\mathbf{z} = \mathbf{x}$. Thus from the above expression we get

$$\phi(\mathbf{x}) = Z + (f(\mathbf{x}^{\mathbf{y}}) - Z) + (f(\mathbf{x}) - f(\mathbf{x}^{\mathbf{y}})) = f(\mathbf{x}).$$

Finally we note that only the terms of the second summation involve variables with indices in S, and whenever $f(\mathbf{z}) - f(\mathbf{z}^{\mathbf{y}}) \neq 0$ we must have an index $j \in S$ such that $z_j \neq y_j$, implying that the corresponding term vanishes when we substitute $x_j = y_j$. □

Remark 5.5 Propositions 5.3 and 5.5 together imply that persistencies for a pseudo-Boolean function are the same as autarkies for its various posiform representations. The above results also show that an appropriately chosen posiform representation can serve as an effective proof that a given partial assignment is indeed a persistency for the PBF under consideration. We use this in an efficient way in the last part of the chapter, when we provide polynomial time algorithms to extract in some sense extremal persistencies for quadratic pseudo-Boolean functions.

Proposition 5.6 *Assume that ϕ is a given posiform, and $\mathbf{x} \in \{0,1\}^V$. Then there exists a unique maximal subset $S \subseteq V$ such that the partial assignment $\mathbf{y} \in \{0,1\}^S$ defined by $y_j = x_j$ for $j \in S$ is an autarky for ϕ, and S can be obtained in time polynomial in the size of ϕ.*

Proof Since the constant term does not play a role in this claim, we can assume w.l.o.g. that $\beta_{\emptyset,\emptyset} = 0$.

Note that if $\phi(\mathbf{x}) = 0$, then $S = V$ and $\mathbf{y} = \mathbf{x}$.

Assume next that $\phi(\mathbf{x}) > 0$. By the definition of an autarky, if for a term of ϕ we have

$$\beta_{P,N} \prod_{j \in P} x_j \prod_{j \in N} \bar{x}_j > 0$$

then we must have $S \cap (P \cup N) = \emptyset$. Let us then remove all occurrences of x_j and $\bar{x}_j$ for all $j \in P \cup N$, remove those terms as well that became empty products, and denote the obtained posiform by ϕ'. Note that if S is a subset for which $\mathbf{x}$ defines an autarky for ϕ, then it is also an autarky for ϕ'. Thus we can repeat the above procedure, until we arrive to a posiform that has value zero when we evaluate it at $\mathbf{x}$. Then defining S as the set of remaining variables completes our proof. □

5.5 Polyhedral Persistency

The third definition view persistency as a polyhedral property. To formally define this notion, let us consider the optimization problems associated to a polytope $Q \subseteq [0,1]^V$ and for a real vector $\mathbf{w} \in \mathbb{R}^V$:

$$\min_{\mathbf{x} \in Q \cap \{0,1\}^V} \mathbf{w}^T \mathbf{x} \tag{IP(Q)}$$

and its continuous relaxation

$$\min_{\mathbf{z} \in Q} \mathbf{w}^T \mathbf{z} \tag{LP(Q)}$$

To a solution $\mathbf{x}$ of (LP(Q)) let us associate two sets $Z, O \subseteq V$ defined by

$$Z(\mathbf{x}) = \{j \in V \mid x_j = 0\} \quad \text{and} \quad O(\mathbf{x}) = \{j \in V \mid x_j = 1\}.$$

Definition 5.3 We say that a polytope $Q \subseteq [0,1]^V$ has the *persistency property* if for every real vector $\mathbf{w} \in \mathbb{R}^V$ and every optimal solution $\mathbf{z}^*$ of (LP(Q)) there exists an optimal solution $\mathbf{x}^*$ of (IP(Q)) such that $x_j^* = z_j^*$ for all $j \in Z(\mathbf{x}^*) \cup O(\mathbf{x}^*)$.

Note that the same notion could be defined for maximization problems, as well, but a standard transformation, switching from x_i to $\bar{x}_i = 1 - x_i$ commutes between these formulations. We are going to recall results form the literature transformed to the minimization case, to follow our treatment of posiform minimization.

This notion, without using the word "persistency" was introduced by Nemhauser and Trotter [20] for the so called stability (or vertex packing) polytope (in the context of maximization). Given a graph $G = (V, E)$ and a weight function $\mathbf{w} \in \mathbb{R}^V$, we are interested in finding a maximum weight independent set in G, that is a subset $I \subseteq V$ such that there are no edges inside I, an $\mathbf{w}(I) = \sum_{i \in I} w_i$ is as large as possible. A standard (sometimes called "edge") formulation of this problem is to consider the polytope

$$Q_{stab}(G) = \{\mathbf{x} \in [0, 1]^V \mid x_i + x_j \leq 1 \text{ for all } (i, j) \in E\}.$$

Balinski [3] observed that all vertices of $Q_{stab}(G)$ are half integral for an arbitrary graph G (integral for bipartite graphs), and Nemhauser and Trotter [20] proved that Q_{stab} has the persistency property. Note that complements of independent set in a graph are vertex covers (some times called transversals of the edge set) and thus stability problems are equivalent with vertex cover problems by the transformation $z_i = 1 - x_i$ for $i \in V$:

$$Q_{vc}(G) = \{\mathbf{z} \in [0, 1]^V \mid z_i + z_j \geq 1 \text{ for all } (i, j) \in E\}.$$

Note that this linear mapping $Q_{stab}(G) \to Q_{vc}(G)$ keeps half-integrality, and thus Balinski's [3] result implies that $Q_{vc}(G)$ has half-integral vertices, too.

We restate now a result of Nemhauser and Trotter [20] for the vertex cover case, and for completeness recall a short proof, based on [16].

Theorem 5.1 ([20]) *The polytope $Q_{vc}(G)$ has the persistency property for all graphs G.*

Proof Consider an arbitrary weight function, $\mathbf{w} \in \mathbb{R}^V$ and let $\mathbf{z}^*$ be a half-integral optimal solution of $LP(Q_{vc}(G))$. Set $Z = Z(\mathbf{z}^*)$, $O = O(\mathbf{z}^*)$ and $H = V \setminus (Z \cup O)$. Note that if $w_i \leq 0$ for some $i \in V$, then in any solution of both $LP(Q_{vc}(G))$ and $I(Q_{vc}(G))$ we can switch to $z_i = 1$ without increasing the objective function. Thus, we can assume w.l.o.g. that $w_i > 0$ for all $i \in V$.

Let us now consider a minimum weight vertex cover S of G. We claim that

$$O \subseteq S \subseteq V \setminus Z.$$

First we show that the right containment implies the left one. For a contradiction, assume that $S \cap Z = \emptyset$ and the there exists a vertex $i \in O \setminus S$. Note that i cannot have a neighbor in Z, since S is a vertex cover. Thus for all edges $(i, j) \in E$ we have $z_j^* > 0$. Consequently, we could decrease z_i^* a little and obtain another feasible solution of $LP(Q_{vc}(G))$ with a strictly smaller objective value, contradicting the optimality of $\mathbf{z}^*$.

To show the right containment, let us now consider a minimum weight vertex cover S for which $|S \cap Z|$ is as small as possible, and assume that the weight of $\mathbf{w}(O \setminus S) < \mathbf{w}(Z \cap S)$. Note that all neighbors of $O \setminus S$ are in $O \cup H \cup (Z \cap S)$ since S is a vertex cover. Thus, for a small $\epsilon > 0$ we could increase z_i^* for $i \in O \setminus S$ by ϵ and decrease z_i^* for $i \in Z \cap S$ by ϵ, and obtain another feasible solution of $LP(Q_{vc}(G))$ with strictly smaller objective value than $\mathbf{z}^*$, contradicting the optimality of $\mathbf{z}^*$. This contradiction proves that the weight of $\mathbf{w}(O \setminus S) \geq \mathbf{w}(Z \cap S)$ implying that $S' = O \cup S \setminus Z$ is another vertex cover with weight not larger than that of S and has $S' \cap Z = \emptyset$. □

This result had numerous applications in approximation algorithms, kernelization methods, branching algorithms, etc., see e.g., [1, 4, 14, 16, 19]. The recent paper [22] shows on the other hand that persistency is not a frequent property of polytopes. They prove that under some mild technical conditions among all polytopes Q with $Q \cap \{0, 1\}^V = Q_{vc} \cap \{0, 1\}^V$ only Q_{vc} and $conv(Q_{vc} \cap \{0, 1\}^V)$ have the persistency property.

Hammer et al. [13] shows that, based on the above result, the standard linearization of QUBO problems also has the persistency property. To a quadratic pseudo-Boolean function

$$f(\mathbf{x}) = \alpha_0 + \sum_{i \in V} \alpha_i x_i + \sum_{(i,j) \in \binom{V}{2}} \alpha_{ij} x_i x_j$$

we can associate its standard linearization (also called *Rhys linearization* due to [21]):

$$\begin{aligned}
\min\ & \alpha_0 + \sum_{i \in V} \alpha_i x_i + \sum_{(i,j) \in \binom{V}{2}} \alpha_{ij} y_{ij} \\
s.t.\ & \\
& \left.\begin{aligned} y_{ij} &\geq x_i + x_j - 1 \\ y_{ij} &\geq 0 \end{aligned}\right\} \quad \text{for all } (i,j) \in \tbinom{V}{2} \text{ with } \alpha_{ij} > 0 \\
& \left.\begin{aligned} y_{ij} &\leq x_i \\ y_{ij} &\leq x_j \end{aligned}\right\} \quad \text{for all } (i,j) \in \tbinom{V}{2} \text{ with } \alpha_{ij} < 0 \\
& \left.\begin{aligned} x_i &\leq 1 \\ x_i &\geq 0 \end{aligned}\right\} \quad \text{for all } i \in V.
\end{aligned} \tag{LP(f)}$$

Theorem 5.2 (Theorem 4.4 in [13]) *If* $(\mathbf{x}, \mathbf{y})$ *is an optimal solution of the Rhys linearization* $LP(f)$*, then there exists a binary minimum* $\mathbf{z} \in \{0, 1\}^V$ *of* f *such that* $z_i = x_i$ *for all* $i \in Z(\mathbf{x}) \cup O(\mathbf{x})$. □

In [13] this result is claimed in slightly different form, but their proof gives an equivalent result, as claimed above. They call this *weak-persistency*, since the claim may not hold for all minima of f.

For the case of quadratic pseudo-Boolea functions, we can claim further specific results, as shown in the next section.

5.6 Persistencies and Autarkies for Quadratic Functions and Posiforms

While persistencies (or equivalently autarkies) are attractive, since potentially we could reduce the size of a hard optimization problem by applying them, for general UBO problems we do not have computationally efficient procedures. For instance, we do not know how to compute $S(\phi)$ for a posiform ϕ in polynomial time in the size of ϕ. Quadratic functions and posiforms are the exception. We already seen that the standard linearization of QUBO problems have the polyhedral persistency property. Here we provide (or recall from the literature) a collection of results that lead to efficient computations of best possible persistencies in some sense, for quadratic functions and posiforms.

To ease the presentation, we introduce simplified notation for the quadratic case. For a finite set S we denote by $\binom{S}{2}$ the set of unordered pairs of distinct elements from S. Recall that $V = [n] = \{1, 2, \ldots, n\}$ is the set of indices of our binary variables and $L = \{x_i, \bar{x}_i \mid i \in V\}$ is the set of $2n$ literals. In this section we assume that a quadratic pseudo-Boolean function, in its unique multilinear polynomial form is given as

$$f(\mathbf{x}) = \alpha_0 + \sum_{i \in V} \alpha_i \cdot x_i + \sum_{(i,j) \in \binom{V}{2}} \alpha_{ij} \cdot x_i \cdot x_j, \tag{5.4}$$

where $\alpha_i, \alpha_{ij} \in \mathbb{R}$ for all $i = 0$, $i \in V$, and $(i, j) \in \binom{V}{2}$. Furthermore, we assume that a quadratic posiform is given in the form

$$\psi(\mathbf{x}) = \beta_0 + \sum_{u \in L} \beta_u \cdot u + \sum_{(u,v) \in \binom{L}{2}} \beta_{uv} \cdot u \cdot v, \tag{5.5}$$

where $\beta_0 \in \mathbb{R}$, and $\beta_u, \beta_{uv} \in \mathbb{R}_+$ for all $u \in L$, and $(u, v) \in \binom{L}{2}$. Note that each literal can be viewed as a function of $\mathbf{x} \in \{0, 1\}^V$. Sometimes we emphasize this by explicitly writing $u = u(\mathbf{x})$, when we talk about the evaluations of these literals at a particular binary vector. When we talk about multiple quadratic posiforms, we can also refer to the coefficients of a particular posiform ψ as e.g., $\beta_0(\psi)$, $\beta_{uv}(\psi)$, etc.

We also assume in the sequel that the quadratic posiforms that we consider are all *simplified* in the sense that the identities $u + \bar{u} = 1$ or $uv + u\bar{v} = u$ cannot be applied any longer. In other words, if $\beta_u > 0$, then $\beta_{\bar{u}} = 0$, and similarly if $\beta_{uv} > 0$ then $\beta_{\bar{u}v} = \beta_{u\bar{v}} = 0$.

First we show that a best possible autarky of a quadratic posiform can be computed in polynomial time.

Proposition 5.7 *Given a quadratic posiform ψ, one can compute the set $S(\psi) \subseteq V$ and a partial assignment $\mathbf{y} \in \{0,1\}^{S(\psi)}$ that is an autarky for ψ in time linear in the size of ψ.*

Proof Assume that ψ is given as in (5.5), and define sets $U = \{u \in L \mid \beta_u > 0\}$ and $T = \{(u,v) \in \binom{L}{2} \mid \beta_{uv} > 0\}$. We associate to ψ its so called *implication graph* (see e.g., [2]) $G_\psi = (L, A)$, the vertices of which are all the literals, and the directed arcs are defined by

$$A = \{u \rightarrow \bar{u} \mid u \in U\} \cup \{u \rightarrow \bar{v},\ v \rightarrow \bar{u} \mid (u,v) \in T\}$$

The meaning of these arcs are related to our desire to make the terms in ψ vanish. Namely, a term involving literals uv is vanishing only if assigning $u = 1$ is coupled with $\bar{v} = 1$, and similarly, setting $v = 1$ makes it necessary to have $\bar{u} = 1$. □

Note also that G_ψ has a strong symmetry, namely (*) there is a directed path from u to v in G_ψ if and only if there exists also a directed path from $\bar{v}$ to $\bar{u}$.

Let us now consider the strong component decomposition of G_ψ, $C_1, C_2, \ldots, C_p$. A strong component is a maximal strongly connected subgraph, and it is known that every directed graph can be decomposed into its maximal strongly connected subgraphs. The arcs between these strong components induce an acyclic graph on vertices C_i, $i = 1, \ldots, p$. By Tarjan [25] such a decomposition can be constructed in time linear in the number of arcs, which is in our case linear in the size of ψ, in such a way that if there is an arc from C_i to C_j, then $i < j$ (so called topological order of the strong components).

Note that by the symmetry (*) if we have $\{u, \bar{u}\} \subseteq C_i$ for a strong component C_i and literal u, then for all literals $v \in C_i$ we must also have $\bar{v} \in C_i$. We call such strong component *self-complementary*. For a non-self-complementary strong component C_i there exists another strong component C_j such that for all $u \in C_i$ we have $\bar{u} \in C_j$. We call C_j the complementary component of C_i, and similarly C_i is the complementary component of C_j.

Let us next note that if $S \subseteq V$ is a set for which there exists a partial assignment $\mathbf{y} \in \{0,1\}^S$ that is an autarky for ψ then no self-complementary component can involve x_j or $\bar{x}_j$ for $j \in S$. To see this claim, assume that C_i is a self-complementary component, $u \in C_i$ and $u = x_j$ or $u = \bar{x}_j$ for some $j \in S$. Then we also have $\bar{u} \in C_i$. Furthermore if $\mathbf{y}$ assigns value 1 to one of u and $\bar{u}$, say to u, then let us consider a directed path $u \rightarrow v_1 \rightarrow \cdots \rightarrow v_k \rightarrow \bar{u}$ inside C_i. Since the term of ϕ corresponding to $u \rightarrow v_1$ must vanish by the definition of an autarky, $\mathbf{y}$ must assign 1 to v_1, too. Repeating the same argument for the terms corresponding to the arcs of this path, one-by-one, we arrive to the conclusion that $\mathbf{y}$ must also assign value 1 to $\bar{u}$, which is impossible, since it assigned value 1 to u. This contradiction proves our claim.

Our main claim is that all other variables, not appearing in self-complementary components form the unique maximal subset $S \subseteq V$ for which there exists a partial assignment $\mathbf{y} \in \{0, 1\}^S$ which is an autarky for ψ. To see this claim, let us first construct $\mathbf{y}$. Consider the non-self-complementary components in their topological order: $C_{i_1}, \ldots, C_{i_q}, i_1 < \ldots i_q$. From right to left we consider these components one-by-one, and if the considered component does not yet have an assignment, assign value 1 to all literals in it, and value 0 to all literals in its complementary component. Notice that every time we assign 1-0 to a pair of complementary components, the one getting the 0 assigned must be to the left of the one getting 1 assigned.

To complete the proof, we partition the literals into there groups: A consisting of all those that got value 1 assigned, B consisting of all those that got 0 assigned, and C containing the left, that is all literals that did not get an assignment in the above procedure (i.e., the literals belonging to self-complementary components).

We show first that we cannot have a directed path connecting a literal in A to a literal in B. Assume for a contradiction that there are literals, $u \in A$ and $v \in B$ such that there exists a path in G_ψ for u to v. Then by the definition of the topological order, the component containing u must be to the left of the component containing v (and they are different components, since u and v got different values. Since the component containing $\bar{u}$ must have been assigned 0 it must be further left from the component of u. Furthermore the component containing $\bar{v}$ must have been assigned 1, and thus it is further to the right from the component of v. Thus, the path from $\bar{v}$ to $\bar{u}$ (which exists in G_ψ by the symmetry (*)) must have some arcs going from a component to another one which is to the left of the first one. This contradicts that components are in their topological order, proving our claim.

It follows now that we cannot have a directed path from A to C, or from C to B. Due to (*), these claims are equivalent, thus we can assume for contradiction that there is a directed path fro m u to v, where $u \in A$ and $v \in C$. By (*) then we also have a directed path from $\bar{v}$ to $\bar{u}$. Since $u \in A$ we must have $\bar{u} \in B$. Furthermore, since $v \in C$ it belongs to a self-complementary component, say C_i, and thus we must have $\bar{v} \in C_i \subseteq C$. Consequently, we also have a directed path from v to $\bar{v}$ since C_i is a strongly connected graph. Chaining these three paths together we obtain a directed path from $u \in A$ to $\bar{u} \in B$, contradicting our previous claim.

We claim finally that $\mathbf{y}$ is an autarky for ψ. To see this, assume that uv is a term with a positive coefficient in ψ, and say $u \in A$ (i.e., $u(\mathbf{y}) = 1$). In G_ψ we have an arc corresponding to this term going from u to $\bar{v}$. By the above arguments, we cannot have $\bar{v}$ inside $B \cup C$, and thus it must belong to A, implying that $v(\mathbf{y}) = 0$. Thus the term uv indeed vanishes. Note that if $u \in B$, then $u(\mathbf{y}) = 0$, and hence the term uv vanishes in this case, too. This shows that $\mathbf{y}$ is indeed an autarky for ψ, completing the proof of our statement. □

Given a quadratic pseudo-Boolean function f (as in (5.4)), we denote by $\mathcal{P}_2(f)$ the set of quadratic posiforms that represent f. Note that since we assume that all these posiforms are simplified, all coefficients appearing in these posiforms are bounded, because they all have f as their unique multilinear polynomial.

To a quadratic posiform ψ given by (5.5) we associate the set $Q(\psi) \subseteq L$ of literals, defined by

$$Q(\psi) \;=\; \{u \in L \mid \beta_u > 0\}.$$

Furthermore, to a quadratic pseudo-Boolean function f we associate

$$C_2(f) \;=\; \max_{\psi \in \mathcal{P}_2(f)} \beta_0(\psi).$$

The analogous quantity for the case of maximization was introduced by Hammer et al. [13], and was called the roof-dual of f. Accordingly, we could call $C_2(f)$ the floor-dual of f. Following [13] let us consider the subset

$$\mathcal{B}(f) \;=\; \{\psi \in \mathcal{P}_2(f) \mid \beta_0(\psi) = C_2(f)\}$$

of the posiforms representing f, and note that convex combinations of posiforms from $\mathcal{B}(f)$ is also a posiform belonging to $\mathcal{B}(f)$. Thus there exists a posiform $\psi* \in \mathcal{B}(f)$ such that

$$Q(\psi^*) \;\supseteq\; Q(\psi) \quad \text{for all} \quad \psi \in \mathcal{B}(f).$$

Theorem 5.3 (Theorem 4.3 in [13]) *Define $S^* \subseteq V$ as the set of indices $j \in V$ such that x_j or $\bar{x}_j$ belongs to $Q(\psi^*)$ and define $\mathbf{y} \in \{0,1\}^{S^*}$ such that $u(\mathbf{y}) = 0$ for all literals $u \in Q(\psi^*)$. Then, $\mathbf{y}$ is a (strong) persistency for f.*

Proof In fact they prove that this is an autarky for ψ^*, without defining this notion. Assume for a contradiction that it is not an autarky. Then we must have a non-vanishing term uv with $\beta_{uv} > 0$ in ψ^* such that it involves a variable x_j with $j \in S^*$. W.l.o.g., $u(\mathbf{y}) = 1$. Since the term uv is not vanishing under the partial assignment $\mathbf{y}$, we have either $v(\mathbf{y}) = 1$ or v corresponds to a variable x_i with $i \notin S^*$.

In the first case, we must have $\beta_{\bar{u}}\bar{u} + \beta_{\bar{v}}\bar{v} + \beta_{uv}uv$ part of ψ^* with all three coefficients positive. Then a for small $\epsilon > 0$ we can write

$$\begin{aligned}\beta_{\bar{u}}\bar{u} + \beta_{\bar{v}}\bar{v} + \beta_{uv}uv &= (\beta_{\bar{u}} - \epsilon)\bar{u} + (\beta_{\bar{v}} - \epsilon)\bar{v} + (\beta_{uv} - \epsilon)uv + \epsilon(\bar{u} + \bar{v} + uv)\\ &= (\beta_{\bar{u}} - \epsilon)\bar{u} + (\beta_{\bar{v}} - \epsilon)\bar{v} + (\beta_{uv} - \epsilon)uv + \epsilon(1 + \bar{u}\bar{v})\end{aligned}$$

and using this identity, we could derive from ψ^* another posiform $\psi^{**} \in \mathcal{P}_2(f)$ with $\beta_0(\psi^{**}) > C_2(f)$, contradicting the definition of $C_2(f)$.

In the second case we have $\beta_{\bar{u}}\bar{u} + \beta_{uv}uv$ is part of ψ^* with both coefficients positive, thus for a small $\epsilon > 0$ we can write

$$\begin{aligned}\beta_{\bar{u}}\bar{u} + \beta_{uv}uv &= (\beta_{\bar{u}} - \epsilon)\bar{u} + (\beta_{uv} - \epsilon)uv + \epsilon(\bar{u} + uv)\\ &= (\beta_{\bar{u}} - \epsilon)\bar{u} + (\beta_{uv} - \epsilon)uv + \epsilon(v + \bar{u}\bar{v})\end{aligned}$$

and using this identity, we could derive again another posiform $\psi^{***} \in \mathcal{P}_2(f)$ from ψ^* such that $Q(\psi^{***}) = Q(\psi^*) \cup \{v\}$, contradicting the maximality of $Q(\psi^*)$, since v in this case corresponds to a variable x_i with $i \notin S^*$.

These contradictions prove that $\mathbf{y}$ is indeed an autarky for ψ^*, that is a persistency of f by Proposition 5.3. Since in ψ^* all literals $u \in Q(\psi^*)$ have a positive coefficient, we get that for all $\mathbf{x} \in \{0,1\}^V$, such that $\mathbf{x} \neq \mathbf{x}^{\mathbf{y}}$ we have $f(\mathbf{x}) = \psi^*(\mathbf{x}) > \psi^*(\mathbf{x}^{\mathbf{y}}) = f(\mathbf{x}^{\mathbf{y}})$, proving that $\mathbf{y}$ is a strong persistency for f. □

Let us call a posiform $\psi \in \mathcal{B}(f)$ a *best quadratic posiform* of f (i.e., if $\beta_0(\psi) = C_2(f)$), and call it *optimal* if $Q(\psi) \supseteq Q(\psi')$ for all $\psi' \in \mathcal{B}(f)$. The proof of the above theorem shows that the linear part of any best quadratic posiform of a given quadratic pseudo-Boolean function f must vanish in all minima of f. In fact such a optimal posiform can be obtained in polynomial time in the size of f. In the paper [13] is shown that an optimal posiform can be obtained from the dual of the linearization (Rhys linearization) of f. Later in [5, 7, 8] several network flow based model was shown for the computation of $C_2(f)$ and an optimal posiform representation of a given quadratic pseudo-Boolean function f.

Let us remark that for a $\psi \in \mathcal{P}_2(f)$ we may have a larger autarky then the one computed above, In fact if $\psi \in \mathcal{B}(f)$ is an optimal representation of f, then we still could have $|S(\psi)| > |Q(\psi)|$. Consider the function $f(x_1, x_2) = x_1 - x_1x_2$. Then $C_2(f) = 0$ and $\mathcal{B}(f) = \{x_1\bar{x}_2\}$. Here $Q(x_1\bar{x}_2) = \emptyset$, while $S(x_1\bar{x}_2) = \{1, 2\}$, since e.g. $x_1 = x_2 = 1$ is an autarky for $x_1\bar{x}_2$. The main reason is that the above technique, focusing on optimal posiforms, provides only strong persistencies.

Since $S(\psi)$ is also computable in polynomial time by Proposition 5.7, it is natural to consider the set

$$S_2(f) = \bigcup_{\psi \in \mathcal{P}_2(f)} S(\psi).$$

Note that by Proposition 5.2 there exists a partial assignment $\mathbf{y} \in \{0,1\}^{S_2(f)}$ that is a persistency of f. It is however not known how to compute $S_2(f)$ and such an assignment efficiently. We do not even know if there exists always a posiform $\psi \in \mathcal{P}_2(f)$ such that $S(\psi) = S_2(f)$, or not.

Let us note that the network flow based approach (e.g., [8]) computes efficiently both strong and weak persistencies for quadratic functions. It maybe the case that it computes actually $S_2(f)$. We just do not have a definitive proof of this, yet.

We close this section with suggesting an efficient heuristics to determine autarkies for higher degree posiforms, based on the efficiency of computing $S(\psi)$ for quadratic posiforms by Proposition 5.7. Note that even quadratic pseudo-Boolean functions have higher degree posiform representations, and those may provide larger autarkies than we can get from quadratic posiforms. However we do not (yet) know an efficient way of computing $S(\phi)$ for a higher degree posiform ϕ.

Since the constant term does not play any role in the problem of minimization and/or autarkies, let us consider a general posiform of the form

$$\phi = \sum_{Q \subseteq L} \delta_Q \prod_{u \in Q} u, \tag{5.6}$$

where $\delta_Q \geq 0$ for all $Q \subseteq L$. Let us also consider a quadratic posiform ψ, as given in (5.4), and assume further that the coefficients β_u, β_{uv} are all zeros or ones and $\beta_0 = 0$. Following [6] let us call ψ a quadratic cover of ϕ if for all $Q \subseteq L$ with $\delta_Q > 0$, $|Q| > 1$, and for all literals $u \in Q$ there exists another literal $v \in Q$ such that $\beta_{uv} = 1$. Furthermore, for all literals $u \in L$, if $\delta_{\{u\}} > 0$ then $\beta_u = 1$. The following claim follows from results in [6]. We include a proof here for completeness, and since this reference is not easily accessible.

Proposition 5.8 *Assume that ψ is a quadratic cover a posiform ϕ, and let $\mathbf{y} \in \{0, 1\}^{S(\psi)}$ be an autarky for ψ. Then $\mathbf{y}$ is an autarky for ϕ, too.*

Proof Assume that ϕ is given as in (5.6). Let us note first that if $\delta_Q > 0$ and $Q = \{u\}$, then u is a linear term of ψ, and thus either it corresponds to a variable x_j with $j \notin S(\psi)$, or we must have $u(\mathbf{y}) = 0$ by the definition of an autarky for ψ. Let us consider next a term of ϕ with $\delta_Q > 0$ and $|Q| > 1$ that involves variables with indices from $S(\psi)$. Say $u \in Q$ is such a literal. If we have $u(\mathbf{y}) = 0$ then this term is vanishing, and there is nothing to prove. If $u(\mathbf{y}) = 1$ then consider such a literal $v \in Q$, $v \neq u$ for which $\beta_{uv} = 1$ in the quadratic cover ψ. By the definition of a quadratic cover, such a v exists. Since $\mathbf{y}$ is an autarky for ψ we must have $v(\mathbf{y}) = 0$ since otherwise the term uv would not vanish in ψ. Thus term Q is vanishing in ϕ, completing our proof. □

This result suggests a simply and fast heuristics for general UBO problems. Let us view such a problem as the minimization of a posiform representation ϕ of its objective function. Generate in linear time in the size of ϕ a quadratic cover ψ and compute in linear time in the size of ψ, which is linear in the size of ϕ, the set $S(\psi)$. We can e.g., randomize the quadratic cover generation, and repeat the above many times. If we obtain some nontrivial autarkies, then we can merge those by Proposition 5.4.

References

1. F.N. Abu-Khzam, M.R. Fellows, M.A. Langston, W.H. Suters, Crown structures for vertex cover kernelization. Theory Comput. Syst. **41**, 411–430 (2007)

2. B. Aspvall, M.F. Plass, R.E. Tarjan, A linear time algorithm for testing the truth of certain quantified Boolean formulas. Inf. Process. Lett. **8**, 121–123 (1979)
3. M.L. Balinski, On maximum matching, minimum covering and their connections, in *Proceedings of the Princeton Symposium on Mathematical Programming*, ed. by H.W. Kuhn (Princeton University Press, Princeton, 1970), pp. 303–312
4. R. Bar-Yehuda, S. Even, A local-ratio theorem for approximating the weighted vetex cover problem. Ann. Discrete Math. **25**, 27–45 (1985)
5. E. Boros, P.L. Hammer, A max-flow approach to improved roof-duality in quadratic $0-1$ minimization. RUTCOR Research Report RRR 15-1989, RUTCOR (1989)
6. E. Boros, P.L. Hammer, A generalization of the pure literal rule for satisfiability problems. RUTCOR Research Report RRR 20-92, RUTCOR (1992). DIMACS Technical Report 92–19
7. E. Boros, P.L. Hammer, Pseudo-Boolean optimization. Discrete Appl. Math. **123**, 155–225 (2002)
8. E. Boros, P.L. Hammer, R. Sun, G. Tavares, A max-flow approach to improved lower bounds for quadratic unconstrained binary optimization (QUBO). Discrete Optim. **5**, 501–529 (2008)
9. J.-M. Bourjolly, P.L. Hammer, W.R. Pulleyblank, B. Simeone, Boolean-combinatorial bounding of maximum 2-satisfiability, in *Computer Science and Operations Research (Williamsburg, VA 1992)* (Pergamon, Oxford, 1992), pp. 23–42
10. A. Fix, A. Gruber, E. Boros, R. Zabih, A graph cut algorithm for higher-order Markov random fields, in *Proceedings of the 2011 IEEE International Conference on Computer Vision (ICCV)* (2011), pp. 1020–1027
11. A. Fix, A. Gruber, E. Boros, R. Zabih, A hypergraph-based reduction for higher-order Markov random fields. IEEE Trans. Pattern Anal. Mach. Intell. **37**, 1387–1395 (2015)
12. P.L. Hammer, S. Rudeanu, *Boolean Methods in Operations Research and Related Areas* (Springer, Berlin, 1968)
13. P.L. Hammer, P. Hansen, B. Simone, Roof duality, complementations, and persistency in quadratic 0-1 optimization. Math. Program. **28**, 121–155 (1984)
14. D.S. Hochbaum, Approximation algorithms for the set covering and vertex cover problems. SIAM J. Comput. **11**, 555–556 (1982)
15. H. Ishikawa, Transformation of general binary MRF minimization to the first-order case. IEEE Trans. Pattern Anal. Mach. Intell. **33**, 1234–1249 (2011)
16. S. Khuller, The vertex cover problem. ACM SIGACT News **33**, 31–33 (2002)
17. V. Kolmogorov, R. Zabih, What energy functions can be minimized via graph cuts? IEEE Trans. Pattern Anal. Mach. Intell. **26**, 147–159 (2004)
18. O. Kullmann, Constraint satisfaction problems in clausal form I: autarkies and deficiency. Fundam. Inf. **109**, 27–81 (2011)
19. D. Lokhstanov, N.S. Narayanaswamy, V. Raman, M.S. Ramanujan, S. Saurabh, Faster parameterized algorithms using linear programming. ACM Trans. Algorithms **11**, 1–31 (2014)
20. G.L. Nemhauser, L.E. Trotter, Jr., Vertex packings: structural properties and algorithms, Math. Program. **8**, 232–248 (1975)
21. J.M.W. Rhys, A selection problem of shared fixed costs and network flows. Manag. Sci. **17**, 200–207 (1970)
22. E. Rodríguez-Heck, K. Stickler, M. Walter, S. Weltge, Persistency of linear programming relaxations for the stable set problem. Math. Program. (2021, to apper)
23. C. Rother, V. Kolmogorov, V. Lempitsky, M. Szummer, Optimizing binary MRFs via extended roof duality, in *IEEE Conference on Computer Vision and Pattern Recognition* (2007)
24. R. Szeliski, R. Zabih, D. Scharstein, O. Veksler, V. Kolmogorov, A. Agarwal, M. Tappen, C. Rother, A comparative study of energy minimization methods for Markov random fields. IEEE Trans. Pattern Anal. Mach. Intell. **30**, 1068–1080 (2008)
25. R.E. Tarjan, Depth-First search and linear graph algorithms. SIAM J. Comput. **1**, 146–160 (1972)

Chapter 6
Mathematical Programming Models and Exact Algorithms

Abraham P. Punnen and Renata Sotirov

Abstract This chapter focusses on exact solution approaches for QUBO. We first discuss various mixed integer linear programming formulations, compare their relative strength in terms of LP relaxations, and resulting upper bounding strategies. Then, new developments based on semidefinite programming approaches are discussed in detail. The mathematical programming formulations discussed here can be used to solve small size problems and the resulting tight upper bounds on the optimal objective function values can be used in developing specialized enumerative algorithms. Capabilities of some promising enumerative algorithms and solvers are also briefly reviewed.

6.1 Introduction

We have seen in Chap. 3 that several special cases of QUBO can be solved in polynomial time. Let us now focus our attention on solving the general problem. Since QUBO is strongly NP-hard, many of the available exact algorithms for the problem are of implicit enumeration type, in one form or other. A straightforward option to solve QUBO to optimality is by making use of 'ready made' general purpose mixed integer quadratic programming solvers such as CPLEX [27], Gurobi [52], SCIP [1], among others. Simple changes in the way we represent the matrix **Q**, as discussed in Chap. 1, can have some effect on the performance of these solvers [108]. The equivalence of QUBO with other optimization problems such as the maximum weight cut problem, the maximum weight stable set problem, bilinear programs, etc., also provide opportunities for solving QUBO using exact algorithms

A. P. Punnen (✉)
Department of Mathematics, Simon Fraser University, Surrey, BC, Canada
e-mail: apunnen@sfu.ca

R. Sotirov
Department of Econometrics & Operations Research, Tilburg School of Economics and Management, Tilburg University, Tilburg, The Netherlands
e-mail: R.Sotirov@tilburguniversity.edu

A. P. Punnen (ed.), *The Quadratic Unconstrained Binary Optimization Problem*,
https://doi.org/10.1007/978-3-031-04520-2_6

developed specifically for these problems. Recent advancements in the capability of semidefinite programming (SDP) solvers [22, 34, 66, 97, 118] provide additional opportunities for solving QUBO by general purpose SDP based solvers or by SDP based solvers developed for equivalent problems such as the maximum weight cut problem [69, 74, 109].

We will start with the discussion of various MILP formulations of QUBO, including some very recent ones and their properties. We then consider SDP formulations of QUBO and other equivalent problems. Discussions on enumerative algorithms that use traditional MILP formulations as well as those based on SDP formulations are also presented.

Recall that QUBO can be stated as the 0-1 quadratic programming problem

$$\text{Maximize} \quad \phi(\mathbf{x}) = \sum_{i=1}^{n} \sum_{j \in R_i} q_{ij} x_i x_j + \sum_{i=1}^{n} c_i x_i$$

$$\text{Subject to:} \quad x_i \in \{0, 1\} \text{ for } i = 1, 2, \ldots, n,$$

where $\mathbf{Q} = (q_{ij})$ is an $n \times n$ symmetric matrix with diagonal entries zero, $\mathbf{c}^{\mathrm{T}} = (c_1, c_2, \ldots, c_n)$ is an n-vector, and $R_i = \{j : q_{ij} \neq 0\}$. When $\mathbf{Q}$, $\mathbf{c}$ and n are given, an instance of QUBO is well defined. Thus, we sometimes use the triplet $(\mathbf{Q}, \mathbf{c}, n)$ to represent an instance of QUBO or simply $(\mathbf{Q}, \mathbf{c})$ since n is implicit in the dimension of $\mathbf{c}$.

6.2 MILP Formulations

Let us now look at some well-known mixed integer linear programming (MILP) formulations of QUBO. Such formulations are useful in a variety of ways. In particular,

1. MILP formulations can be used to solve smaller size QUBO to optimality using general purpose MILP solvers.
2. The linear programming relaxations of these MILP formulations provide upper bounds on the optimal objective function value, which in turn can be used to design special purpose exact algorithms for QUBO.
3. The MILP formulations can be used as quick heuristic procedures by running the model on powerful MILP solvers with a time limit restriction and collect the best solution obtained.
4. MILP formulation can also be used to design effective matheuristics [90] to solve QUBO.

In addition, there are many different ways to strengthen the MILP formulations yielding tighter upper bounds and improved problem solving capabilities. These include adding strong valid inequalities (see Chap. 4), effective reformulations (see Chap. 7), constructing higher level RLT formulations [112] and using resulting strong cutting planes, applying formulations based on disjunctive programming [13], and using SDP based cuts.

Early use of binary variables in translating logical conditions to a set linear inequalities (in binary and continuous variables) can be traced back to [28, 36, 43, 121, 123]. The product $x_i x_j$ of binary variables x_i and x_j takes value one, if $x_i = x_j = 1$ and zero, otherwise. The process of replacing the product $x_i x_j$ by a single $0-1$ variable, say y_{ij}, along with additional linear constraints such that $y_{ij} = 1$ if and only if $x_i = x_j = 1$ is called *linearization of the product* $x_i x_j$. The theorem below summarizes two basic linearization techniques for the product $x_i x_j$ considered respectively by Watters [121] and Glover and Woolsey [43]. It may be noted that earlier works of other authors also considered linearization of the product of binary variables, either implicitly or explicitly. For example, Goldman [47] mentioned the works of Fortet [36, 37] and Dantzig [28] as earlier examples of linearizations.

Theorem 6.1 *The product $x_i x_j$ of binary variables can be linearized using the variable y_{ij} by introducing any of the following set of constraints.*

$$
\begin{array}{llll}
(a) & x_i + x_j - y_{ij} \le 1 & (b) & x_i + x_j - y_{ij} \le 1 \\
 & 2y_{ij} - x_i - x_j \le 0 & & y_{ij} \le x_i,\, y_{ij} \le x_j \\
 & y_{ij} \in \{0, 1\}. & & y_{ij} \ge 0.
\end{array}
$$

It is easy to verify that conditions (a) or (b) guarantee that $y_{ij} = 1$ precisely when $x_i = x_j = 1$. This leads to the following MILP formulations of QUBO.

$$\text{WT:}\quad \text{Maximize} \quad \sum_{i=1}^{n} \sum_{j \in R_i} q_{ij} y_{ij} + \sum_{i=1}^{n} c_i x_i$$

$$\text{Subject to:}\quad x_i + x_j - y_{ij} \le 1 \text{ for } j \in R_i, i = 1, 2, \ldots, n \tag{6.1}$$

$$2y_{ij} - x_i - x_j \le 0 \text{ for } j \in R_i, i = 1, 2, \ldots, n \tag{6.2}$$

$$x_i \in \{0, 1\} \text{ for } i = 1, 2, \ldots, n \tag{6.3}$$

$$y_{ij} \in \{0, 1\} \text{ for } j \in R_i, i = 1, 2, \ldots, n. \tag{6.4}$$

and

$$\text{GW:}\quad \text{Maximize} \quad \sum_{i=1}^{n} \sum_{j \in R_i} q_{ij} y_{ij} + \sum_{i=1}^{n} c_i x_i$$

$$\text{Subject to:}\quad x_i + x_j - y_{ij} \le 1 \text{ for } j \in R_i, i = 1, 2, \ldots, n \tag{6.5}$$

$$y_{ij} - x_i \le 0 \text{ for } j \in R_i, i = 1, 2, \ldots, n \tag{6.6}$$

$$y_{ji} - x_i \leq 0 \text{ for } j \in R_i, i = 1, 2, \ldots, n \tag{6.7}$$

$$x_i \in \{0, 1\} \text{ for } i = 1, 2 \ldots, n \tag{6.8}$$

$$y_{ij} \geq 0 \text{ for } j \in R_i, i = 1, 2, \ldots, n. \tag{6.9}$$

It may be noted that the constraints (6.6) and (6.7) are presented in a slightly different form than that is given in Theorem 6.1 (b). However, they are equivalent when considering all the indices $i = 1, 2, \ldots, n, j \in R_i$. As indicated in Chap. 1, GW is normally called *the standard linearization* or the *Glover-Woolsey linearization*. Both WT and GW belong to the class of *explicit linearizations* [105] as each of these formulations uses a variable y_{ij} to represent the product $x_i x_j$. Explicit linearizations are of considerable interest since their liner programming relaxations (LP relaxations) can be strengthened by directly using facet defining inequalities of the Boolean quadric polytope [99]. See Chap. 4 for a detailed discussion of this polytope.

We denote the objective function value of a mathematical programming formulation P by OBJ(P). Also, the LP relaxation of an MILP formulation P is denoted by P*.

Lemma 6.1 ([105]) *OBJ(GW*) $\leq$ OBJ(WT*). Further, OBJ(WT*) could be arbitrarily worse than OBJ(GW*).*

Proof Note that each constraint from (6.2) is a linear combination of appropriate constraints from (6.6) and (6.7) of GW. Thus, any feasible solution of GW* is also a feasible solution to WT* and hence OBJ(GW*) $\leq$ OBJ(WT*). To prove the second part of the lemma, consider the following example [105]. Choose

$$\mathbf{Q} = \begin{pmatrix} 0 & \alpha \\ \alpha & 0 \end{pmatrix} \text{ and } \mathbf{c} = (-2\alpha, 1), \text{ where } \alpha > 0.$$

Now, $x_1 = y_{12} = y_{21} = 0, x_2 = 1$ is an optimal solution to GW* with OBJ(GW*) = 1 whereas for WT*, $x_1 = 0, x_2 = 1, y_{12} = y_{21} = 0.5$ is an optimal solution with OBJ(WT*)= $1 + 2\alpha$. □

The value OBJ(GW*) is equal to the upper bound value obtained from the best posiform representation of $\phi(\mathbf{x})$ and is also equal to the roof duality bound [2, 58] (See also, Chaps. 1 and 5). Despite the negative result established in Lemma 6.1 for WT, it may be noted that by using appropriate coefficients for the variables y_{ij}, x_i and x_j in (6.2), WT can be strengthened yielding the corresponding LP relaxation value matches with OBJ(GW*) [105]. Computational results reported in [105] show that this modified WT model works well in practice. The formulation GW plays a fundamental role in various MILP formulations of QUBO and the strength of such formulations is often compared to the value OBJ(GW*).

Let us now discuss ways to simplify the formulations WT and GW. For $i = 1, 2, \ldots, n$, let $R_i^+ = \{j : q_{ij} > 0\}$ and $R_i^- = \{j : q_{ij} < 0\}$. Note that the constraints (6.1) and (6.2) can be stated as [105]

$$x_i + x_j - 1 \leq y_{ij} \leq \frac{x_i + x_j}{2}.$$

Therefore, at optimality, constraint (6.1) is not active if $j \in R_i^+$ and constraint (6.2) is not active if $j \in R_i^-$. Thus, WT can be re-stated as

$$\text{RWT:} \quad \text{Maximize} \quad \sum_{i=1}^{n}\sum_{j\in R_i} q_{ij}y_{ij} + \sum_{i=1}^{n} c_i x_i$$

$$\text{Subject to:} \quad x_i + x_j - y_{ij} \leq 1 \text{ for all } j \in R_i^-, i = 1, 2, \ldots, n$$

$$2y_{ij} - x_i - x_j \leq 0 \text{ for all } j \in R_i^+, i = 1, 2, \ldots, n$$

$$x_i \in \{0, 1\} \text{ for all } i = 1, 2, \ldots, n$$

$$y_{ij} \in \{0, 1\} \text{ for all } j \in R_i, i = 1, 2, \ldots, n.$$

The number of general constraints in RWT is reduced by half, compared to that of WT. It may be noted that, for RWT, the definition $y_{ij} = x_i x_j$ is guaranteed only at optimality, and not necessarily for an arbitrary feasible solution. An MILP model with such a property is called *optimality restricted models* [105]. Explicit MILP models that guarantee $y_{ij} = x_i x_j$ for all feasible solutions are called *complete explicit models* [105]. GW and WT are complete explicit models whereas RWT is an optimality restricted model. A complete model sometimes have better heuristic value in the sense that if we terminate the solver before reaching optimality, the best solution produced can be taken as a heuristic solution, without additional calculations. But, for optimality restricted models, if we terminate the solver prematurely, the objective function value reported by the solver for the best solution obtained need not match with the QUBO objective function value of the solution defined by the corresponding **x** variables. The objective function value of such a solution could be very different from the value reported by the solver [105] and therefore it is important to recompute the objective function value using the **x** variables and it will require an additional $O(n^2)$ time. However, optimality restricted models are interesting because of the reduced problem size. It may be noted that OBJ(WT*) = OBJ(RWT*).

For given i and j, the constraints (6.5), (6.6), (6.7), and (6.9) of GW can be stated as

$$\max\{0, x_i + x_j - 1\} \leq y_{ij} \leq \min\{x_i, x_j\}.$$

Thus, constraints (6.5) and (6.9) are redundant at optimality if $j \in R_i^+$ and constraints (6.6) and (6.7) are redundant at optimality if $j \in R_i^-$. Removing these redundant constraints, an optimality restricted version of GW [43, 44, 59] can be obtained as given below.

$$\begin{aligned}
\text{RGW:}\quad \text{Maximize}\quad & \sum_{i=1}^{n}\sum_{j\in R_i} q_{ij}y_{ij} + \sum_{i=1}^{n} c_i x_i \\
\text{Subject to:}\quad & x_i + x_j - y_{ij} \le 1 \text{ for all } j \in R_i^-, i = 1, 2, \ldots, n \\
& y_{ij} - x_i \le 0 \text{ for all } j \in R_i^+, i = 1, 2, \ldots, n \\
& y_{ji} - x_i \le 0 \text{ for all } j \in R_i^+, i = 1, 2, \ldots, n \\
& x_i \in \{0, 1\} \text{ for all } i = 1, 2, \ldots, n \\
& y_{ij} \ge 0 \text{ for all } j \in R_i^-, i = 1, 2, \ldots, n.
\end{aligned}$$

Note that OBJ(GW*) = OBJ(RGW*).

Weighted aggregation of selected constraints of WT and GW can generate alternative MILP formulations of QUBO with various properties [105]. In particular, this class of explicit formulations have less number of constraints compared to WT and GW. For $\theta_{ij} > 0$, consider the inequality

$$\sum_{j\in R_i} \theta_{ij}(x_j - y_{ij}) \le \left(\sum_{j\in R_i} \theta_{ij}\right)(1 - x_i), \text{ for } i = 1, 2, \ldots, n. \tag{6.10}$$

Let AWT(θ) be the model obtained from WT by replacing constraint (6.1) with (6.10) and AGW(θ) be the model obtained from GW by replacing constraints (6.5) with (6.10).

Theorem 6.2 ([105]) *AWT(θ) and AGW(θ) are valid MILP models for QUBO.*

Proof Since (6.10) is a positively weighted linear combination of some constraints of WT, every feasible solution of WT is also a feasible solution of AWT. Now, choose a solution $(\mathbf{x}, \mathbf{y})$ of AWT(θ) and let x_i be an arbitrary component of $\mathbf{x}$. If $x_i = 0$ then constraint (6.10) is redundant and in this case (6.1) is clearly satisfied. If $x_i = 1$, then from (6.10)

$$\sum_{j\in R_i} \theta_{ij}(x_j - y_{ij}) \le 0. \tag{6.11}$$

If $x_j - y_{ij} < 0$ for some $j \in R_i$ then $x_j = 0$ and $y_{ij} = 1$ and this violates (6.2). Thus, the only possibility is that $x_j - y_{ij} = 0$ establishing that (6.1) is satisfied. This shows that, if $(\mathbf{x}, \mathbf{y})$ is a feasible solution to AWT(θ) then it is also feasible solution to WT establishing that AWT(θ) and WT have the same set of feasible solutions. Proving the validity of AGW(θ) is not difficult and the proof is omitted. □

When $\theta_{ij} = 1$ for all i and j, the constraint (6.10) becomes

$$\sum_{j \in R_i} (x_j - y_{ij}) \leq |R_i|(1 - x_i), \text{ for } i = 1, 2, \ldots, n. \tag{6.12}$$

This leads to the unit-weight special cases of AWT(θ) and AGW(θ).

Theorem 6.3 ([105]) *OBJ(AWT(θ)*) $\geq$ OBJ(WT*) and OBJ(AGW(θ)*) $\geq$ OBJ (GW*). Further, if θ_{ij} is taken as an optimal dual variable associated with the corresponding constraint in WT* and GW* then OBJ(AWT(θ)*) $=$ OBJ(WT*) and OBJ(AGW(θ)*) $=$ OBJ(GW*).*

Proof Since every feasible solution of WT* (GW*) is also a feasible solution to AWT(θ))* (AGW(θ)*), OBJ(AWT(θ)*) $\geq$ OBJ(WT*) (AGW(θ)*$\geq$OBJ(GW*)). The second part of the theorem follows from Theorem 6.20 in Appendix. □

As noted in [105], choosing θ_{ij} as optimal dual variables in Theorem 6.3 needs to be made with caution. Some of these dual variables could be zero but the validity of the model AWT(θ) (AGW(θ)) assumes $\theta_{ij} > 0$. Thus, if any of the optimal dual variables that we consider turned out to be zero, we can set its value to a small positive real number without significantly deteriorating the performance guarantee offered by Theorem 6.3, in general.

Let us now consider another weighted aggregation model where constraints (6.6) and (6.7) of GW are aggregated. Let ζ_{ij} and η_{ij} be positive real numbers. Consider the inequality

$$\sum_{j \in R_i} \zeta_{ij} y_{ij} + \sum_{j \in R_i} \eta_{ij} y_{ji} \leq \sum_{j \in R_i} (\zeta_{ij} + \eta_{ij}) x_i, \text{ for } i = 1, 2, \ldots, n. \tag{6.13}$$

The MILP model obtained from GW by replacing the class of inequalities (6.6) and (6.7) with (6.13) and adding the new upper bound constraints $y_{ij} \leq 1$ for $i = 1, 2, \ldots, n$, $j \in R_i$ is denoted by AGW(ζ, η).

Theorem 6.4 ([105]) *AGW(ζ, η) is a valid MILP model for QUBO and OBJ$\big($AGW(ζ, η)$^*\big)$ $\geq$ OBJ(GW*). Further, when the multipliers ζ_{ij} and η_{ij} are selected as optimal dual variables associated with corresponding constraints in GW*, OBJ$\big($AGW(ζ, η)$^*\big)$ $=$ OBJ(GW*)*

Proof Constraint (6.13) guarantees that if $x_i = 0$ for any i, then $y_{ij} = y_{ji} = 0$ for $j \in R_i$ and if $x_i = 1$, constraint (6.13) is redundant. When $x_i = 1$, constraint (6.5) together with the upper bound restriction on y_{ij} in AGW(ζ, η) we have $x_j \leq y_{ij} \leq 1$. Thus if x_j is also 1, then $y_{ij} = 1$. If $x_j = 0$ then we can use (6.13) and by choosing this j in place i, we get $y_{ij} = y_{ji} = 0$. This completes the proof of the first part of the theorem. The last part of the theorem can be proved using Theorem 6.20 from Appendix. □

When $\eta_{ij} = \zeta_{ij} = 1$ constraint (6.13) becomes

$$\sum_{j \in R_i} \zeta_{ij} y_{ij} + \sum_{j \in R_i} \eta_{ij} y_{ji} \leq 2|R_i|x_i, \text{ for } i = 1, 2, \ldots, n. \tag{6.14}$$

In this case AGW(ζ, η) becomes the Glover-Woolsey aggregation model [44].

Now, consider the model AGW(θ, ζ, η) obtained by weighted aggregations constraints (6.6) and (6.7) as well as weighted aggregation of (6.5).

$$\begin{aligned}
\text{Maximize} \quad & \sum_{(i,j) \in S} q_{ij} y_{ij} + \sum_{i=1}^{n} c_i x_i \\
\text{Subject to:} \quad & \sum_{j \in R_i} \theta_{ij}(x_j - y_{ij}) \leq \left(\sum_{j \in R_i} \theta_{ij} \right) (1 - x_i), \text{ for } i = 1, 2, \ldots, n \\
& \sum_{j \in R_i} \zeta_{ij} y_{ij} + \sum_{j \in S_i} \eta_{ij} y_{ji} \leq \left(\sum_{j \in R_i} \zeta_{ij} + \sum_{j \in S_i} \eta_{ij} \right) x_i, \text{ for } i = 1, 2, \ldots, n \\
& x_i \in \{0, 1\} \text{ for all } i = 1, 2, \ldots, n \\
& 0 \leq y_{ij} \leq 1 \text{ for all } j \in R_i, i = 1, 2, \ldots, n.
\end{aligned}$$

where $S_i = \{j : q_{ji} \neq 0\}$. Since we assume that $\mathbf{Q}$ is symmetric, $R_i = S_i$ for all i. But we retained the separate notation of S_i for clarity.

Theorem 6.5 *AGW(θ, ζ, η) is a valid model for QUBO and* $\text{OBJ}(AGW(\theta, \zeta, \eta)^*) \geq \text{OBJ}(GW^*)$*. Further, when the multipliers θ_{ij}, ζ_{ij} and η_{ij} are selected as optimal dual variables associated with the corresponding constraints in GW^*,* $\text{OBJ}(AGW(\theta, \zeta, \eta)^*) = \text{OBJ}(GW^*)$

For the proof of the above theorem, we refer to [105]. Among all of the complete and explicit MILP models of QUBO, AGW(θ, ζ, η) have the least number of general constraints. When $\theta_{ij} = \zeta_{ij} = \eta_{ij} = 1$ we have the unit-weight special case of AGW(θ, ζ, η) which offers a nice and simple constraint set.

Using weighted aggregation in various forms and combinations, many more explicit MILP formulations of QUBO can be constructed. Despite the interesting theoretical properties, the computational behaviour of models using aggregation constraints (6.10) is not very good [105]. On the other hand, some of the other aggregation models that does not use (6.10) performed better than GW, on average. In particular, the modified version of WT mentioned earlier in this section is one of the better MILP formulations of QUBO, along with some other aggregation based models. For details on the experimental behaviour of various explicit linearization models, we refer to [35, 59, 105]. Optimality restricted versions also allow weighted aggregations, but with some level of restrictions [105].

6.2.1 Compact MILP Formulations

The explicit MILP formulations discussed so far have $O(n^2)$ variables and $O(n)$ to $O(n^2)$ constraints. In compact formulations, the number variables and constraints is $O(n)$. Most of the compact formulations reported in the literature use a linearization technique proposed by Glover [41] in some way or other.

Theorem 6.6 *(Glover's Linearization Theorem) Let x be a binary variable and y be a real number such that $\ell \leq y \leq u$. Then, $w = xy$ if*

$$\ell x \leq w \leq ux \tag{6.15}$$

$$y - u(1-x) \leq w \leq y - \ell(1-x). \tag{6.16}$$

Proof Note that w is either 0 or y. When $x = 0$ condition (6.15) guarantees that $w = 0$ and (6.16) is redundant. Similarly, when $x = 1$ the condition (6.16) guarantees that $w = y$ and the condition (6.15) is redundant. □

A more general set of conditions approximating product of two real variables is presented in [92] and the conditions of Theorem 6.6 follow from it (see Chap. 1). Let $g_i(\mathbf{x}) = \sum_{j \in R_i} q_{ij} x_j$. Then, the objective function $\phi(\mathbf{x})$ of QUBO can be written as

$$\phi(\mathbf{x}) = \sum_{i=1}^{n} x_i g_i(\mathbf{x}) + \sum_{i=1}^{n} c_i x_i. \tag{6.17}$$

For any $\mathbf{x} \in \{0, 1\}^n$, $g_i(\mathbf{x})$ is a real number satisfying

$$\ell_i = \sum_{j \in R_i} \min\{0, q_{ij}\} \leq g_i(\mathbf{x}) \leq u_i = \sum_{j \in R_i} \max\{0, q_{ij}\}. \tag{6.18}$$

Applying Glover's linearization theorem on the product $x_i g_i(\mathbf{x})$ in (6.17), we get the compact formulation [41]

$$\text{GL:} \quad \text{Maximize} \quad \sum_{i=1}^{n} w_i + \sum_{i=1}^{n} c_i x_i$$

$$\text{Subject to:} \quad w_i \leq u_i x_i, \text{ for } i = 1, 2, \ldots, n \tag{6.19}$$

$$w_i \leq \ell_i x_i + \sum_{j \in R_i} q_{ij} x_j - \ell_i, \text{ for } i = 1, 2, \ldots, n \tag{6.20}$$

$$w_i \geq \ell_i x_i, \text{ for } i = 1, 2, \ldots, n \tag{6.21}$$

$$w_i \geq u_i x_i + \sum_{j \in R_i} q_{ij} x_j - u_i, \text{ for } i = 1, 2, \ldots, n \tag{6.22}$$

$$x_i \in \{0, 1\} \text{ for } i = 1, 2, \ldots, n. \tag{6.23}$$

The MILP formulation GL is known as *Glover's compact linearization* and it is a complete formulation. An optimality restricted version of GL can also be obtained. Note that the general constraints of GL can be written as

$$\max\{u_ix_i + \sum_{j\in R_i} q_{ij}x_j - u_i, \ell_ix_i\} \le w_i \le \min\{u_ix_i, \ell_ix_i + \sum_{j\in R_i} q_{ij}x_j - \ell_i\}, i = 1, 2, \ldots, n. \tag{6.24}$$

Since we have a maximization objective with the coefficient of w_i in the objective function is one for all i, the inequalities (6.21) and (6.22) can be discarded from GL without affecting optimality [3] (see also inequality (6.24)). This leads to the optimality restricted version, GLO, of GL considered in [5]:

$$\text{GLO:} \quad \text{Maximize} \quad \sum_{i=1}^{n} w_i + \sum_{i=1}^{n} c_ix_i$$

$$\text{Subject to:} \quad w_i \le u_ix_i, \ \text{for } i = 1, 2, \ldots, n \tag{6.25}$$

$$w_i \le \ell_ix_i + \sum_{j\in R_i} q_{ij}x_j - \ell_i, \ \text{for } i = 1, 2, \ldots, n \tag{6.26}$$

$$x_i \in \{0, 1\} \text{ for } i = 1, 2, \ldots, n. \tag{6.27}$$

The model GLO has $2n$ general constraints compared to the 4n general constraints for GL. Let us now discuss another optimality restricted model, proposed in [106] which is a variation of the model by Oral and Kettami [98] who developed it for a minimization problem.

$$\text{PK:} \quad \text{Maximize} \quad \sum_{i=1}^{n}(c_i + u_i)x_i - \sum_{i=1}^{n} z_i$$

$$\text{Subject to:} \quad (u_i - \ell_i)x_i - z_i - \sum_{j\in R_i} q_{ij}x_j \le -\ell_i, \ \text{for } i = 1, 2, \ldots, n \tag{6.28}$$

$$x_i \in \{0, 1\} \text{ for } i = 1, 2, \ldots, n \tag{6.29}$$

$$z_i \ge 0 \text{ for } i = 1, 2, \ldots, n. \tag{6.30}$$

Theorem 6.7 ([106]) *PK is a valid optimality restricted MILP model for QUBO.*

Proof When $x_i = 1$, we have $z_i \ge u_ix_i - \sum_{j\in R_i} q_{ij}x_j$ and since the objective function value is to be maximized and the coefficient of z_i is -1, at optimality $z_i = u_ix_i - \sum_{j\in R_i} q_{ij}x_j$. Similarly, when $x_i = 0$, we have $z_i \ge -\sum_{j\in R_i} q_{ij}x_j + \ell_i$ and $z_i \ge 0$. Thus, at optimality, when $x_i = 0$, $z_i = 0$ and the result follows. □

PK can be derived from GLO by eliminating the unrestricted variables $w_i, i = 1, 2, \ldots, n$ using inequality (6.25) and the associated slack variable. Adams and

Forrester [3] made a similar comparison for the minimization version of QUBO and the model of [98]. Note that PK has only n general constraints. Eliminating $w_i, i = 1, 2, \ldots, n$ from GLO using inequality (6.26) and the associated slack variable, another MILP formulation with only n constraints can be obtained [106].

Lemma 6.2 *OBJ(GL*) = OBJ(GLO*) = OBJ(PK*) ≥ OBJ(GW*). Further, there exist QUBO instances such that OBJ(GL*) > OBJ(GW*).*

Proof The equality OBJ(GL*) = OBJ(GLO*) follows from the fact that the optimality restriction argument extends to LP relaxations as well. The equality OBJ(GLO*) = OBJ(PK*) follows since PK can be obtained by eliminating unrestricted variables from GLO. Let $(\mathbf{x}, \mathbf{y})$ be any feasible solution to GW*. Define $w_i = \sum_{j \in R_i} q_{ij} x_j$. Then $(\mathbf{x}, \mathbf{w})$ is a feasible solution to GL*. Further the objective function value of $(\mathbf{x}, \mathbf{y})$ in GW* is the same as the objective function value of $(\mathbf{x}, \mathbf{w})$ in GL*. Thus OBJ(GL*) ≥ OBJ(GW*).

It is easy to construct examples establishing the strict inequality OBJ(GL*) > OBJ(GW*). □

The inequality OBJ(GL*) ≥ OBJ(GW*) is proved in [5]. The compact formulations that we have discussed so far possess weak LP relaxation bounds. Let us now explore how to obtain compact MILP formulations with the corresponding LP relaxation bound is at least as good as OBJ(GW*).

Let y and z be real numbers and $x \in \{0, 1\}$. Then, $w = xy + (1 - x)z$ is called the *sum of complementary products* [106]. When $x = 0, w = z$ and when $x = 1, w = y$.

Theorem 6.8 ([59, 106]) *Let x be a binary variable and y and z be real numbers such that $\ell \leq g(y) - h(y) \leq u$. Then, $w = xy + (1 - x)z$ if the following conditions are satisfied.*

$$\ell x + z \leq w \leq z + ux \tag{6.31}$$

$$y - u(1 - x) \leq w \leq y - \ell(1 - x). \tag{6.32}$$

Proof Since, $w = xy + (1 - x)z$ we have $w - z = x(y - z)$. The result now follows by applying Glover's linearization theorem on $(w - z) = x(y - z)$. □

Let us now look at a special representation of the objective function $\phi(\mathbf{x})$ of QUBO. The function $\phi(\mathbf{x})$ is said to be in *decomposition form* [18, 59, 106] if there exist non-negative linear functions $L(\mathbf{x}), f_i(\mathbf{x}), g_i(\mathbf{x}), i = 1, 2, \ldots, n$ defined over binary variables such that

$$\phi(\mathbf{x}) = Z - \left(L(\mathbf{x}) + \sum_{i=1}^{n} \Big[x_i f_i(\mathbf{x}) + (1 - x_i) g_i(\mathbf{x}) \Big] \right). \tag{6.33}$$

Theorem 6.9 ([18, 106]) *From any dual feasible solution of GW* with dual objective function value Z, we can construct linear functions $L(\mathbf{x}), f_i(\mathbf{x}), g_i(\mathbf{x})$,* *for* $i = 1, 2, \ldots, n$ *that are non-negative over the unit cube in* $\mathbb{R}^n$ *such that*

$$\phi(\mathbf{x}) = Z - \left(L(\mathbf{x}) + \sum_{i=1}^{n} \Big[x_i f_i(\mathbf{x}) + (1 - x_i) g_i(\mathbf{x}) \Big] \right).$$

Proof Let u^1_{ij} be the dual variable associated with (i, j)th constraint from (6.5), u^2_{ij} be the dual variable associated with (i, j)th constraint from (6.6), u^3_{ij} be the dual variable associated with (i, j)th constraint from (6.7), u^4_i be the dual variable associated with ith constraint $x_i \le 1$ in GW*. Also, let Z be the dual objective function value of GW* corresponding to the dual solution $(\mathbf{u}^1, \mathbf{u}^2, \mathbf{u}^3, \mathbf{u}^4)$. Further, let s_i be the slack variable in the dual constraint of GW* corresponding to the primal variable x_i and s_{ij} be the slack variable in the dual constraint corresponding to the primal variable y_{ij}. Then, from Theorem 6.21 in Appendix, for any feasible solution $(\mathbf{x}, \mathbf{y})$ of GW*, we have

$$\begin{aligned} F(\mathbf{x}, \mathbf{y}) =& Z - \sum_{i=1}^{n} s_i x_i - \sum_{i=1}^{n} \sum_{j \in R_i} s_{ij} y_{ij} - \sum_{i=1}^{n} \sum_{j \in R_i} u^1_{ij} \left(1 - x_i - x_j + y_{ij}\right) - \\ & \sum_{i=1}^{n} \sum_{j \in R_i} u^2_{ij} \left(-y_{ij} + x_i\right) - \sum_{i=1}^{n} \sum_{j \in R_i} u^3_{ij} \left(-y_{ji} + x_i\right) - \sum_{i=1}^{n} u^4_{ij} \left(1 - x_i\right). \end{aligned} \tag{6.34}$$

For any solution $\mathbf{x}$ of QUBO, there exists a unique $\mathbf{y}$ defined by $y_{ij} = x_i x_j$. Thus, in view of (6.34), for any $\mathbf{x} \in \{0, 1\}^n$,

$$\begin{aligned} F(\mathbf{x}) =& Z - \sum_{i=1}^{n} s_i x_i - \sum_{i=1}^{n} \sum_{j \in R_i} s_{ij} x_i x_j - \sum_{i=1}^{n} \sum_{j \in R_i} u^1_{ij} \left(1 - x_i\right) \left(1 - x_j\right) - \\ & \sum_{i=1}^{n} \sum_{j \in R_i} u^2_{ij} x_i \left(1 - x_j\right) - \sum_{i=1}^{n} \sum_{j \in R_i} u^3_{ij} x_i \left(1 - x_i\right) - \sum_{i=1}^{n} u^4_{ij} \left(1 - x_i\right). \end{aligned} \tag{6.35}$$

Choose $L(x) = \sum_{i=1}^{n} s_i x_i + \sum_{i=1}^{n} u^4_{ij} (1 - x_i)$, $g_i(x) = \sum_{j \in R_i} u^3_{ij} (1 - x_j)$, and $f_i(x) = \sum_{j \in R_i} \left(s_{ij} x_j + u^1_{ij} (1 - x_j) + u^2_{ij} (1 - x_j) \right)$, the result follows. □

Corollary 6.1 *If* $(\mathbf{u}^1, \mathbf{u}^2, \mathbf{u}^3, \mathbf{u}^4)$ *is an optimal solution to the dual of* GW^**, then* $Z = \text{OBJ}(GW^*)$.

If the decomposition form of $\phi(\mathbf{x})$ constructed in Theorem 6.9 uses optimal dual variables $(\mathbf{u}^1, \mathbf{u}^2, \mathbf{u}^3, \mathbf{u}^4)$, we say that (6.33) is in *optimal decomposition form* [18, 59, 106]. In view of Theorem 6.9, for any instance of QUBO, there exists an equivalent instance where the objective function is in the decomposition form (optimal decomposition form). The decomposition form discussed above is a best posiform representation of $\phi(\mathbf{x})$.

Applying Theorem 6.8 on $\phi(\mathbf{x})$ given in (6.33), we get the following compact linearization [106] which is a variation of an analogous model given in [18] for the minimization version of QUBO.

$$\text{D1:}\quad \text{Maximize}\quad Z - L(\mathbf{x}) - \sum_{i=1}^{n} w_i$$

$$\text{Subject to:}\quad w_i \leq g_i(\mathbf{x}) + u_i x_i,\ \text{ for } i = 1, 2, \ldots, n \tag{6.36}$$

$$w_i \leq \ell_i x_i + f_i(\mathbf{x}) - \ell_i,\ \text{ for } i = 1, 2, \ldots, n \tag{6.37}$$

$$w_i \geq g_i(\mathbf{x}) + \ell_i x_i,\ \text{ for } i = 1, 2, \ldots, n \tag{6.38}$$

$$w_i \geq u_i x_i + f_i(\mathbf{x}) - u_i,\ \text{ for } i = 1, 2, \ldots, n \tag{6.39}$$

$$w_i \geq 0,\ \text{ for } i = 1, 2, \ldots, n \tag{6.40}$$

$$x_i \in \{0, 1\} \text{ for } i = 1, 2, \ldots, n. \tag{6.41}$$

D1 have $4n$ general constraints, n binary variables, and n continuous variables.

Lemma 6.3 ([106]) *When $\phi(\mathbf{x})$ is in optimal decomposition form,* OBJ*($D1^*$) $\leq$* OBJ*(GW^*).*

Proof When $\phi(\mathbf{x})$ is in optimal decomposition form, $Z =$ OBJ(GW^*) [2, 18, 58, 59]. Since $L(\mathbf{x})$ and w_i are non-negative, OBJ$(D1^*) \leq Z$ and the result follows. □

In D1, the constraint (6.40) is essentially redundant and can be removed. However, the resulting model could have an LP relaxation value more than OBJ(GW*) [59, 106]. The constraint (6.40) in D1 can also be replaced by

$$\sum_{i=1}^{n} w_i \geq 0$$

and it can be verified that the resulting model will have an LP relaxation value no more than OBJ(GW*) [59, 106]. There are several variations of D1. For a comprehensive review of them along with computational analysis, an interested reader is referred to [106]. The model D1 is closely related to a similar model developed by Hansen and Meyar [59] and also by Adams et al. [7].

Other compact models of interest include [5, 7, 17, 25, 35, 42, 50, 61, 84, 103] and some of these models are developed for the minimization version of QUBO. Detailed computational analysis and relationships between these models are discussed in [106]. An extended MILP model for QUBO using more variables was proposed in [39] where the corresponding LP-relaxation bound matches OBJ(GW*).

Let us now look at another MILP modelling strategy that is applicable when q_{ij} is an integer for all i, j. Note that, in this case $\sum_{j\in R_i} q_{ij}x_j$ must be an integer. Let

ℓ_i and u_i be upper and lower bounds on $\sum_{j\in R_i} q_{ij}x_j$. Then $\sum_{j\in R_i} q_{ij}x_j - l_i$ is a non-negative integer bounded above by $u_i - \ell_i$. Let $t_i = \lfloor \log_2(u_i - \ell_i)\rfloor + 1$. Then, $\sum_{j\in R_i} q_{ij}x_j - l_i$ can be written as

$$\sum_{j\in R_i} q_{ij}x_j - l_i = \sum_{k=1}^{t_i} 2^{k-1}u_{ik}$$

and we have the 0-1 programming formulation of QUBO as

$$\begin{aligned}
\text{Maximize} \quad & \sum_{i=1}^{n}\sum_{k=1}^{t_i} 2^{k-1}u_{ik}x_i + \sum_{i=1}^{n}(c_i + l_i)x_i \\
\text{Subject to:} \quad & \sum_{j\in R_i} q_{ij}x_j = l_i + \sum_{k=1}^{t_i} 2^{k-1}u_{ik}, \text{ for } i = 1, 2, \dots, n \\
& x_i \in \{0, 1\} \text{ for all } i = 1, 2, \dots, n \\
& u_{ik} \in \{0, 1\} \text{ for all } i = 1, 2, \dots, n, k = 1, 2, \dots, t_i.
\end{aligned}$$

Now, we can linearize the product $u_{ik}x_i$ of binary variables using any of the strategies discussed in Theorem 6.1 to get another linearization of QUBO. Similar modelling strategy was used by others for several 0-1 quadratic optimization problems [4]. For variations of this type of linearization and applications to low-rank QUBO, we refer to [107, 108].

The linearizations discussed above can be strengthened using different techniques. This include, adding valid inequalities, applying reformulation linearization techniques of Adams and Sherali [6] and of Boros et al. [19], and disjunctive programming techniques introduced by Balas [13].

6.3 QUBO and Semidefinite Programming

Semidefinite programming belongs to the class of convex conic programming, in which the objective is linear and the constraint set is given by the intersection of an affine space with a cone of positive semidefinite matrices. Moreover, SDP can be viewed as an extension of linear programming where the vector variables are replaced by matrix variables and the nonnegativity constraints are replaced by positive semidefiniteness constraints. Semidefinite programming plays an important role in combinatorial optimization where it is used to solve convex relaxations of NP-hard problems. Semidefinite programming problems are solvable in polynomial time, with fixed precision, by interior point methods. For more details on the development of semidefinite optimization and recent approaches the interested reader is refereed to [10, 122].

SDP has been studied since 1940 under different names. One of the first papers on the theory of semidefinite programming appeared in 1963 by Bellman and Fan [15]. An explicit use of semidefinite programming in combinatorial optimization appeared in the seminal work of Lovász [86] on the ϑ-function, in the late 1970s. The development of semidefinite optimization has grown tremendously after the extension of interior point methods from linear programming to semidefinite programming by Nesterov and Nemirovski [96] in the early 1990s. Interior point methods are the most prominent algorithms for solving semidefinite optimization problems. Another important result that contributed to the development of semidefinite programming is the Goemans-Williamson randomized approximation algorithm for the maximum cut problem [46]. That paper demonstrates the strength of semidefinite relaxations in combinatorial optimization.

SDP solvers based on interior point methods exhibit difficulties in terms of running time as well as memory for solving even medium-size SDP relaxations. However, SDP is a viable approach in solving problems like QUBO due to recent developments in efficient SDP algorithms [22, 34, 63, 68, 97, 118]. With this motivation, let us explore SDP in the context of QUBO, Ising QUBO, and the stable set problem.

We use the following notation through the section. The set of all $n \times n$ real symmetric matrices is denoted by $\mathcal{S}^n$. We denote by $\mathcal{S}^n_+$ the set of positive semidefinite matrices of order n i.e., $\mathcal{S}^n_+ := \{\mathbf{X} \in \mathcal{S}^n : \mathbf{X} \succeq \mathbf{0}\}$. A symmetric matrix $\mathbf{X}$ of order n is positive semidefinite if and only if $\mathbf{z}^{\mathrm{T}}\mathbf{X}\mathbf{z} \geq 0$ for all $\mathbf{z} \in \mathbb{R}^n$. We will sometimes also use the notation $\mathbf{X} \succeq \mathbf{0}$ instead of $\mathbf{X} \in \mathcal{S}^n_+$, if the order of the matrix is clear from the context. The space of symmetric matrices is considered with the trace inner product

$$\langle \mathbf{A}, \mathbf{B} \rangle = \mathrm{tr}(\mathbf{AB}) = \sum_{i=1}^{n}\sum_{j=1}^{n} a_{ij} b_{ij}, \quad \mathbf{A}, \mathbf{B} \in \mathcal{S}^n,$$

and its associated norm is the Frobenius norm, denoted by $||\mathbf{A}|| = \sqrt{\mathrm{tr}(\mathbf{AA})}$.

For two matrices $\mathbf{X}, \mathbf{Y} \in \mathbb{R}^{n \times m}$, $\mathbf{X} \geq \mathbf{Y}$ means $x_{ij} \geq y_{ij}$ for all i, j. The 'diag' operator maps an $n \times n$ matrix to the n-vector given by its diagonal. We denote by 'Diag' the adjoint operator of 'diag'. The rank of matrix $\mathbf{X}$ is denoted by $\mathrm{rank}(\mathbf{X})$.

We denote by $\mathbf{J}_n$ and $\mathbf{I}_n$ the $n \times n$ matrix of all ones and the $n \times n$ identity matrix, respectively. Further, we denote by $\mathbf{e}_n \in \mathbb{R}^n$ the vector of all ones. In case the order of these matrices is clear, we omit the subscript to simplify notation. We denote the set $\{1, \ldots, n\}$ by $[n]$.

6.3.1 The QUBO Formulation and SDP

In this section, we consider the following QUBO:

$$\begin{aligned} \text{QB:} \quad & \text{Maximize} \quad \mathbf{x}^{\mathrm{T}}\mathbf{Q}\mathbf{x} \\ & \text{Subject to:} \quad \mathbf{x} \in \{0, 1\}^n, \end{aligned}$$

where $\mathbf{Q} \in \mathcal{S}^n$. Recall that one may assume without loss of generality that $\mathbf{Q}$ is symmetric, see Chap. 1. Furthermore, we assume that liner costs are modeled on the main diagonal of $\mathbf{Q}$ since $x_i^2 = x_i$ for $i \in [n]$.

In order to derive an SDP relaxation of QB, we proceed as follows. The objective function can be rewritten as $\mathbf{x}^{\mathrm{T}}\mathbf{Q}\mathbf{x} = \langle \mathbf{Q}, \mathbf{x}\mathbf{x}^{\mathrm{T}} \rangle$. To linearize the objective function, we replace $\mathbf{x}\mathbf{x}^{\mathrm{T}}$ by a matrix variable $\mathbf{X} \in \mathcal{S}^n$, and relax the nonconvex constraint $\mathbf{X} - \mathbf{x}\mathbf{x}^{\mathrm{T}} = \mathbf{0}$ to the convex constraint $\mathbf{X} - \mathbf{x}\mathbf{x}^{\mathrm{T}} \succeq \mathbf{0}$. The following lemma shows that the latter constraint can be rewritten as $\begin{pmatrix} 1 & \mathbf{x}^{\mathrm{T}} \\ \mathbf{x} & \mathbf{X} \end{pmatrix} \succeq \mathbf{0}$.

Lemma 6.4 (Schur Complement) *Let* $\mathbf{S}$ *be a symmetric matrix partitioned in blocks*

$$\mathbf{S} = \begin{pmatrix} \mathbf{A} & \mathbf{B} \\ \mathbf{B}^{\mathrm{T}} & \mathbf{C} \end{pmatrix},$$

where $\mathbf{A} \in \mathcal{S}^m$, $\mathbf{C} \in \mathcal{S}^n$, *and* $\mathbf{B} \in \mathbb{R}^{m \times n}$. *Assume that* $\mathbf{A}$ *is invertible. Then* $\mathbf{S} \succeq \mathbf{0}$ *if and only if* $\mathbf{A} \succeq \mathbf{0}$ *and* $\mathbf{C} - \mathbf{B}^{\mathrm{T}}\mathbf{A}^{-1}\mathbf{B} \succeq \mathbf{0}$.

Proof We define the following matrix $\mathbf{X} := \begin{pmatrix} \mathbf{I}_m & -\mathbf{A}^{-1}\mathbf{B} \\ \mathbf{0} & \mathbf{I}_n \end{pmatrix}$. The result follows from

$$\mathbf{X}^{\mathrm{T}}\mathbf{S}\mathbf{X} = \begin{pmatrix} \mathbf{A} & \mathbf{0} \\ \mathbf{0} & \mathbf{C} - \mathbf{B}^{\mathrm{T}}\mathbf{A}^{-1}\mathbf{B} \end{pmatrix},$$

and the fact that positive semidefiniteness is invariant under basis transformations.

□

Note that $\operatorname{diag}(\mathbf{x}\mathbf{x}^{\mathrm{T}}) = \mathbf{x}$ for a binary vector $\mathbf{x}$. For completeness, we prove the following result.

Lemma 6.5 *The optimization problem QB is equivalent to*

$$\begin{aligned} \textit{QB1:} \quad & \textit{Maximize} \quad \langle \mathbf{Q}, \mathbf{X} \rangle \\ & \textit{Subject to:} \quad \widetilde{\mathbf{X}} = \begin{pmatrix} 1 & \operatorname{diag}(\mathbf{X})^{\mathrm{T}} \\ \operatorname{diag}(\mathbf{X}) & \mathbf{X} \end{pmatrix} \succeq \mathbf{0} \\ & \qquad \operatorname{rank}(\widetilde{\mathbf{X}}) = 1. \end{aligned}$$

Proof Let $\widetilde{\mathbf{X}}$ be feasible for QB1. We index the first row and first column of that matrix as row zero and column zero, respectively. Since rank$(\widetilde{\mathbf{X}}) = 1$ and $\widetilde{\mathbf{X}} \succeq \mathbf{0}$, there exists $\mathbf{z} \in \mathbb{R}^n$ such that $\widetilde{\mathbf{X}} = \widetilde{\mathbf{z}}\widetilde{\mathbf{z}}^T$ where $\widetilde{\mathbf{z}} = [1, \mathbf{z}^{\mathrm{T}}]^T$. From the positive semidefiniteness of order two principal submatrices of $\widetilde{\mathbf{X}}$ it follows that $0 \le \widetilde{x}_{ii} \le 1$ for $i \in [n]$. From this and $\widetilde{x}_{ii} = z_i z_i$ for $i \in [n]$ it follows that $\mathbf{z} \in \{0, 1\}^n$.

The converse inclusion is trivial. Since the objectives of QB and QB1 coincide, we conclude that the two problems are equivalent. □

Dropping the rank one constraint in the QUBO formulation from Lemma 6.5, yields the following SDP relaxation of QB:

$$\begin{aligned} &\text{Maximize} \quad \langle \mathbf{Q}, \mathbf{X} \rangle \\ &\text{Subject to:} \quad \begin{pmatrix} 1 & \operatorname{diag}(\mathbf{X})^{\mathrm{T}} \\ \operatorname{diag}(\mathbf{X}) & \mathbf{X} \end{pmatrix} \succeq \mathbf{0}. \end{aligned}$$

We say that this is the *basic* SDP relaxation of QB. In order to strengthen the basic SDP relaxation of QB, one can add the following facet defining inequalities of the Boolean quadric polytope:

$$\begin{aligned} 0 \le x_{ij} &\le x_{ii}, \\ x_{ii} + x_{jj} &\le 1 + x_{ij}, \\ x_{ik} + x_{jk} &\le x_{kk} + x_{ij}, \\ x_{ii} + x_{jj} + x_{kk} &\le x_{ij} + x_{ik} + x_{jk} + 1, \end{aligned} \tag{6.42}$$

for $1 \le i, j, k \le n$, see [99] and Chap. 4 for details.

6.3.2 The Ising QUBO Formulation and SDP

The equivalence of QUBO and the maximum cut problem is explained in Chap. 1. We present here the basic SDP relaxation for the maximum cut problem and show how to improve it. The SDP relaxations presented in this section are used in the state of the art SDP-based solvers for solving the maximum cut problem.

Let $G = (V, E)$ be an undirected graph on $n = |V|$ vertices and $m = |E|$ edges with edge weights w_{ij} for $(i, j) \in E$. Let $\mathbf{A} = (a_{ij}) \in \mathcal{S}^n$ be the weighted adjacency matrix of G where

$$a_{ij} = \begin{cases} w_{ij} & \text{if } (i, j) \in E \\ 0 & \text{if } (i, j) \notin E. \end{cases}$$

We let $\mathbf{L}$ denote the Laplacian matrix associated with $\mathbf{A}$, i.e.,

$$\mathbf{L} = \mathrm{Diag}(\mathbf{Ae}) - \mathbf{A}.$$

For $S \subseteq V$, the cut is the set of edges $(i, j) \in E$ that have one endpoint in S and the other in $V \backslash S$. The weight of the cut given by $S \subseteq V$ is

$$\mathrm{cut}(S, V \backslash S) = \sum_{i \in S, j \in V \backslash S, (i,j) \in E} w_{ij}.$$

If $S = \emptyset$ or $S = V$, the weight of the cut is defined to be zero. The maximum cut problem asks to partition the vertex set V into S and $V \backslash S$ in such a way that the total weight of the cut is maximized. Thus, the maximum cut problem is

$$\max_{S \subseteq V} \sum_{i \in S, j \in V \backslash S, (i,j) \in E} w_{ij}.$$

For $S \subseteq V$, we introduce the cut vector $\mathbf{y} \in \{-1, 1\}^n$ such that $y_i = 1$ for $i \in S$ and $y_i = -1$ for $i \in V \backslash S$. Now, one can verify (see also Chap. 1, Sect. 1.4.2) that the weight of the cut given by $S \subseteq V$ is

$$\mathrm{cut}(S, V \backslash S) = \frac{1}{4} \mathbf{y}^{\mathrm{T}} \mathbf{L} \mathbf{y},$$

where $\mathbf{L}$ is the Laplacian matrix of the graph associated with $\mathbf{A}$. Thus, finding a maximum cut in a graph is equivalent to solving the following quadratic optimization problem

$$\begin{aligned} \text{MC:} \quad & \text{Maximize} \quad \mathbf{y}^{\mathrm{T}} \mathbf{L} \mathbf{y} \\ & \text{Subject to:} \quad \mathbf{y} \in \{-1, 1\}^n, \end{aligned}$$

that corresponds to the Ising QUBO($\mathbf{L}, \mathbf{0}$), see Chap. 1, Sect. 1.4.2.

Note that the objective function in MC can be rewritten as $\mathbf{y}^{\mathrm{T}} \mathbf{L} \mathbf{y} = \langle \mathbf{L}, \mathbf{y}\mathbf{y}^{\mathrm{T}} \rangle$. Define $\mathbf{Y} := \mathbf{y}\mathbf{y}^{\mathrm{T}}$ and observe that $\mathrm{diag}(\mathbf{Y}) = \mathbf{e}_n$ since $\mathbf{y} \in \{-1, 1\}^n$. This brings us to the following optimization problem:

$$\begin{aligned} \text{MC1:} \quad & \text{Maximize} \quad \langle \mathbf{L}, \mathbf{Y} \rangle \\ & \text{Subject to:} \quad \mathrm{diag}(\mathbf{Y}) = \mathbf{e} \\ & \qquad \mathrm{rank}(\mathbf{Y}) = 1 \\ & \qquad \mathbf{Y} \succeq \mathbf{0}. \end{aligned}$$

Let us relate MC and MC1.

Lemma 6.6 *The optimization problems MC and MC1 are equivalent.*

Proof Let $\mathbf{Y}$ be feasible for MC1. Since $\text{rank}(\mathbf{Y}) = 1$ and $\mathbf{Y} \succeq \mathbf{0}$, there exists $\mathbf{z} \in \mathbb{R}^n$ such that $\mathbf{Y} = \mathbf{z}\mathbf{z}^{\mathrm{T}}$. Then from $y_{ii} = z_i z_i = 1$ for $i \in [n]$ it follows that $\mathbf{z} \in \{-1, 1\}^n$.

The converse inclusion is trivial. Since the objectives of MC and MC1 coincide, we conclude that the two problems are equivalent. □

We present one more equivalent formulation for the maximum cut problem, that is based on the following result.

Theorem 6.10 ([11]) *Let* $\mathbf{Y}$ *be an* $n \times n$ *symmetric matrix. Then*

$$\mathbf{Y} \succeq \mathbf{0},\ \mathbf{Y} \in \{-1, 1\}^{n \times n} \quad \textit{if and only if} \quad \mathbf{Y} = \mathbf{y}\mathbf{y}^{\mathrm{T}},\ \textit{for some}\ \ \mathbf{y} \in \{-1, 1\}^n.$$

Proof Observe that $\mathbf{Y} \succeq \mathbf{0},\ \mathbf{Y} \in \{-1, 1\}^{n \times n}$ implies that $\text{diag}(\mathbf{Y}) = \mathbf{e}_n$. Therefore, $\mathbf{Y} = \begin{pmatrix} 1 & \mathbf{y}^{\mathrm{T}} \\ \mathbf{y} & \widetilde{\mathbf{Y}} \end{pmatrix}$ for some $\mathbf{y} \in \{-1, 1\}^{n-1}$, $\widetilde{\mathbf{Y}} \in \mathcal{S}^{n-1}$. It follows from the Schur complement lemma that

$$\mathbf{Y} \succeq \mathbf{0} \quad \Leftrightarrow \quad \widetilde{\mathbf{Y}} - \mathbf{y}\mathbf{y}^{\mathrm{T}} \succeq \mathbf{0}.$$

From $\widetilde{\mathbf{Y}} - \mathbf{y}\mathbf{y}^{\mathrm{T}} \succeq \mathbf{0}$ and $\text{diag}(\widetilde{\mathbf{Y}} - \mathbf{y}\mathbf{y}^{\mathrm{T}}) = \mathbf{0}$, it follows that $\widetilde{\mathbf{Y}} = \mathbf{y}\mathbf{y}^{\mathrm{T}}$. The converse inclusion is trivial. □

The proof of Theorem 6.10 is from [9]. In the view of Theorem 6.10, the maximum cut problem can be formulated as follows:

$$\begin{aligned} \text{MC2:}\quad & \text{Maximize} \quad \langle \mathbf{L}, \mathbf{Y} \rangle \\ & \text{Subject to:} \quad \text{diag}(\mathbf{Y}) = \mathbf{e} \\ & \qquad\qquad\quad \mathbf{Y} \in \{-1, 1\}^{n \times n} \\ & \qquad\qquad\quad \mathbf{Y} \succeq \mathbf{0}. \end{aligned}$$

Thus, we can replace the rank one constraint in MC1 by $\mathbf{Y} \in \{-1, 1\}^{n \times n}$ to obtain an equivalent formulation of the maximum cut problem MC2. Different problem formulations might lead to the different solving approaches.

Dropping the rank one constraint in MC1, or the constraint $\mathbf{Y} \in \{-1, 1\}^{n \times n}$ in MC2, yields the *basic* SDP relaxation for the maximum cut problem:

$$\begin{aligned} \text{SDP}_{basic}:\quad & \text{Maximize} \quad \langle \mathbf{L}, \mathbf{Y} \rangle \\ & \text{Subject to:} \quad \text{diag}(\mathbf{Y}) = \mathbf{e} \\ & \qquad\qquad\quad \mathbf{Y} \succeq \mathbf{0}. \end{aligned}$$

The set

$$\mathcal{E}_n = \{\mathbf{Y} \in \mathcal{S}^n : \text{diag}(\mathbf{Y}) = \mathbf{e},\ \mathbf{Y} \succeq \mathbf{0}\}$$

is called *elliptope* [81]. The basic SDP relaxation for MC is studied extensively, see e.g., [11, 29, 65, 81, 122]. It is known that the basic SDP relaxations of QB and MC are equivalent see [64].

The quality of the basic SDP bound for the maximum cut problem is analyzed by Goemans and Williamson [46]. They prove the following bound for the integrality ratio of any graph with nonnegative edge weights:

$$\frac{\text{OBJ}(\text{MC})}{\text{OBJ}(\text{SDP}_{basic})} \geq \alpha,$$

where OBJ(MC) is the optimal value of MC, $\text{OBJ}(\text{SDP}_{basic})$ the corresponding basic SDP bound, and $0.87856 < \alpha < 0.87857$. Goemans and Williamson [46] also present a celebrated randomized algorithm that provides a cut whose expected weight is at least $\alpha \cdot \text{OBJ}(\text{SDP}_{basic})$.

The basic SDP bound for the maximum cut problem is introduced by Delorme and Poljak [29] as an eigenvalue bound for the problem. Namely, to derive the bound for the maximum cut problem Delorme and Poljak exploit the following result:

$$\max_{\mathbf{y}\in\{-1,1\}^n} \mathbf{y}^{\mathrm{T}}\mathbf{L}\mathbf{y} \leq \max_{\mathbf{y}^{\mathrm{T}}\mathbf{y}=n} \mathbf{y}^{\mathrm{T}}\mathbf{L}\mathbf{y} = n\lambda_{\max}(\mathbf{L}),$$

that is provided by Mohar and Poljak [93]. Here $\lambda_{\max}(\mathbf{L})$ denotes the largest eigenvalue of $\mathbf{L}$, and the equality follows from the Rayleigh-Ritz theorem. In order to improve this bound, Delorme and Poljak exploit the fact that $\mathbf{y}^{\mathrm{T}}\,\text{Diag}(\mathbf{u})\mathbf{y} = 0$ for all $\mathbf{y} \in \{-1, 1\}^n$ and $\mathbf{u} \in \mathbb{R}^n$ such that $\mathbf{e}^{\mathrm{T}}\mathbf{u} = 0$. Notice that these diagonal perturbations when added to the cost matrix $\mathbf{L}$ do not change the optimal value of the maximum cut problem but have an influence on eigenvalues of $\mathbf{L}$. The resulting eigenvalue bound from [29] is:

$$\text{MC}_{eig}: \quad \min_{\mathbf{u}\in\mathbb{R}^n,\ \mathbf{e}^{\mathrm{T}}\mathbf{u}=0} \quad n\lambda_{\max}(\mathbf{L} + \text{Diag}(\mathbf{u})).$$

This eigenvalue bound is related to the dual problem of SDP_{basic}, which is the following optimization problem:

$$\begin{aligned} \text{DSDP}_{basic}: \quad & \text{Minimize} \quad \mathbf{e}^{\mathrm{T}}\mathbf{z} \\ & \text{Subject to:} \quad \text{Diag}(\mathbf{z}) - \mathbf{L} \succeq \mathbf{0}. \end{aligned}$$

In particular, Poljak and Rendl prove the following result.

Lemma 6.7 ([104]) *DSDP_{basic} and MC_{eig} are equivalent optimization problems.*

Proof The eigenvalue problem MC_{eig} can be reformulated as the following SDP problem

$$\begin{aligned} \text{Minimize} \quad & n\lambda \\ \text{Subject to:} \quad & \lambda \mathbf{I}_n - (\mathbf{L} + \text{Diag}(\mathbf{u})) \succeq \mathbf{0} \\ & \mathbf{e}^{\mathrm{T}}\mathbf{u} = 0. \end{aligned}$$

We rewrite $\lambda \mathbf{I}_n - \mathbf{L} - \text{Diag}(\mathbf{u}) = \text{Diag}(\lambda \mathbf{e}_n - \mathbf{u}) - \mathbf{L} = \text{Diag}(\mathbf{z}) - \mathbf{L}$ where $\mathbf{z} = \lambda \mathbf{e}_n - \mathbf{u}$. Since $\mathbf{e}^{\mathrm{T}}\mathbf{z} = \lambda n$, the result follows. □

It is well known that one can strengthen semidefinite programming relaxations by adding polyhedral information, see e.g., [82]. Generic inequalities for $\{-1, 1\}$ problems are known as the *hypermetric inequalities* [30, 31]. Those inequalities are based on the fact that for any $\mathbf{y} \in \{-1, 1\}^n$, the inequality $|\mathbf{b}^{\mathrm{T}}\mathbf{y}| \geq 1$ is valid if $\mathbf{b}^{\mathrm{T}}\mathbf{e}$ is odd and $\mathbf{b}$ integer. In particular, for any $\mathbf{b}$ with the mentioned properties we have:

$$|\mathbf{b}^{\mathrm{T}}\mathbf{y}| \geq 1 \;\Leftrightarrow\; \mathbf{b}^{\mathrm{T}}\mathbf{y}\mathbf{y}^{\mathrm{T}}\mathbf{b} \geq 1 \;\Leftrightarrow\; \langle \mathbf{b}\mathbf{b}^{\mathrm{T}}, \mathbf{y}\mathbf{y}^{\mathrm{T}} \rangle \geq 1. \tag{6.43}$$

Thus, any integer vector $\mathbf{b}$ for which $\mathbf{b}^{\mathrm{T}}\mathbf{e}$ is odd defines a hypermetric inequality. The hypermetric inequalities are valid for any symmetric matrix $\mathbf{Y}$ from the convex hull of rank-one matrices $\mathbf{y}\mathbf{y}^{\mathrm{T}}$ where $\mathbf{y} \in \{-1, 1\}^n$.

Well known hypermetric inequalities are the *triangle inequalities*. Each triangle inequality is generated by $\mathbf{b}$ that has three non-zero entries of value -1 or 1. In particular, the triangle inequalities are as follows:

$$y_{ij} + y_{ik} + y_{jk} \geq -1, \tag{6.44}$$

$$y_{ij} - y_{ik} - y_{jk} \geq -1, \tag{6.45}$$

$$-y_{ij} + y_{ik} - y_{jk} \geq -1, \tag{6.46}$$

$$-y_{ij} - y_{ik} + y_{jk} \geq -1, \quad \forall i < j < k. \tag{6.47}$$

To derive those inequalities we use that $\mathbf{Y}$ is symmetric and $\text{diag}(\mathbf{Y}) = \mathbf{e}$. We say that $\mathbf{Y} \in \text{MET}$ if and only if $\mathbf{Y}$ satisfies (6.44)–(6.47). The triangle inequalities impose that for every 3-cycle of vertices one can cut either zero or two of the edges. The polyhedron defined by these inequalities is called the *metric polytope*, i.e.,

$$\mathcal{M}_n = \left\{ \mathbf{Y} \in \mathcal{S}^n : \text{diag}(Y) = \mathbf{e}, \; \mathbf{Y} \in \text{MET} \right\}.$$

Several papers investigate quality of the bounds that are obtained after adding (subsets of) the triangle inequalities to the basic SDP relaxation, see e.g., [34, 65–67]. The numerical results show that the resulting bounds may improve significantly over the basic SDP bound. Therefore the following SDP relaxation for the maximum cut problem:

$$\begin{aligned} \text{SDP}_{met}: \quad & \text{Maximize} \quad \langle \mathbf{L}, \mathbf{Y} \rangle \\ & \text{Subject to:} \quad \mathbf{Y} \in \mathcal{E}_n \cap \mathcal{M}_n, \end{aligned}$$

is exploited in the SDP-based exact solvers for solving the maximum cut problem [54, 65, 69, 75, 109]. Since the SDP program SDP$_{met}$ has $4\binom{n}{3}$ triangle inequalities, only a subset of those valid inequalities is added to SDP$_{basic}$ in each node of a branch and bound tree. Various approaches, including an interior point algorithm [65], bundle method [34], spectral bundle method [66], quasi-Newton method [74], alternating direction method of multipliers [69], are implemented for solving SDP$_{met}$.

In general, if $|\mathbf{b}|$ in (6.43) is a characteristic vector of a clique of odd order k, i.e., $\mathbf{b}$ has k non-zero entries of value -1 or 1, then the resulting hypermetric inequality is known as a clique inequality. The numerical results in [65] show that adding clique inequalities of order $k > 3$ to SDP$_{met}$, in each node of a tree, improves the performance of a branch and bound algorithm over an algorithm that implements only SDP$_{met}$. In [54, 69], the authors strengthen SDP$_{met}$ by adding pentagonal and heptagonal inequalities. In particular, pentagonal (resp., heptagonal) inequalities are 5-clique (resp., 7-clique) inequalities where $\mathbf{b}$ has five (resp., seven) non-zero entries of value -1 or 1.

6.3.3 Linearly Constrained Quadratic Problems and the Maximum Cut Problem

Lasserre [79] showed how to reformulate a quadratic (or linear) binary program with linear constraints into a maximum cut problem by exploiting semidefinite programming. The reformulated problem can be then solved by any of the available approaches for solving the maximum cut problem. Reformulations of constrained binary optimization problems into QUBO are also discussed in Chap. 7.

Consider the following quadratic optimization problem with linear constraints

$$\begin{aligned} \text{QP:} \quad & \text{Minimize} \quad \mathbf{y}^T\widetilde{\mathbf{Q}}\mathbf{y} + \mathbf{c}^T\mathbf{y} \\ & \text{subject to:} \quad \mathbf{A}\mathbf{y} = \mathbf{b} \\ & \qquad\qquad\quad \mathbf{y} \in \{-1, 1\}^n, \end{aligned}$$

where $\widetilde{\mathbf{Q}} \in \mathcal{S}^n$, $\mathbf{c} \in \mathbb{R}^n$, $\mathbf{A} \in \mathbb{Z}^{m\times n}$ and $\mathbf{b} \in \mathbb{Z}^m$. For a penalty parameter $\sigma > 0$, consider the following minimization problem

$$\min_{\mathbf{y}\in\{-1,1\}^n} \mathbf{y}^T\widetilde{\mathbf{Q}}\mathbf{y} + \mathbf{c}^T\mathbf{y} + \sigma||\mathbf{A}\mathbf{y} - \mathbf{b}||^2.$$

Let us rewrite the objective function as follows

$$\begin{aligned}\mathbf{y}^{\mathrm{T}}\widetilde{\mathbf{Q}}\mathbf{y} + \mathbf{c}^{\mathrm{T}}\mathbf{y} + \sigma\|\mathbf{A}\mathbf{y} - \mathbf{b}\|^2 &= \mathbf{y}^{\mathrm{T}}(\widetilde{\mathbf{Q}} + \sigma\mathbf{A}^{\mathrm{T}}\mathbf{A})\mathbf{y} + (\mathbf{c} - 2\sigma\mathbf{A}^{\mathrm{T}}\mathbf{b})^{\mathrm{T}}\mathbf{y} + \sigma\mathbf{b}^{\mathrm{T}}\mathbf{b}\\ &= \tilde{\mathbf{y}}^{\mathrm{T}}\mathbf{Q}\tilde{\mathbf{y}},\end{aligned}$$

where $\tilde{\mathbf{y}}^{\mathrm{T}} = (1, \mathbf{y}^{\mathrm{T}}) \in \{-1, 1\}^{n+1}$ and

$$\mathbf{Q} := \begin{pmatrix} \sigma\mathbf{b}^{\mathrm{T}}\mathbf{b} & (\mathbf{c} - 2\sigma\mathbf{A}^{\mathrm{T}}\mathbf{b})^{\mathrm{T}}/2 \\ (\mathbf{c} - 2\sigma\mathbf{A}^{\mathrm{T}}\mathbf{b})/2 & \widetilde{\mathbf{Q}} + \sigma\mathbf{A}^{\mathrm{T}}\mathbf{A} \end{pmatrix} \in \mathcal{S}^{n+1}. \tag{6.48}$$

This leads to the following optimization problem

$$\begin{aligned} \mathrm{MC}_{n+1} \quad & \text{Minimize} \quad \mathbf{y}^{\mathrm{T}}\mathbf{Q}\mathbf{y} \\ & \text{subject to:} \quad y_0 = 1,\ \mathbf{y} \in \{-1, 1\}^{n+1}, \end{aligned}$$

where we indexed by zero the first element in the vector of variables. Note that if $(y_0, \mathbf{y}^{\mathrm{T}})^{\mathrm{T}} \in \{-1, 1\}^{n+1}$ is an optimal solution for this optimization problem, then $(-y_0, -\mathbf{y}^{\mathrm{T}})^{\mathrm{T}} \in \{-1, 1\}^{n+1}$ is also an optimal solution for the same optimization problem. Therefore, one may w.l.o.g. fix $y_0 = 1$. Thus MC_{n+1} is the maximum cut problem on a graph with $n + 1$ vertices.

To obtain an equivalent maximum cut formulation MC_{n+1} of QP one has to determine an appropriate parameter σ in the definition of $\mathbf{Q}$, see (6.48). Lasserre [79] defines

$$\sigma := 2\rho + 1, \tag{6.49}$$

where

$$\rho := \max\{|\rho^1|, |\rho^2|\} \tag{6.50}$$

and

$$\rho^1 = \min\left\{\langle\widetilde{\mathbf{Q}}, \mathbf{Y}\rangle + \mathbf{c}^{\mathrm{T}}\mathbf{y} : \begin{pmatrix} 1 & \mathbf{y}^{\mathrm{T}} \\ \mathbf{y} & \mathbf{Y} \end{pmatrix} \succeq \mathbf{0},\ \mathrm{diag}(\mathbf{Y}) = \mathbf{e}_n\right\},$$

$$\rho^2 = \max\left\{\langle\widetilde{\mathbf{Q}}, \mathbf{Y}\rangle + \mathbf{c}^{\mathrm{T}}\mathbf{y} : \begin{pmatrix} 1 & \mathbf{y}^{\mathrm{T}} \\ \mathbf{y} & \mathbf{Y} \end{pmatrix} \succeq \mathbf{0},\ \mathrm{diag}(\mathbf{Y}) = \mathbf{e}_n\right\}.$$

Thus, to compute σ one has to solve two semidefinite programming problems that are derived from the quadratic problem QP. Note that

$$\max\{\mathbf{y}^{\mathrm{T}}\widetilde{\mathbf{Q}}\mathbf{y} + \mathbf{c}^{\mathrm{T}}\mathbf{y} :\ \mathbf{y} \in \{-1, 1\}^n\} \leq \rho.$$

Theorem 6.11 ([79]) *Let q^* be the optimal value of QP and m^* the optimal value of MC_{n+1}. Let $\sigma = 2\rho + 1$ where ρ is defined as in* (6.50).

If $q^* < +\infty$, then $q^* = m^*$, i.e., q^* is the optimal value of the maximum cut problem MC_{n+1}. Moreover $q^* = +\infty$ if and only if $m^* > \rho$.

Proof Let $\mathcal{F} := \{\mathbf{y} \in \{-1, 1\}^n : \mathbf{Ay} = \mathbf{b}\} \neq \emptyset$ be the feasible set of QP, and

$$f(\mathbf{y}) = \mathbf{y}^{\mathrm{T}}\widetilde{\mathbf{Q}}\mathbf{y} + \mathbf{c}^{\mathrm{T}}\mathbf{y} + \sigma||\mathbf{Ay} - \mathbf{b}||^2.$$

There are two cases to consider. The first case is

$$f(\mathbf{y}) = \mathbf{y}^{\mathrm{T}}\widetilde{\mathbf{Q}}\mathbf{y} + \mathbf{c}^{\mathrm{T}}\mathbf{y} \leq \rho \quad \text{for} \quad \mathbf{y} \in \mathcal{F},$$

and the second case is

$$f(\mathbf{y}) = \mathbf{y}^{\mathrm{T}}\widetilde{\mathbf{Q}}\mathbf{y} + \mathbf{c}^{\mathrm{T}}\mathbf{y} + \sigma||\mathbf{Ay} - \mathbf{b}||^2 > \rho \quad \text{for} \quad \mathbf{y} \in \{-1, 1\}^n \setminus \mathcal{F}.$$

In the latter case we use that $||\mathbf{Ay} - \mathbf{b}||^2 \geq 1$ for all $\mathbf{y} \in \{-1, 1\}^n \setminus \mathcal{F}$, because $\mathbf{A} \in \mathbb{Z}^{m \times n}$ and $\mathbf{b} \in \mathbb{Z}^m$. Thus,

$$q^* = \min_{\mathbf{y} \in \{-1,1\}^n} \mathbf{y}^{\mathrm{T}}\widetilde{\mathbf{Q}}\mathbf{y} + \mathbf{c}^{\mathrm{T}}\mathbf{y} + \sigma||\mathbf{Ay} - \mathbf{b}||^2.$$

By using (6.48), we have

$$\min_{\mathbf{y} \in \{-1,1\}^n} \mathbf{y}^{\mathrm{T}}\widetilde{\mathbf{Q}}\mathbf{y} + \mathbf{c}^{\mathrm{T}}\mathbf{y} + \sigma||\mathbf{Ay} - \mathbf{b}||^2 = \min_{\mathbf{y} \in \{-1,1\}^{n+1}} \mathbf{y}^{\mathrm{T}}\mathbf{Q}\mathbf{y} = m^*.$$

It is also clear that $m^* > \rho$ if and only if $\mathcal{F} = \emptyset$. □

Thus, solving the constrained optimization problem QP is equivalent to solving the maximum cut problem MC_{n+1}. The described approach can also be applied to quadratic binary optimization problems with linear inequality constraints. Lasserre [79] also investigates lower bounds obtained from the basic SDP relaxation of the (reformulated) maximum cut problem. Preliminary tests show that the resulting bounds are mostly better than the bounds obtained from the standard linear programming relaxation of the original problem.

The idea from [79] is further exploited in [53, 54] to develop algorithms for solving binary quadratic problems with linear equality constraints. EXPEDIS [53] uses BiqMac [109] to solve the (reformulated) maximum cut problems.

6.3.4 The Stable Set Problem

It is well known that QUBO is equivalent to the maximum weight stable set problem and the maximum clique problem, see also Chap. 1. Nevertheless, there exist bounding and exact approaches that are developed specifically for each of the problems. We present here SDP approaches for solving the stable set problem. Section 6.3.5 presents SDP approaches for solving the maximum k-colorable subgraph problem, that can be seen as a generalization of the stable set problem.

Let $G = (V, E)$ be an undirected graph with $n = |V|$ vertices and $m = |E|$ edges. A stable set S in G is a subset of vertices such that no two vertices in S are adjacent in G. Note that, S forms a complete subgraph in the complement of G. The maximum stable set problem is the problem of finding a stable set of maximum cardinality. If there are weights associated with vertices, then the maximum weight stable set problem is to find a stable set in G such that the sum of weights of the vertices in the stable set is maximized. We consider the stable set problem in an unweighted graph.

Let $\alpha(G)$ denote the *stability number* of G, that is the maximum cardinality of a stable set in G. Let $\omega(G)$ denote the *clique number* of G, that is the maximum cardinality of a clique in G. A clique C in a graph is a subset of vertices such that every two distinct vertices in C are adjacent. The problem of determining $\omega(G)$ is called the maximum clique problem. Clearly, $\alpha(G) = \omega(\overline{G})$ where $\overline{G}$ denotes the complement of G.

The maximum stable set problem can be modeled as an integer programming problem. Let $\mathbf{x} \in \{0, 1\}^n$ be the characteristic vector of a (nonempty) stable set. Then the stability number of G is the solution of the following optimization problem:

$$\begin{aligned} \alpha(G) := \text{Maximize} \quad & \mathbf{x}^{\mathrm{T}}\mathbf{x} \\ \text{subject to:} \quad & x_i x_j = 0 \ \text{ for } (i, j) \in E \\ & \mathbf{x} \in \{0, 1\}^n. \end{aligned}$$

There exist several equivalent formulations of the stable set problem. The constraint $x_i x_j = 0$ for $(i, j) \in E$, that at most one of the nodes i and j can be in a stable set, may be replaced by $x_i + x_j \leq 1$ for $(i, j) \in E$. Further, one can replace $\mathbf{x} \in \{0, 1\}^n$ by $x_i^2 = x_i$ for all i. Note also that $\mathbf{x}^{\mathrm{T}}\mathbf{x} = \mathbf{e}^{\mathrm{T}}\mathbf{x}$.

To derive an SDP relaxation for the stable set problem assume again $\mathbf{x} \in \{0, 1\}^n$ is the characteristic vector of a (nonempty) stable set, and define

$$\mathbf{X} := \frac{1}{\mathbf{x}^{\mathrm{T}}\mathbf{x}}\mathbf{x}\mathbf{x}^{\mathrm{T}}.$$

Then $\mathbf{X}$ is positive semidefinite matrix and $x_{ij} = 0$ for $(i, j) \in E$. Moreover,

$$\mathrm{tr}(\mathbf{X}) = \frac{1}{\mathbf{x}^{\mathrm{T}}\mathbf{x}}\mathbf{e}^{\mathrm{T}}\mathbf{x} = 1$$

and

$$\langle \mathbf{J}, \mathbf{X} \rangle = \frac{1}{\mathbf{x}^{\mathrm{T}}\mathbf{x}}(\mathbf{e}^{\mathrm{T}}\mathbf{x})^2 = \mathbf{e}^{\mathrm{T}}\mathbf{x}.$$

These observations give rise to the well-known SDP relaxation for the stable set problem:

$$\begin{aligned} \vartheta(G) := \text{Maximize} \quad & \langle \mathbf{J}, \mathbf{X} \rangle \\ \text{subject to:} \quad & x_{ij} = 0 \text{ for } (i,j) \in E \\ & \operatorname{tr}(\mathbf{X}) = 1 \\ & \mathbf{X} \succeq \mathbf{0}. \end{aligned} \tag{6.51}$$

This relaxation was proposed by Lovász [86]. The graph parameter $\vartheta(G)$ is known as the *ϑ-number* or *Lovász theta number*. The Lovász theta number provides bounds for both the clique number and the chromatic number of a graph. The *chromatic number* of G, denoted by $\chi(G)$, is the smallest number of colors needed to (properly) color G. The well known result that establishes the following relation

$$\alpha(G) \le \vartheta(G) \le \chi(\overline{G}),$$

or equivalently

$$\omega(G) \le \vartheta(\overline{G}) \le \chi(G),$$

is called the *sandwich theorem* [87].

To obtain a stronger upper bound for the stable set problem than $\vartheta(G)$, Schrijver [111] proposed to add non-negativity constraints on the matrix variable. The resulting relaxation is:

$$\begin{aligned} \vartheta'(G) := \text{Maximize} \quad & \langle \mathbf{J}, \mathbf{X} \rangle \\ \text{subject to:} \quad & x_{ij} = 0 \text{ for } (i,j) \in E \\ & x_{ij} \ge 0 \text{ for } (i,j) \notin E \\ & \operatorname{tr}(\mathbf{X}) = 1 \\ & \mathbf{X} \succeq \mathbf{0}. \end{aligned} \tag{6.52}$$

The graph parameter $\vartheta'(G)$ is known as the *Schrijver's number*. Djukanović and Rendl [32] further strengthen the above relaxation by adding the following constraints:

$$x_{ij} \le x_{ii} \tag{6.53}$$

$$x_{ik} + x_{jk} \le x_{ij} + x_{kk}, \tag{6.54}$$

for all $i, j, k \in V$, see also [88]. The mentioned upper bounds for $\alpha(G)$ are tested on various highly symmetric graphs in [32]. The numerical results show that $\vartheta'(G)$ improves upon $\vartheta(G)$ however the additional inclusion of constraints (6.53)–(6.54) often does not improve upon $\vartheta'(G)$. Moreover, adding cutting planes to the SDP relaxation for the Schrijver's number significantly increases computational cost.

Another formulation for the ϑ-number is given by Grötschel et al. [48]. Suppose that $\mathbf{x} \in \{0, 1\}^n$ is the characteristic vector of a (nonempty) stable set and let $\mathbf{X} = \mathbf{x}\mathbf{x}^{\mathrm{T}}$. We relax the nonconvex constraint $\mathbf{X} - \mathbf{x}\mathbf{x}^{\mathrm{T}} = \mathbf{0}$ to the convex constraint $\mathbf{X} - \mathbf{x}\mathbf{x}^{\mathrm{T}} \succeq \mathbf{0}$ and observe that $\mathrm{diag}(\mathbf{X}) = \mathbf{x}$. This leads to the following formulation of the ϑ-number:

$$
\begin{aligned}
\vartheta(G) = \text{Maximize} \quad & \mathbf{e}^{\mathrm{T}}\mathbf{x} \\
\text{subject to:} \quad & x_{ij} = 0 \ \text{ for } \ (i, j) \in E \\
& \mathrm{diag}(\mathbf{X}) = \mathbf{x} \\
& \begin{pmatrix} 1 & \mathbf{x}^{\mathrm{T}} \\ \mathbf{x} & \mathbf{X} \end{pmatrix} \succeq \mathbf{0}.
\end{aligned}
$$

This formulation of the ϑ-number can be further strengthened towards $\alpha(G)$ by adding valid constraints such as constraints of type (6.42) or the odd-cycle constraints, see e.g., [22, 49]. Galli and Letchford [40] compare quality of bounds obtained after adding different kind of valid inequalities to both, here presented, SDP formulations of the ϑ-number. The results show that the equivalence of the bounds breaks down after adding cuts, and that the Grötschel et al. [48] formulation yields mostly stronger bounds than the Lovász formulation [86].

Connections between semidefinite programming relaxations of the maximum cut problem and the stable set problem are studied in detail in [82]. The BiqCrunch solver [75] exploits the SDP relaxation for the ϑ'-number that is strengthened with valid inequalities to find maximum stable sets in graphs.

More intricate approaches for obtaining hierarchies of semidefinite programming bounds for the stability number one can find in the work of Lovász and Schrijver [88], Lasserre [77, 78], and de Klerk and Pasechnik [62]. The hierarchy of Lasserre is known to converge in $\alpha(G)$ steps and it refines the hierarchy of Lovász and Schrijver and the hierarchy of de Klerk and Pasechnik, for details see [55, 80]. In general, it is difficult to compute bounds resulting from these hierarchies since SDP relaxations already in the second level of a hierarchy turn out to be too large for standard SDP solvers.

6.3.5 The Maximum k-Colorable Subgraph Problem

The maximum k-colorable subgraph (MkCS) problem is to find the largest induced subgraph in a given graph that can be colored with k colors such that no two adjacent vertices have the same color. The MkCS problem is also known as the maximum k-partite induced subgraph problem. The MkCS problem has a number of applications, such scheduling [16], VLSI design [38], channel assignment in spectrum sharing networks [73, 117], genetic research [85]. The maximum k-colorable subgraph problem reduces to the maximum stable set problem when $k = 1$. Moreover, the MkCS problem on a graph can be considered as the stable set problem on the Cartesian product of the complete graph on k vertices and the graph under consideration.

Let $G = (V, E)$ be an undirected graph with $n = |V|$ vertices and $m = |E|$ edges. A graph G is k-colorable ($1 \leq k \leq n - 1$) if one can assign to each vertex in G one of the k colors such that adjacent vertices do not have the same color. An induced subgraph of a graph $G = (V, E)$ is a graph $G' = (V', E')$ such that $V' \subseteq V$ and $E' \subseteq E$ is the set of all edges in E connecting the vertices in V'.

The maximum k-colorable subgraph problem can be modeled as an integer programming problem. Let $\mathbf{X} \in \{0, 1\}^{n \times k}$ be the matrix with one in the entry (i, r) if vertex $i \in [n]$ is colored with color $r \in [k]$ and zero otherwise. The number of vertices in a maximum k-colorable subgraph of G, denoted by $\alpha_k(G)$, is the solution of the MkCS problem:

$$
\begin{aligned}
\alpha_k(G) := \text{Maximize} \quad & \sum_{i \in [n],\, r \in [k]} x_{ir} \\
\text{subject to:} \quad & x_{ir} x_{jr} = 0 \ \text{ for } \ (i, j) \in E,\ r \in [k] \\
& \sum_{r \in [k]} x_{ir} \leq 1 \ \text{ for } \ i \in [n] \\
& \mathbf{X} \in \{0, 1\}^{n \times k}.
\end{aligned}
$$

Note that for $k = 1$, we have that $\alpha_1(G) = \alpha(G)$ where $\alpha(G)$ denotes the stability number of G. The first set of constraints in the model ensures that two adjacent vertices are not colored with the same color, and the second set of constraints takes care that each vertex is colored with at most one color. There are several equivalent formulations of the MkCS problem. For example, constraints $x_{ir} x_{jr} = 0$ for $(i, j) \in E,\ r \in [k]$ may be replaced by

$$
x_{ir} + x_{jr} \leq 1 \ \text{ for } \ (i, j) \in E,\ r \in [k].
$$

The resulting problem formulation was studied in [70].

The MkCS problem on a graph G can be modeled as the stable set problem on the Cartesian product of the complete graph on k vertices and G. We denote by $K_k = (V_k, E_k)$ the complete graph on k vertices. The Cartesian product $K_k \Box G$ of graphs K_k and $G = (V, E)$ is the graph with vertex set $V_k \times V$ and edge set $E_\Box$ where two vertices (u, i) and (v, j) are adjacent if $u = v$ and $(i, j) \in E$ or $i = j$ and $(u, v) \in E_k$. The following result relates the MkCS problem on G and the stable set problem on $K_k \Box G$.

Theorem 6.12 ([94]) *Let $G = (V, E)$ be a given graph, and K_k the complete graph on k vertices. Then $\alpha_k(G) = \alpha(K_k \Box G)$.*

Proof Let $S_1, \ldots, S_k$ be disjoint stable sets in G, then $\{1\} \times S_1, \ldots, \{k\} \times S_k$ is a stable set in $K_k \square G$. Assume now that S is a stable set in $K_k \square G$ of the largest cardinality. Then S can be partitioned into $S_1, \ldots, S_k$ such that $S_1 = \{1\} \times \hat{S}_1, \ldots, S_k = \{k\} \times \hat{S}_k$, $\hat{S}_1 \subseteq V, \ldots, \hat{S}_k \subseteq V$. Moreover, $\hat{S}_1, \ldots, \hat{S}_k$ are disjoint since $u \in \hat{S}_l \cap \hat{S}_p$ for some $l, p \in [k]$ with $l \neq p$ implies that there is an edge between (l, u) and (p, u) that is also in the stable set S. Hence $\hat{S}_1, \ldots, \hat{S}_k$ are disjoint stable sets in G. □

This short proof is from [76]. Thus, computing $\alpha_k(G)$ for a given k and graph G corresponds to computing the stability number of the larger graph $K_k \square G$. However, computing the stability number of a medium size graph is already difficult, and thus computing $\alpha(K_k \square G)$ is limited to small k and n. Therefore approaches developed specifically for the MkCS problem are beneficial for solving the problem [23, 71].

Interestingly, many concepts developed for the maximum stable set problem extend to the maximum k-colorable subgraph problem. Let $\omega_k(G)$ denote the size of a largest induced subgraph of G that can be covered with k cliques. Then $\alpha_k(G) = \omega_k(\overline{G})$ where $\overline{G}$ denotes the complement of G. The *generalized ϑ-number* for a given k, denoted by $\vartheta_k(G)$, is introduced by Narasimhan and Manber [95] as an eigenvalue bound for $\alpha_k(G)$, i.e.,

$$\alpha_k(G) \leq \vartheta_k(G) := \min_{\mathbf{A} \in \mathcal{S}^n} \left\{ \sum_{i=1}^{k} \lambda_i(\mathbf{A}) : a_{ij} = 1 \text{ for } (i, j) \notin E \text{ or } i = j \right\},$$

where $\lambda_1(\mathbf{A}) \geq \lambda_2(\mathbf{A}) \geq \cdots \geq \lambda_n(\mathbf{A})$ are eigenvalues of $\mathbf{A} \in \mathcal{S}^n$. This eigenvalue bound is reformulated by Alizadeh [8] to the following SDP relaxation:

$$\begin{aligned} \vartheta_k(G) = \text{Maximize} \quad & \langle \mathbf{J}, \mathbf{X} \rangle \\ \text{subject to:} \quad & x_{ij} = 0 \text{ for } (i, j) \in E \\ & \text{tr}(\mathbf{X}) = k \\ & \mathbf{X} \succeq \mathbf{0} \\ & \mathbf{I} - \mathbf{X} \succeq \mathbf{0}. \end{aligned}$$

The constraint $\mathbf{I} - \mathbf{X} \succeq \mathbf{0}$ is redundant when $k = 1$ since $\mathbf{X}$ is positive semidefinite and its eigenvalues sum up to one. Thus for $k = 1$ the generalized ϑ-number reduces to the ϑ-number (6.51).

An in depth analysis of the generalized ϑ-number is given by Sinjorgo and Sotirov [115]. The authors provide closed form expressions for the generalized ϑ-number on several classes of graphs including the Kneser graphs, cycle graphs, strongly regular graphs and orthogonality graphs. The following properties of $\vartheta_k(G)$, among many others, are proven in [115]:

- $\vartheta_k(G) \leq |V|$ for all k;
- $\vartheta_k(G) \leq \vartheta_{k+1}(G)$ with equality if and only if $\vartheta_k(G) = |V|$;
- Let C_n be an odd cycle. Then $\vartheta_2(C_n) = 2\vartheta(C_n)$.

Narasimhan and Manber [95] prove the generalized sandwich theorem:

$$\alpha_k(G) \leq \vartheta_k(G) \leq \chi_k(\overline{G}),$$

or equivalently

$$\omega_k(G) \leq \vartheta_k(\overline{G}) \leq \chi_k(G),$$

where $\chi_k(G)$ is the minimum number of colors needed for a valid k-multicoloring of G. The k-multicoloring of a graph is to assign k distinct colors to each vertex in the graph such that two adjacent vertices are assigned disjoint sets of colors.

The generalized ϑ-number can be strengthened by adding non-negativity constraint to the matrix variable. The resulting graph parameter

$$\begin{aligned}
\vartheta_k'(G) := \text{Maximize} \quad & \langle \mathbf{J}, \mathbf{X} \rangle \\
\text{subject to:} \quad & x_{ij} = 0 \text{ for } (i, j) \in E \\
& x_{ij} \geq 0 \text{ for } (i, j) \notin E \\
& \text{tr}(\mathbf{X}) = k \\
& \mathbf{X} \succeq \mathbf{0} \\
& \mathbf{I} - \mathbf{X} \succeq \mathbf{0},
\end{aligned}$$

is called the *generalized* ϑ'*-number* [76]. For $k = 1$, the graph parameter $\vartheta_k'(G)$ reduces to the Schrijver's number (6.52).

The following result characterises a family of graphs for which $\vartheta_k(G)$ and $\vartheta_k'(G)$ provide tight bounds.

Theorem 6.13 ([76]) *For a given* k $(1 \leq k \leq n-1)$*, let* G *be a graph such that* $\alpha_k(G) = k\vartheta(G)$*. Then*

$$\vartheta_k(G) = \vartheta_k'(G) = k\alpha(G) = \alpha_k(G).$$

The set of vertex-transitive graphs contains a number of non-trivial examples for Theorem 6.13. We say that a graph is vertex-transitive if its automorphism group acts transitively on vertices, i.e., if for every two vertices there is an automorphism that maps one to the other one.

Various semidefinite programming relaxations for the MkCS problem, with increasing complexity are derived in [76]. In order to reduce the sizes of these SDP relaxations, the authors exploit the fact that the MkCS problem is invariant under permutations of the colors. They also show how to further reduce the SDP relaxations for highly symmetric graphs. Currently, the are no specialised SDP-based exact algorithms for the MkCS problem.

6.4 Upper Bounds by Decomposition

In the previous sections, we have seen various MILP formulations and SDP formulations of QUBO. Relaxations of these formulations offer upper bounds on the optimal objective function value of QUBO, at varying levels of strength. Stronger bounds generally require more computational effort. These upper bounds can be embedded within effective enumerative algorithms to solve QUBO. Let us now discuss some additional upper bounding strategies.

We first give a simple Gilmore-Lawler type upper bound for QUBO. Let

$$u_i = c_i + \sum_{j \in R_i}^{n} \max(q_{ij}, 0).$$

Then $\text{UB1} = \sum_{i=1}^{n} \max(u_i, 0)$ is an upper bound on the optimal objective function value of QUBO. A similar lower bound for the minimization QUBO was considered by Li et al. [83] and Pardalos and Rodgers [101]. The upper bound UB1 could be arbitrarily bad. Let us now see how to improve this bound.

Consider the bilinear program

$$\text{BLP}(i)\text{:} \quad \text{Maximize} \quad c_i x_i + \sum_{j=1, j \neq i}^{n} q_{ij} x_i y_j + \sum_{j=1, j \neq i}^{n} c_j y_j \tag{6.55}$$

$$\text{Subject to:} \quad x_i \in \{0, 1\} \tag{6.56}$$

$$y_j \in \{0, 1\} \text{ for all } j = 1, 2, \ldots, n, j \neq i. \tag{6.57}$$

Then, the bound UB1 is precisely $\sum_{i=1}^{n} \text{OBJ}(BLP(i))$. This bilinear programming interpretation of UB1 leads us to an improved upper bounding scheme, where the bound obtained matches $\text{OBJ}(\text{GW}^*)$.

For each variable x_j, introduce its $n-1$ copies, say $y_{ij}, i = 1, 2, \ldots, n, j \neq i$. Then, QUBO can be written as

$$\text{Maximize} \quad \sum_{i=1}^{n} \sum_{j=1, j \neq i}^{n} q_{ij} x_i y_{ij} + \sum_{i=1}^{n} c_i x_i$$

$$\text{Subject to:} \quad x_j = y_{ij}, i, j = 1, 2, \ldots, n, j \neq i \tag{6.58}$$

$$x_i y_{ij} = x_j y_{ji}, i, j = 1, 2, \ldots, n, j \neq i \tag{6.59}$$

$$x_i \in \{0, 1\}, y_{ij} \in \{0, 1\} \text{ for all } i, j = 1, 2, \ldots, n, , i \neq j.$$

Let λ_{ij} and μ_{ij} respectively be the Lagrange multipliers associated with the constraint corresponding to the pair i, j in equation (6.58) and equation (6.59). Also, we denote $y_i = \{y_{ij} : j = 1, 2, \ldots, n, j \neq i\}$. Using these multipliers, construct the Lagrangian objective function

$$L(\mathbf{x}, \mathbf{y}, \boldsymbol{\lambda}, \boldsymbol{\mu}) = \sum_{i=1}^{n} L_i(x_i, y_i, \boldsymbol{\lambda}, \boldsymbol{\mu})$$

where $L_i(x_i, y_i, \boldsymbol{\lambda}, \boldsymbol{\mu})$ is equal to

$$\sum_{i=1}^{n}\left(\Big(c_i + \sum_{k=1,k\neq i}^{n} \lambda_{ki}\Big)x_i - \sum_{j=1,j\neq i}^{n} \lambda_{ij} y_{ij} + \sum_{j=1,j\neq i}^{n} (q_{ij} + \mu_{ij} - \mu_{ji}) x_i y_{ij}\right).$$

and $\boldsymbol{\lambda}$, $\boldsymbol{\mu}$ are the matrices formed by the corresponding Lagrange multipliers λ_{ij} and μ_{ij}. For a given $\boldsymbol{\lambda}$ and $\boldsymbol{\mu}$ values, consider the bilinear program

$$\begin{aligned} P_i(\boldsymbol{\lambda}, \boldsymbol{\mu}):\quad & \text{Maximize} && L_i(x_i, y_i, \boldsymbol{\lambda}, \boldsymbol{\mu}) \\ & \text{Subject to:} && x_i \in \{0, 1\} && (6.60) \\ & && y_{ij} \in \{0, 1\} \text{ for all } j = 1, 2, \ldots, n,\ j \neq i. && (6.61) \end{aligned}$$

Then, $\sum_{i=1}^{n} \text{OBJ}(P_i(\boldsymbol{\lambda}, \boldsymbol{\mu}))$ is an upper bound for QUBO for any given $\boldsymbol{\lambda}$ and $\boldsymbol{\mu}$. Compare this bound with UB1. The best such upper bound, denoted by OBJ(LD), can be obtained by solving the Lagrangian dual problem

$$\text{LD:}\quad \text{Minimize}\quad \sum_{i=1}^{n} \text{OBJ}(P_i(\boldsymbol{\lambda}, \boldsymbol{\mu})).$$

This bound was first proposed by Chardaire and Sutter [26] and later for a more general problem by Elloumi et al. [33]. Let us call it the *simple decomposition bound* or the SD bound. Interestingly, the SD bound matches with OBJ(GW*), as shown below.

Theorem 6.14 ([26, 33]) *The SD bound is equal to* OBJ*(GW*)*

Proof Consider the linear program

$$\begin{aligned} \text{P:}\quad & \text{Maximize} && \sum_{i=1}^{n} \sum_{j=1,j\neq i}^{n} q_{ij} x_i y_{ij} + \sum_{i=1}^{n} c_i x_i \\ & \text{Subject to:} && x_j = y_{ij},\ i, j = 1, 2, \ldots, n,\ j \neq i && (6.62) \\ & && w_{ij} = w_{ji},\ i, j = 1, 2, \ldots, n,\ j \neq i \\ & && w_{ij} \leq x_i,\ i, j = 1, 2, \ldots, n,\ j \neq i \\ & && w_{ij} \leq y_{ij},\ i, j = 1, 2, \ldots, n,\ j \neq i \end{aligned}$$

$$x_i + y_{ij} - w_{ij} \leq 1, i, j = 1, 2, \ldots, n, j \neq i$$

$$w_{ij} \geq 0, i, j = 1, 2, \ldots, n, j \neq i.$$

We will now show that OBJ(P) = OBJ(GW*) and also OBJ(P) = SD bound, leading to the proof of the theorem. Eliminating y_{ij} from P using the equality (6.62) and noting that $w_{ij} = w_{ji}$, we can see that OBJ(P)=OBJ(GW*). To establish the second equality, construct the linearization $P_i(\boldsymbol{\lambda}, \boldsymbol{\mu})^*$ of $P_i(\boldsymbol{\lambda}, \boldsymbol{\mu})$ as follows:

$$\text{Maximize } \sum_{i=1}^{n}\left(\Big(c_i + \sum_{k=1,k\neq i}^{n} \lambda_{ki}\Big)x_i - \sum_{j=1,j\neq i}^{n} \lambda_{ij}y_{ij} + \sum_{j=1,j\neq i}^{n} \Big(q_{ij} + \mu_{ij} - \mu_{ji}\Big)x_i y_{ij}\right)$$

$$\begin{aligned}
\text{Subject to: } & w_{ij} \leq, x_i, j = 1, 2, \ldots, n, j \neq i \\
& w_{ij} \leq y_{ij}, j = 1, 2, \ldots, n, j \neq i \\
& x_i + y_{ij} - w_{ij} \leq 1, j = 1, 2, \ldots, n, j \neq i \\
& w_{ij} \geq 0, j = 1, 2, \ldots, n, j \neq i \\
& x_i \in \{0, 1\}, j = 1, 2, \ldots, n, j \neq i \\
& y_{ij} \in \{0, 1\}, j = 1, 2, \ldots, n, j \neq i.
\end{aligned}$$

By Theorem 6.22 in Appendix, OBJ$(P_i(\boldsymbol{\lambda}, \boldsymbol{\mu}))$ = OBJ$(P_i(\boldsymbol{\lambda}, \boldsymbol{\mu})^*)$. Thus, LD can be viewed as the Langrangian dual problem of P when relaxing the first two constraints from P in the Lagrangian way. Thus, by linear programming duality, OBJ(P) = OBJ(LD) and this completes the proof. □

We now discuss a more general decomposition scheme [26, 33, 91] to improve the SD bound. Here, we partition and decompose QUBO into several pseudo-bilinear programs [26] using Lagrangian decomposition. Consider the partition $V_1, V_2, \ldots, V_p$ of $N = \{1, 2, \ldots, n\}$ where $V_i \cap V_j = \emptyset$ for $i \neq j$, $\cup_{k=1}^{p} V_k = N$ and denote $\bar{V}_k = N - V_k$. For each $k = 1, 2, \ldots, p$, introduce a copy y_{kj} of the variable x_j for each $j \in \bar{V}_k$. Recall that $q_{ii} = 0$ for all $i \in \{1, 2, \ldots, n\}$. Let

$$F_k(\mathbf{x}, \mathbf{y}) = \sum_{j\in V_k} c_j x_j + \sum_{i\in V_k}\sum_{j\in V_k} q_{ij}x_i x_j + \sum_{i\in V_k}\sum_{j\in \bar{V}_k} q_{ij}x_i y_{kj}$$

and consider the pseudo-bilinear program [26]

$$\begin{aligned}
\text{PBLP(k):}\quad & \text{Maximize} && F_k(\mathbf{x}, \mathbf{y}) \\
& \text{Subject to:} && x_j \in \{0, 1\}, \text{ for } j \in V_k \\
& && y_{kj} \in \{0, 1\} \text{ for } j \in \bar{V}_k.
\end{aligned}$$

Then, $\sum_{k=1}^{p} PBLP(k)^*$ is an upper bound for QUBO. When $|V_k|$ is small, PBLP(k) can be solved by enumerating the values of the $\mathbf{x}$ variables and choosing the corresponding optimal $\mathbf{y}$ variables. When $|V_k| = 1$, this bound is equal to UB1. To improve this upper bound, we link the variables y_{kj} to x_j and use Lagrangian decomposition. Note that QUBO is equivalent to

$$\text{EQB:} \quad \text{Maximize} \quad \sum_{k=1}^{p}\left(\sum_{j\in V_k} c_j x_j + \sum_{i\in V_k}\sum_{j\in V_k} q_{ij}x_i x_j + \sum_{i\in V_k}\sum_{j\in \bar{V}_k} q_{ij}x_i y_{kj}\right)$$

$$\text{Subject to:} \quad x_j = y_{kj},\ j = 1,2,\ldots,n,\ k = 1,2,\ldots,p,\ j \in \bar{V}_k \tag{6.63}$$

$$x_i y_{kj} = x_j y_{ri},\ k,r = 1,2,\ldots,p;\ i \in V_k,\ j \in V_r;\ k \neq r, \tag{6.64}$$

$$x_i \in \{0,1\},\ i = 1,2,\ldots,n$$

$$y_{kj} \in \{0,1\} \text{ for all } k = 1,2,\ldots,p,\ j \in \bar{V}_k.$$

Relaxing constraints (6.63) and (6.64) and taking them into the objective function using Lagrange multipliers we get p independent subproblems of the type PBLP(k) (but with different data, which dependents on the value of the Lagrange multipliers). These subproblems can be solved either exactly or by appropriate LP relaxation of equivalent linearizations. The resulting Lagrangian dual problem can be solved using subgradient optimization and the optimal objective function value gives an upper bound. Different implementations of this general scheme are reported by Chardaire and Sutter [26], Elloumi et al. [33], and Mauri and Lorena [91].

6.5 Variable Fixing

A priori knowledge of the value of a variable x_j in an optimal solution can be exploited as a useful preprocessing tool for both exact and heuristic algorithms for QUBO. An important result of this type is the persistency property of GW* discussed in Chap. 5. For convenience, we restate this here.

Theorem 6.15 ([58, 99]) *If a variable x_i is zero or one in an optimal solution of GW*, then there exists an optimal solution to QUBO where x_i assumes the same values.*

Let us now look at some simple conditions to fix variables [45, 56, 119]. For more involved persistency results and logical conditions we refer to Chap. 5. Suppose that a variable x_i is known to be equal to zero in an optimal solution. Without loss of generality assume that $i = n$. Then, we can reduce the problem dimension by 1. Delete row n and column n of $\mathbf{Q}$ to obtain an $(n-1)\times(n-1)$ matrix $\mathbf{Q}'$ and delete the last component of $\mathbf{c}$ to obtain the $(n-1)$-vector $\mathbf{c}'$. Then the QUBO

$(\mathbf{Q}, \mathbf{c}, n)$ can be solved by solving the QUBO $(\mathbf{Q}', \mathbf{c}', n-1)$. Similarly, assume without loss of generality that $x_n = 1$ in an optimal solution. Then again we can reduce the problem size by 1. Recall that the diagonal elements of $\mathbf{Q}$ are zeros and $\mathbf{Q}$ is symmetric. Delete row n and column n from $\mathbf{Q}$ to obtain the matrix $\mathbf{Q}''$. Add twice row n to $\mathbf{c}$ and delete the last component of the resulting vector to obtain $\mathbf{c}''$. Then the QUBO $(\mathbf{Q}, \mathbf{c}, n)$ can be solved by solving the QUBO $(\mathbf{Q}'', \mathbf{c}'', n-1)$ and the optimal objective function value of QUBO $(\mathbf{Q}, \mathbf{c}, n)$ is c_n plus the optimal objective function value of QUBO $(\mathbf{Q}'', \mathbf{c}'', n-1)$.

We now look at some very simple data dependent conditions that permits variable fixing. These conditions (some are for the minimization version of QUBO) were obtained independently by many researchers [12, 20, 24, 45, 56–58, 83, 102, 120]. For each $i = 1, 2, \ldots, n$ define $\ell_i = c_i + \sum_{j\in R_i} \min\{0, 2q_{ij}\}$ and $u_i = c_i + \sum_{j\in R_i} \max\{0, 2q_{ij}\}$.

Theorem 6.16 *If $\ell_i = 0$ then $x_i = 1$ in some optimal solution. If $\ell_i > 0$ then $x_i = 1$ in every optimal solution. Likewise, if $u_i = 0$ then $x_i = 0$ in some optimal solution. If $u_i < 0$ then $x_i = 0$ in every optimal solution.*

Proof Without loss of generality assume $i = n$. Then the objective function of QUBO can be written as

$$\phi(\mathbf{x}) = \left(\sum_{j=1}^{n-1} c_j x_j + \sum_{i=1}^{n-1}\sum_{j=1}^{n-1} q_{ij} x_i x_j\right) + \left(x_n\left(c_n + \sum_{j=1}^{n} 2q_{nj} x_j\right)\right). \tag{6.65}$$

The first bracket in equation (6.65) is independent of x_n. The minimum contribution from the second bracket is ℓ_n and if this is positive, then x_n must be at value 1 in every optimal solution. Likewise, the maximum contribution from the last bracket is u_n and if this is negative, then x_n must be zero in every optimal solution. □

It is possible that after fixing a variable at 0 or 1, new candidates for fixing the value may emerge.

Let us now discuss another very simple variable fixing rule obtained from a posiform that is particularly useful in branch and bound algorithms. Recall that $\phi(\mathbf{x})$ can be written as $\phi(\mathbf{x}) = \psi_0 - \psi(\mathbf{x}, \bar{\mathbf{x}})$ where $\psi(\mathbf{x}, \bar{\mathbf{x}}) = \sum_{j-1}^{n}(d_j x_j + \bar{d}_j \bar{x}_j) + \sum_{i=1}^{t} \alpha_i Ti$ and each T_i is a product of two literals from $\{x, \bar{x}_i, i = 1, 2, \ldots, n\}$, $\alpha_i \geq 0$ for $i = 1, 2, \ldots, t$ and $d_i, \bar{d}_i \geq 0$ [60]. The decomposition form of $\phi(\mathbf{x})$ constructed in Theorem 6.9 is an example of such a posiform representation. Note that ψ_0 is an upper bound on the optimal objective function value of QUBO. We assume that $d_j \bar{d}_j = 0$ since if both are positive, we can use the transformation $x_i = 1 - \bar{x}_i$ which results in a posiform with a larger constant value. Let $\mathbf{x}^0$ be a given solution.

Lemma 6.8 ([60]) *If $\psi_0 - d_i \leq \phi(\mathbf{x}^0)$ then there exists an optimal solution $\mathbf{x}$ with $x_i = 0$. If $\psi_0 - \bar{d}_i \leq \phi(\mathbf{x}^0)$ then there exists an optimal solution $\mathbf{x}$ with $x_i = 1$.*

Proof If $\psi_0 - d_i \le \phi(\mathbf{x}^0)$ then, no solution with $x_i = 1$ can have a better objective function value than $\mathbf{x}^0$ and hence there exists an optimal solution $\mathbf{x}$ with $x_i = 0$. The proof of the second part follows analogously. □

Additional rules for fixing variables or establishing logical relationships between variables are discussed in detail in [45, 56].

6.6 Algorithms and Solvers

The MILP formulations discussed in the previous sections or the natural quadratic 0-1 formulation of QUBO can be used to solve the problem with the aid of general purpose solvers such as Gurobi [52], CPLEX [27], SCIP [1] etc. There are several algorithms available to solve non-linear 0-1 programming problems with linear or quadratic constraints (for example, [89, 114, 116]) and these algorithms can also be used to solve QUBO. Further, algorithms to solve the maximization problem of convex quadratic functions in continuous bounded variables can be used to solve QUBO. SDP based algorithms for the maximum cut problem or QUBO offer yet another class of viable alternatives to solve QUBO. Exact algorithms specifically designed for QUBO are mostly of enumerative type. Let us now discuss some of the simple special purpose algorithms to solve QUBO that can be implemented easily, followed by a brief summary of other available algorithms.

Gulati et al. [51] developed a branch and search algorithm that enumerates promising local optima with respect to the flip neighborhood for the minimization version of QUBO. Let us discuss the algorithm in the context of the maximization version of QUBO with some minor modifications. The *flip neighborhood* of an $\mathbf{x} \in \{0, 1\}^n$, denoted by $\mathbb{F}(\mathbf{x})$, is the collection of all solutions obtained by replacing one component of $\mathbf{x}$, say x_k by its complement $1 - x_k$ (flipping x_k). Let $\mathbf{x}^k = (x_1^k, x_2^k, \ldots, x_n^k)$ be the solution obtained from $\mathbf{x}$ by flipping x_k. Then, $x_i^k = x_i$ if $i \ne k$ and $x_k^k = 1 - x_k$. A solution $\mathbf{x}$ is locally optimal with respect to the flip neighborhood if and only is $\phi(\mathbf{x}) \ge \phi(\mathbf{x}^k)$ for all $k = 1, 2, \ldots, n$.

Theorem 6.17 *The QUBO solution $\mathbf{x} \in \{0, 1\}^n$ is a local optimum with respect to the flip neighborhood if and only if $c_k + \sum_{j \in R_k} 2q_{kj}x_j \ge 0$ if $x_k = 1$ and $c_k + \sum_{j \in R_k} 2q_{kj}x_j \le 0$ if $x_k = 0$ for all $k = 1, 2, \ldots, n$.*

Proof The objective function $\phi(\mathbf{x})$ can be written as

$$\phi(\mathbf{x}) = \left(\sum_{j=1, j\ne k}^{n} c_j x_j + \sum_{i=1, i\ne k}^{n} \sum_{j=1, j\ne k}^{n} q_{ij} x_i x_j \right) + x_k \Big(c_k + \sum_{j=1, j\ne k}^{n} 2q_{kj} x_j \Big). \tag{6.66}$$

Note that the first bracket is independent of x_k. Thus, the condition of the theorem is not satisfied for any k, by flipping x_k we can find an improved solution and if the condition is satisfied, the solution obtained by flipping any component of $\mathbf{x}$ leads to a worse (or at best an equivalent) solution. □

Let S^0, S^1, S^f be a partition of $\{1, 2, \ldots, n\}$ where $x_i = 1$ for $i \in S^1$, $x_i = 0$ for $i \in S^0$, and x_i is a free variable for any $i \in S^f$. This partition generates a partial solution $\mathbf{x}$ where $x_i = 1$ for $i \in S^1$ and $x_i = 0$ for $i \in S^0$. The values x_i for $i \in S^f$ is not yet fixed. A solution obtained from the partial solution $\mathbf{x}$ by assigning $0-1$ values to the free variables is called a *completion* of $\mathbf{x}$. Given a partial solution $\mathbf{x}$, does there exist a completion of $\mathbf{x}$ which is a local optimum with respect to the flip neighborhood? We now discuss some sufficient conditions for non-existence of completions which are local optima.

Theorem 6.18 *Let* $\mathbf{x}$ *be a partial solution defined by the partition* (S^0, S^1, S^f)*. If*

$$c_k + \sum_{j\in S^1} 2q_{kj} + \sum_{j\in S^f, j\neq k} \max\{0, q_{kj}\} < 0 \tag{6.67}$$

then a completion with $x_k = 1$ *will not lead to a local optimum with respect to the flip neighborhood. Similarly, if*

$$c_k + \sum_{j\in S^1} 2q_{kj} + \sum_{j\in S^f, j\neq k} \min\{0, q_{kj}\} > 0 \tag{6.68}$$

then a completion with $x_k = 0$ *will not lead to a local optimum with respect to the flip neighborhood.*

Proof Suppose inequality (6.67) holds for some $k \in S^f$ and let $\mathbf{x}$ be a completion with $x_k = 1$ for some $k \in S^f$. Then

$$c_k + \sum_{j\in R_k} 2q_{kj}x_j \le c_k + \sum_{j\in S^1} 2q_{kj} + \sum_{j\in S^f, j\neq k} \max\{0, q_{kj}\} < 0,$$

and hence by Theorem 6.17, $\mathbf{x}$ cannot be a local optimum with respect to the flip neighborhood. The proof of the second part of the theorem follows analogously. □

Note that the conditions of Theorem 6.18 is precisely the variable fixing rule of Theorem 6.16 restricted to the free variables.

The local optima enumeration algorithm maintains a binary search tree. Each node represents a variable x_k and the two outgoing branches represent a branching decision whether $x_k = 1$ or $x_k = 0$. The unique path from x_k to the root node defines a partial solution. Theorem 6.18 can be used to possibly fix values of more free variables at the node x_k. Let QUBO(S^0, S^1, S^f) be the reduced QUBO obtained after fixing variables in S^0 and S^1 at their corresponding values. Apply a local search algorithm with flip neighborhood on QUBO(S^0, S^1, S^f) to identify a local optimum and extend it using the fixed variables to a solution $\mathbf{x}$ of QUBO. Then, $\mathbf{x}$ may or may not be local optimum with respect to the flip neighborhood. If it is, record this local optimum. If all free variables are fixed, prune node x_k. Otherwise a branching operation is performed. The algorithm terminates when no free variables are left at any of the nodes and also produce a global optimum.

The algorithm can be enhanced by making use of the objective function values of the local optima (and other solutions) identified, along with simple upper bound values that can be calculated efficiently. Pardalos and Rodgers [102] used the simple upper bound

$$UB(S^0, S^1, S^f) = \sum_{i \in U^1} \left(c_i + \sum_{j \in U^1} q_{ij} \right) + \sum_{i \in U^f} \max\{0, c_i + \sum_{j \in U^f} \max\{0, q_{ij}\}\}$$

in their branch and bound algorithm, which also maintains a binary tree. At any node of the search tree, if the best solution obtained have an objective function value greater than or qual to the upper bound at the node, the node is pruned.

Combining the ideas discussed above, a simple branch and bound algorithm can easily be developed to solve QUBO (for example, [51, 102]). The performance of the algorithm can likely be improved by using strong heuristics and/or metaheuristics (see Chap. 9) in place of the flip based local search and stronger upper bounding mechanisms, particularly at the root node. The effects of all these possibilities need to be accessed using experimental analysis. Let us briefly summarize some additional exact algorithms developed specifically for QUBO.

Kalantari and Bagchi [72] developed a branch and bound algorithm to solve the minimization version of QUBO by viewing it as a continuous optimization problem, with the assumption that **Q** is positive semidefinite. As discussed in Chap. 1, such an assumption can be made without loss of generality. Since our version of QUBO is of maximization type, the algorithm of Kalantari and Bagchi [72] can be modified, assuming **Q** is negative semidefinite. We can also use the algorithm of [72] directly to solve the maximization problem by negating the objective function.

Li et al. [83] developed a branch and bound algorithm to solve the minimization version of QUBO, exploiting geometrical properties. Their algorithm can be modified to solve the maximization version of QUBO, or by converting QUBO into an equivalent minimization version and directly use the algorithm. They also proposed some variable fixing rules and upper bound calculations.

Barahona et al. [14] developed a branch and cut algorithm for the minimization version of QUBO. They first convert the problem into a maximum cut problem (see Chap. 1) and then use partial description of the cut polytope (see Chap. 4) to generate an LP relaxation and additional cutting planes. The algorithm also uses variable fixing strategies and when no cutting plane is identified, the algorithm performs a branching operation and new cutting planes are explored from a selected node.

Hansen et al. [60] developed a branch and bound algorithm to solve the minimization version of QUBO, using roof dual upper bound (see Chap. 1), which is equivalent to the LP relaxation of GW [2]. They work with a posiform representation and the roof dual bound is calculated using maximum flow computations. Variable fixing strategies generated by persistency properties (see Chap. 5) and those given in Theorem 6.16 are also used.

Some other exact algorithms for QUBO include Pan et al. [100] which uses continuous optimization methods, Shimizu et al. [113] that solves maximum edge weighted clique problem by branch and bound.

In early days of development, SDP based algorithms were not widely adopted to solve QUBO. However, a breakthrough in solving dense QUBO instances is due to semidefinite programming, in the late 1990s. One of the reasons for not widely using SDP within exact algorithms is due to the computational effort required to solve SDP relaxations via interior point methods. Namely, SDP solvers based on interior point algorithms exhibit problems in terms of both time and memory for solving even medium-size SDP relaxations. However, recent developments of efficient algorithms for solving semidefinite programming problems [22, 34, 63, 68, 97, 118] are changing the trend of designing exact solvers for combinatorial optimization problems. We list below SDP-based exact solves for QUBO in chronological order.

One of the first results on solving QUBO using semidefinite programming is the work of Helmberg and Rendl [65]. More specifically, [65] presents a branch and bound algorithm for solving QUBO that combines semidefinite programming and polyhedral approaches. To solve semidefinite programming relaxations, the authors use an interior point algorithm which is the main drawback of the approach. The computational experiments show that the algorithm from [65] is robust.

In 2007, Billionnet and Elloumi [17] show how to use general-purpose mixed integer quadratic programming solvers for solving QUBO, see also Chap. 7. The main idea in [17] is to convexify the objective function of QUBO by using semidefinite programming. The resulting convex problem is then handled by CPLEX [27].

A dynamic version of the bundle method that provides an efficient machinery to solve SDP problems, including SDP relaxations for the maximum cut problem, with a nearly arbitrary number of inequalities is developed in [34, 110]. Another efficient approach for solving SDP relaxations of the maximum cut problem is the spectral bundle method of Helmberg and Rendl [66]. Those bundle approaches are exploited within a branch and bound framework and resulted in the publicly usable BiqMac solver by Rendl et al. [109]. The BiqMac solver solves any test instance up to 100 vertices, and problems of special structure as well as sparse problems up to 300 vertices. BiqMac was superior to other available methods for solving the maximum cut problem, at the time it was designed. Another SDP-based exact solver for binary quadratic optimization problems appeared a few years after BiqMac. BiqCrunch [74, 75] is a branch and bound solver that uses adjustable semidefinite programming bounds that are computed by a quasi-Newton method. The numerical results by Krislock et al. [74] show that BiqCrunch is often faster and more robust for solving the maximum cut instances than BiqMac. Moreover, the BiqCrunch solver has been successfully tested on a variety of combinatorial optimization problems, including the maximum stable set problem. BiqCrunch is available as a free and open-source software.

BiqBin [54] is a branch and bound solver for binary quadratic problems with linear constraints. The BiqBin solver exploits the fact that any linearly constrained quadratic binary problem can be reformulated into an equivalent maximum cut problem [79]. The sequential version of BiqBin outperforms BiqCrunch on the maximum cut instances, while the two solvers perform similarly on general QUBO instances. Gurobi [52] and SCIP [1] fail to solve the same QUBO test instances in a reasonable time frame. Numerical experiments also show that Gurobi is competitive with BiqBin on test instances with linear constraints. The parallel version of BiqBin solves QUBO instances up to 300 vertices. BiqBin solver is available as a web application.

Hrga and Povh [69] have recently introduced MADAM, a parallel exact solver for the maximum cut problem, that is based on the alternating direction method of multipliers [21]. The numerical results show that the serial version of MADAM outperforms BiqMac, BiqCrunch, and BiqBin on dense instances up to 100 vertices. The authors present numerical results computed by the parallel version of MADAM for graphs with at most 180 vertices.

Appendix: Linear Programming and Duality

In this appendix, we will discuss various results from linear programming that are used in this chapters.

Consider the linear program

$$\begin{aligned} \text{P:}\quad \text{Maximize}\quad & \mathbf{c}^{\mathrm{T}}\mathbf{x} \\ \text{Subject to:}\quad & \mathbf{A}\mathbf{x} \leq \mathbf{b}, \\ & \mathbf{x} \geq \mathbf{0}, \end{aligned}$$

where the matrix $\mathbf{A}$ and vectors $\mathbf{b}$, $\mathbf{c}$ and $\mathbf{x}$ are of appropriate dimensions and shapes. Let $\mathbf{w}^*$ be a given non-negative row vector. Now consider the continuous knapsack problem obtained from P as

$$\begin{aligned} \text{CKP}(\mathbf{w}^*):\quad \text{Maximize}\quad & \mathbf{c}^{\mathrm{T}}\mathbf{x} \\ \text{Subject to:}\quad & \mathbf{w}^*\mathbf{A}\mathbf{x} \leq \mathbf{w}^*\mathbf{b} \\ & \mathbf{x} \geq \mathbf{0}. \end{aligned}$$

Theorem 6.19 *If P has an optimal solution with an optimal dual solution* $\mathbf{w}^0$, *then CKP(*$\mathbf{w}^0$*) also has an optimal solution. Further, the optimal objective function values of P and CKP(*$\mathbf{w}^0$*) are the same.*

Now, consider the linear program

$$
\begin{aligned}
P^*:\quad & \text{Maximize} \quad \mathbf{c}^T\mathbf{x} \\
& \text{Subject to:} \quad \mathbf{A}\mathbf{x} \le \mathbf{b}, \\
& \qquad \mathbf{A}^1\mathbf{x} \le \mathbf{b}^1, \mathbf{A}^2\mathbf{x} \le \mathbf{b}^2, \ldots, \mathbf{A}^p\mathbf{x} \le \mathbf{b}^p, \\
& \qquad \mathbf{x} \ge \mathbf{0}
\end{aligned}
$$

where $\mathbf{b}^i \in R^{m_i}$ and $\mathbf{A}^i$ is an $m_i \times n$ matrix, for $i = 1, 2, \ldots, p$. Let $\mathbf{w}^i \in R^{m_i}$ be given non-negative row vectors in R^{m_i}, for $i = 1, 2, \ldots, p$. Now consider the new linear program P^w obtained from P^* using the weighted aggregation of constraints $\mathbf{A}^i\mathbf{x} \le \mathbf{b}^i$, $i = 1, 2, \ldots, p$. Then P^w can be written as

$$
\begin{aligned}
P^w:\quad & \text{Maximize} \quad \mathbf{c}^T\mathbf{x} \\
& \text{Subject to:} \quad \mathbf{A}\mathbf{x} \le \mathbf{b}, \\
& \qquad \mathbf{w}^1\mathbf{A}^1X \le \mathbf{w}^1\mathbf{b}^1, \mathbf{w}^2\mathbf{A}^2\mathbf{x} \le \mathbf{w}^2\mathbf{b}^2, \ldots, \mathbf{w}^p\mathbf{A}^p\mathbf{x} \le \mathbf{W}^p\mathbf{b}^p, \\
& \qquad \mathbf{x} \ge 0.
\end{aligned}
$$

Theorem 6.20 *When $\mathbf{w}^i$ is the part of an optimal dual solution $\mathbf{w}$ of P^* that is associated with the constraint block $\mathbf{A}^i\mathbf{x} \le \mathbf{b}^i$, $i = 1, 2, \ldots, p$, the optimal objective function values of P^* and P^w are the same.*

For the proof of Theorems 6.19 and 6.20, we refer to [105].

Theorem 6.21 ([18, 106]) *Let $\mathbf{w}^0$ be a given solution to dual D of P and $\mathbf{s}^0$ be the associated surplus vector in D. Then, $\mathbf{c}^T\mathbf{x} = \mathbf{w}^0\mathbf{b} - \mathbf{w}^0(\mathbf{b} - \mathbf{A}\mathbf{x}) - (\mathbf{s}^0)^T\mathbf{x}$.*

Note that $\mathbf{w}^0(\mathbf{b} - \mathbf{A}\mathbf{x}) \ge 0$ and $(\mathbf{s}^0)^T\mathbf{x} \ge \mathbf{0}$. For the proof of Theorem 6.21, we refer to [106].

Consider the bilinear program

$$
\begin{aligned}
\text{BLP:}\quad & \text{Maximize} \quad ax + \sum_{j=1}^{n} b_j x y_j + \sum_{j=1}^{n} c_j y_j \\
& \text{Subject to:} \quad x \in \{0, 1\} && (6.69) \\
& \qquad y_j \in \{0, 1\} \text{ for all } j = 1, 2, \ldots, n. && (6.70)
\end{aligned}
$$

Let BLP′ be the problem obtained from BLP by relaxing (6.70) to $0 \le y_j \le 1$ and BLP″ be the problem obtained from BLP′ by relaxing (6.69) to $0 \le x \le 1$. Now consider the following LP which is the LP relaxation of the linearization of BLP′

$$
\begin{aligned}
\text{LP:}\quad \text{Maximize}\quad & ax + \sum_{j=1}^{n} b_j w_j + \sum_{j=1}^{n} c_j y_j \\
\text{Subject to:}\quad & x + y_j w_j \le 1, \text{ for } j = 1, 2, \ldots, n, \\
& w_j \le x, \\
& w_j \le y_j, \text{ for } j = 1, 2, \ldots, n, \\
& w_j \ge 0 \text{ for } j = 1, 2, \ldots, n.
\end{aligned}
$$

Theorem 6.22 *The optimal objective function values of BLP, BLP′, BLP″, and LP are the same.*

Proof The equivalence of the optimal objective function values of BLP, BLP′, and BLP″ follows from the extreme point optimality property of bilinear programs. Let us now show that the objective function value of BLP′ is equal to that of LP.

Note that the constraints on LP guarantees that $0 \le x \le 1$. Thus, it is sufficient to show that there exists an optimal solution to LP where $x \in \{0, 1\}$. For this, we show that for any extreme point of LP, $x \in \{0, 1\}$. Suppose this is not true. Then, there exists an extreme point feasible solution $(\mathbf{x}^*, \mathbf{y}^*, \mathbf{w}^*)$ to LP such that $0 < x < 1$. Now consider $\tilde{\mathbf{y}}$ where $\tilde{y}_i = (y_i^* - w_i^*)/(1 - x), i = 1, 2, \ldots, n$ and $\hat{\mathbf{y}}$ where $\hat{y}_j = w_j^*/x, j = 1, 2, \ldots, n$. Then $\tilde{\mathbf{z}} = (0, \tilde{\mathbf{y}}, \mathbf{0})$ and $\hat{\mathbf{z}} = (1, \hat{\mathbf{y}}, \hat{\mathbf{y}})$ are feasible solutions of LP and $(x, \mathbf{y}^*, \mathbf{w}^*) = (1-x)\tilde{\mathbf{z}} + x\hat{\mathbf{z}}$. Thus, $(x, \mathbf{y}^*, \mathbf{w}^*)$ is not an extreme point solution and the result follows. □

The last equality also follows directly from a necessary and sufficient condition proved in [69] and follows from a result proved in a more general context in [33].

References

1. T. Achterberg. SCIP: Solving constraint integer programs. Math. Program. Comput. **1**, 1–41 (2009)
2. W.P. Adams, P.M. Dearing, On the equivalance between roof duality and Lagrangian duality for unconstrained 0-1 quadratic programming problems. Discrete Appl. Math. **48**, 1–20 (1994)
3. W.P. Adams, R.J. Forrester, A simple recipe for concise mixed 0-1 linearizations. Oper. Res. Lett. **33**, 55–61 (2005)
4. W.P. Adams, S.M. Henry, Base-2 expansions for linearizing products of functions of discrete variables. Oper. Res. **60**, 1477–1490 (2012)
5. W.P. Adams, H.D. Sherali, A tight linearization and an algorithm for 0–1 quadratic programming problems. Manag. Sci. **32**, 1274–1290 (1986)
6. W.P. Adams, H.D. Sherali, A hierarchy of relaxations leading to the convex hull representation for general discrete optimization problems. Annals Oper. Res. **140**, 21–47 (2005)

7. W.P. Adams, R. Forrester, F. Glover, Comparisons and enhancement strategies for linearizing mixed 0-1 quadratic programs. Discrete Optim. **1**, 99–120 (2004)
8. F. Alizadeh, Interior point methods in semidefinite programming with applications to combinatorial optimization. SIAM J. Optim. **5**(1), 13–51 (1995)
9. M.F. Anjos, New convex relaxations for the maximum cut and VLSI layout problems, PhD thesis, The University of Waterloo, 2001
10. M.F. Anjos, J. Lasserre (ed.), *Handbook of Semidefinnite, Conic and Polynomial Optimization: Theory, Algorithms, Software and Applications*, volume 166 of International Series in Operational Research and Management Science (Springer, 2012)
11. M.F. Anjos, H. Wolkowicz, Strengthened semidefinite relaxations via a second lifting for the Max-Cut problem. Discrete Appl. Math. **119**, 79–106 (2002)
12. D. Axehill, A. Hansson, A preprocessing algorithm for MIQP solvers with applications to MPC, in *43rd IEEE Conference on Decision and Control*, December 14–17, 2004, Atlantis (Paradise Island, Bahamas, 2004)
13. E. Balas, Disjunctive programming and a hierarchy of relaxations for discrete optimization. SIAM J. Algebraic Discrete Methods **6**, 466–486 (1985)
14. F. Barahona, M. Jünger, G. Reinelt, Experiments in quadratic 0-1 programming. Math. Program. **44**, 127–137 (1989)
15. R. Bellman, K. Fan, On systems of linear inequalities in hermitian matrix variables, in *Proceedings of Symposia in Pure Mathematics*, ed. by V.L. Klee (AMS, Providence, 1963), pp. 1–11
16. M. Bentert, R. van Bevern, R. Niedermeier, Inductive k-independent graphs and c-colorable subgraphs in scheduling: a review. J. Scheduling **22**, 3–20 (2019)
17. A. Billionnet, S. Elloumi, Using mixed integer quadratic programming solver for the unconstrained quadratic 0-1 problem. Math. Program. A **109**, 55–68 (2007)
18. A. Billionnet, S. Elloumi, M.C. Plateau, Quadratic 0-1 programming: Tightening linear or quadratic convex reformulation by use of relaxations. RAIRO Oper. Res. **42**, 103–121 (2008)
19. E. Boros, Y. Crama, P.L. Hammer, Upper bounds for quadratic 0-1 maximization. Oper. Res. Lett. **9**, 73–79 (1990)
20. E. Boros, P.L. Hammer, G. Tavares, Preprocessing of unconstrained quadratic binary optimization. RUTCOR Research Report RRR 10-2006, April 2006, Rutgers University, USA (2006)
21. S. Boyd, N. Parikh, E. Chu, B. Peleato, J. Eckstein, Distributed optimization and statistical learning via the alternating direction method of multipliers. Found. Trends® Mach. Learn. **3**(1), 1–122 (2011)
22. S. Burer, D. Vandenbussche, Solving lift-and-project relaxations of binary integer programs. SIAM J. Optim. **16**, 726–750 (2006)
23. M. Campêlo, R.C. Corrêa, A combined parallel Lagrangian decomposition and cutting-plane generation for maximum stable set problems. Electron. Notes Discrete Math. **36**, 503–510 (2010)
24. M.W. Carter, The indefinite zero-one quadratic problem. Discrete Appl. Math. **7**, 23–44 (1984)
25. W. Chaovalitwongse, P. Pardalos, O.A. Prokopyev, A new linearization technique for multi-quadratic 0–1 programming problems. Oper. Res. Lett. **32**, 517–522 (2004)
26. P. Chardaire, A. Sutter, A decomposition method for quadratic 0-1 programming. Manag. Sci. **41**, 704–712 (1995)
27. Cplex II. V12. 1: User's Manual for CPLEX, International Business Machines Corporation 46(53):157 (2009)
28. G.B. Dantzig, On the significance of solving linear programming problems with some integer variables. Econometrica **28**, 30–44 (1960)
29. C. Delorme, S. Poljak, Laplacian eigenvalues and the maximum cut problem. Math. Program. **62**, 557–574 (1993)
30. M. Deza, M. Laurent, Applications of cut polyhedra I. J. Comput. Appl. Math. **55**, 191–216 (1994)

31. M. Deza, M. Laurent, Applications of cut polyhedra II. J. Comput. Appl. Math. **55**, 217–247 (1994)
32. I. Dukanović, F. Rendl, Semidefinite programming relaxations for graph coloring and maximal clique problems. Math. Program. **109**, 345–365 (2007)
33. S. Elloumi, A. Faye, E. Soutif, Decomposition and linearization for 0-1 quadratic programming. Ann. Oper. Res. **99**, 79–93 (2000)
34. I. Fischer, G. Gruber, F. Rendl, R. Sotirov, Computational experience with a bundle approach for semidefinite cutting plane relaxations of Max-Cut and equipartition. Math. Program. B **105**, 451–469 (2006)
35. R.J. Forrester, N. Hunt-Isaak, Computational comparison of exact solution methods for 0-1 quadratic programs: Recommendations for practitioners. J. Appl. Math. **2020**, Article ID 5974820 (2020)
36. R. Fortet, Applications de l'algèbre de boole en recherche opérationelle. Rev. Francaise Recherche Opér. **4**, 5–36 (1959)
37. R. Fortet, L'algèbre de boole et ses applications en recherche opérationnelle. Cahiers du Centre d'Etudes de Recherche Opér. **4**, 17–26 (1960)
38. P. Fouilhoux, A.R. Mahjoub. Solving VLSI design and DNA sequencing problems using bipartization of graphs. Comput. Optim. Appl. **51**(2), 749–781 (2012)
39. F. Furini, E. Traversi, Extended linear formulation for binary quadratic problems. Optim. Online (2013)
40. L. Galli, A.N. Letchford. On the Lovász theta function and some variants. Discrete Optim. **25**, 159–174 (2017)
41. F. Glover, Improved linear integer programming formulations of nonlinear integer problems. Manag. Sci. **22**, 455–460 (1975)
42. F. Glover, An improved MIP formulation for products of discrete and continuous variables. J. Inf. Optim. Sci. **5**, 469–471 (1984)
43. F. Glover, E. Woolsey, Further reduction of zero-one polynomial programming problems to zero-one linear programming problems. Oper. Res. **21**, 141–161 (1973)
44. F. Glover, E. Woolsey, Converting the 0–1 polynomial programming problem to a 0–1 linear program. Oper. Res. **22**, 180–182 (1974)
45. F. Glover, M. Lewis, G. Kochenberger, Logical inequality implications for reducing the size and difficulty of quadratic unconstrained binary optimization problems. Eur. J. Oper. Res. **265**, 829–842 (2018)
46. M.X. Goemans, D.P. Williamson, Improved approximation algorithms for maximum cut and satisfiability problems using semidefinite programming. J. ACM **42**(6), 1115–1145 (1995)
47. A. J. Goldman, Linearization in 0-1 variables: a clarification. Oper. Res. **31**, 946–947 (1983)
48. M. Grötschel, L. Lovász, A.J. Schrijver, *Geometric Algorithms and Combinatorial Optimization* (Wiley, New York, 1988)
49. G. Gruber, F. Rendl, Computational experience with stable set relaxations. SIAM J. Optim. **13**, 1014–1028 (2003)
50. S. Gueye, P. Michelon, A linearization framework for unconstrained quadratic (0-1) problems. Discrete Appl. Math. **157**, 1255–1266 (2009)
51. V.P. Gulati, S.K. Gupta, A.K. Mittal, Unconstrained quadratic bivalent programming problem. Eur. J. Oper. Res. **15**, 121–125 (1984)
52. Gurobi Optimization, LLC, Gurobi Optimizer Reference Manual, 2021
53. N. Gusmeroli, A. Wiegele, EXPEDIS: An exact penalty method over discrete sets. Discrete Optim. (2021). https://doi.org/10.1016/j.disopt.2021.100622
54. N. Gusmeroli, T. Hrga, B. Lužar, J. Povh, M. Siebenhofer, A. Wiegele, BiqBin: a parallel branch-and-bound solver for binary quadratic problems with linear constraints. Preprint (2021). https://arxiv.org/pdf/2009.06240.pdf
55. N. Gvozdenović, M. Laurent, Semidefinite bounds for the stability number of a graph via sums of squares of polynomials. Math. Program. **110**, 145–173 (2007)
56. P.L. Hammer, P. Hansen, Logical relations in quadratic 0-1 programming. Rev. Roumaine Math. Pures Appl. **26**, 421–429 (1981)

57. P.L. Hammer, S. Rudeanu, *Boolean Methods in Operations Research and Related Areas* (Springer, Berlin, 1968)
58. P.L. Hammer, P. Hansen, B. Simone, Roof duality, complementations, and persistency in quadratic 0-1 optimization. Math. Program. **28**, 121–155 (1984)
59. P. Hansen, C. Meyer, Improved compact linearizations for the unconstrained quadratic 0-1 minimization problem. Discrete Appl. Math. **157**, 1267–1290 (2009)
60. P. Hansen, B. Jaumard, C. Meyer, A simple enumerative algorithm for unconstrained 0-1 quadratic programming, Cahier du GERAD G-2000-59, GERAD, November (2000)
61. X. He, A. Chen, W.A. Chaovalitwongse, H.X. Liu, An improved linearization technique for a class of quadratic 0-1 programming problems. Optim. Lett. **6**, 31–41 (2012)
62. E. de Klerk, D.V. Pasechnik, Approximating of the stability number of a graph via copositive programming. SIAM J. Optim. **12**(4), 875–892 (2002)
63. F. de Meijer, R. Sotirov. SDP-based bounds for the Quadratic Cycle Cover Problem via cutting plane augmented Lagrangian methods and reinforcement learning. INFORMS J. Comput. **33**(4), 1259–1684 (2021)
64. C. Helmberg, Fixing variables in semidefinite relaxations. SIAM J. Matrix Anal. Appl. **21**, 952–969 (1996)
65. C. Helmberg, F. Rendl, Solving quadratic (0,1)-problems by semidefinite programs and cutting planes. Math. Program. **82**, 291–315 (1998)
66. C. Helmberg, F. Rendl, A spectral bundle method for semidefinite programming. SIAM J. Optim. **10**(3), 673–696 (2000)
67. C. Helmberg, K.C. Kiwiel, F. Rendl, Incorporating inequality constraints in the spectral bundle method, in *Integer Programming and Combinatorial Optimization*, ed. by E.A. Boyd R.E. Bixby, R.Z. Rios-Mercado, Springer Lecture Notes 1412 (1998), pp. 423–435
68. C. Helmberg, M.L. Overton, F. Rendl, The spectral bundle method with second-order information. Optim. Methods Softw. **29**(4), 855–876 (2014)
69. T. Hrga, J. Povh, MADAM: a parallel exact solver for max-cut based on semidefinite programming and ADMM. Comput. Optim. Appl. **80**(2), 347–375 (2021)
70. T. Januschowski, M.E. Pfetsch, The maximum k-colorable subgraph problem and orbitopes. Discrete Optim. **8**, 478–494 (2011)
71. T. Januschowski, M.E. Pfetsch. Branch-cut-and-propagate for the maximum k-colorable subgraph problem with symmetry, in *Integration of AI and OR Techniques in Constraint Programming for Combinatorial Optimization Problems*, ed. by T. Achterberg, J. Christopher Beck (Springer, Berlin Heidelberg, 2011), pp. 99–116
72. B. Kalantari, A. Bagchi, An algorithm for quadratic zero-one programs. Naval Res. Logist. **37**, 527–538 (1990)
73. A.M.C.A. Koster, M. Scheffel, A routing and network dimensioning strategy to reduce wavelength continuity conflicts in all-optical networks, in *Proceedings of INOC 2007, International Network Optimization Conference, Spa, April 22–25* (2007)
74. N. Krislock, J. Malick, F. Roupin, Improved semidefinite bounding procedure for solving max-cut problems to optimality. Math. Program. A **143**, 61–86 (2014)
75. N. Krislock, J. Malick, F. Roupin, BiqCrunch: A semidefinite branch-and-bound method for solving binary quadratic problems. ACM Trans. Math. Softw. (TOMS) **43**(4), 1–23 (2017)
76. O. Kuryatnikova, R. Sotirov, J. Vera, The maximum k-colorable subgraph problem and related problems. INFORMS J. Comput. **34**(1), 656–669 (2021)
77. J.B. Lasserre, Global optimization with polynomials and the problem of moments. SIAM J. Optim. **11**, 796–817 (2001)
78. J.B. Lasserre, An explicit exact SDP relaxation for nonlinear 0–1 programs, in *Lecture Notes in Computer Science*, ed. by K. Aardal, A.M.H. Gerards, vol. 2081 (2001), pp. 293–303
79. J.B. Lasserre, A MAX-CUT formulation of 0/1 programs. Oper. Res. Lett. **44**(2), 158–164 (2016)
80. M. Laurent, A comparison of the Sherali-Adams, Lovász–Schrijver, and Lasserre relaxations for 0–1 programming. Math. Oper. Res. **28**, 470–496 (2003)

81. M. Laurent, S. Poljak, On a positive semidefinite relaxation of the cut polytope. Linear Algebra Appl. **223/224**, 439–461 (1995)
82. M. Laurent, S. Poljak, F. Rendl, Connections between semidefinite relaxations of the max-cut and stable set problems. Math. Program. **77**, 225–246 (1997)
83. D. Li, X.L. Sun, C.L. Liu, An exact solution method for the unconstrained quadratic 0-1 programming: a geometric approach. J. Glob. Optim. **52**, 797–829 (2012)
84. L. Liberti, Compact linearization for binary quadratic problems. 4OR **5**(3), 231–245 (2007)
85. R. Lippert, R. Schwartz, G. Lancia, S. Istrail, Algorithmic strategies for the single nucleotide polymorphism haplotype assembly problem. Brief. Bioinform. **3**(1), 23–31 (2002)
86. L. Lovász, On the Shannon capacity of a graph. IEEE Trans. Inf. Theory **25**(1), 1–7 (1979)
87. L. Lovász, *An Algorithmic Theory of Numbers, Graphs, and Convexity* (SIAM, Philadelphia, 1986)
88. L. Lovász, A. Schrijver, Cones of matrices and set-functions and 0-1 optimization. SIAM J. Optim. **1**, 166–190 (1991)
89. S. H. Lu, An improved enumerative algorithm for solving quadratic zero-one programs. Eur. J. Oper. Res. **15**, 110–120 (1984)
90. V. Maniezzo, T. Stützle, S. Voß, *Matheuristics - Hybridizing Metaheuristics and Math. Program.*, Annals of Information Systems, vol. 10, (Springer, 2010), ISBN 978-1-4419-1305-0
91. G.R. Mauri, L.A.N. Lorena, Improving a Lagrangian decomposition for the unconstrained binary quadratic programming problem. Comput. Oper. Res. **39**, 1577–1581 (2012)
92. G.P. McCormick, Computability of global solutions to factorable nonconvex programs: Part I - Convex underestimating problems. Math. Program. **10**, 147–175 (1976)
93. B. Mohar, S. Poljak, Eigenvalues and the max-cut problem. Czechoslovak Math. J. **40**, 343–352 (1990)
94. G. Narasimhan, The maximum $k-$colorable subgraph problem, PhD thesis, University of Wisconsin-Madison, 1989
95. G. Narasimhan, R. Manber, *A Generalization of Lovász ϑ Function*, DIMACS Series in Discrete Mathematics and Computer Science (1990), pp. 19–27
96. Y. Nesterov, A. Nemirovski, *Interior-point Polynomial Algorithms in Convex Programming* (SIAM Studies in Applied Mathematics, Philadelphia, 1994)
97. D.E. Oliveira, H. Wolkowicz, Y. Xu, ADMM for the SDP relaxation of the QAP. Math. Program. Comput. **10**, 631–658 (2018)
98. M. Oral, O. Kettani, A linearization procedure for the quadratic and cubic mixed integer programs. Oper. Res. **40**(supplement-1), S109–S116 (1992)
99. M. Padberg, The Boolean quadric polytope: Some characteristics, facets and relatives. Math. Program. **45**, 134–172 (1989)
100. S. Pan, T. Tan, Y. Jiang, A global continuation algorithm for solving binary quadratic programming problems. Comput. Optim. Appl. **41**, 349–362 (2008)
101. P.M. Pardalos, G.P. Rodgers, Parallel branch and bound algorithms for unconstrained quadratic zero-one programming, in *Impacts of Recent Computer Advances on Operations Research*, ed. by R. Sharda et al. (North-Holland, 1989), pp. 131–143
102. P.M. Pardalos, G.P. Rodgers, Computational aspects of a branch and bound algorithm for quadratic zero-one programming. Computing **45**, 131–144 (1990)
103. P.M. Pardalos, W. Chaovalitwongse, L.D. Iasemidis, J.C. Sackellares, D.S. Shiau, P.R. Carney, O. A. Prokopyev, V.A. Yatsenko, Seizure warning algorithm based on optimization and nonlinear dynamics. Math. Program. B **101**, 365–385 (2004)
104. S. Poljak, F. Rendl, Nonpolyhedral relaxations of graph-bisection problems. SIAM J. Optim. **5**, 467–487 (1995)
105. A.P. Punnen, N. Kaur, Revisiting some classical explicit linearizations for the quadratic binary optimization problem, Research Report, Department of Mathematics, Simon Fraser Univerity, 2021
106. A.P. Punnen, N. Kaur, On compact linearizations of the quadratic binary optimization problem, Research Report, Department of Mathematics, Simon Fraser University, 2021

107. A.P. Punnen, N. Kaur, Low rank quadratic unconstrained binary optimization problem, Research Report, Department of Mathematics, Simon Fraser University, 2021
108. A.P. Punnen, P. Pandey, M. Friesen, Representations of quadratic combinatorial optimization problems: A case study using the quadratic set covering problem. Comput. Oper. Res. **112**, 104769 (2019)
109. F. Rendl, G. Rinaldi, A. Wiegele, Solving Max-Cut to optimality by intersecting semidefinite and polyhedral relaxations. Math. Program. **121**, 307–335 (2010)
110. F. Rendl, R. Sotirov, Bounds for the quadratic assignment problem using the bundle method. Math. Program. **109**, 505–524 (2007)
111. A. Schrijver, A comparison of the Delsarte and Lovász bounds. IEEE Trans. Inf. Theory **25**(4), 425–429 (1979)
112. H. Sherali, W. Adams, A hierarchy of relaxations between the continuous and convexhull representations for zero-one programming problems. SIAM J. Discrete Math. **3**(3), 411–430 (1990)
113. S. Shimizu, K. Yamaguchi, S. Masuda, A maximum edge-weight clique extraction algorithm based on branch-and-bound. Discrete Optim. **37**, 100583 (2020)
114. N.Z. Shor, A.S. Davydov, On a bounding method in quadratic extremal problems with 0-1 variables. Kibernetika **54**, 48–50 (1985)
115. L. Sinjorgo, R. Sotirov. On the generalized ϑ-number and related problems for highly symmetric graphs. Preprint (2021). https://arxiv.org/abs/2104.11910
116. R.A. Stubbs, S. Mehrotra, A branch-and-cut method for 0-1 mixed convex programming. Math. Program. **86**, 515–532 (1999)
117. A.P. Subramanian, H. Gupta, S.R. Das, M.M. Buddhikot, Fast spectrum allocation in coordinated dynamic spectrum access based cellular networks, in *2007 2nd IEEE International Symposium on New Frontiers in Dynamic Spectrum Access Networks* (2007), pp. 320–330
118. D. Sun, K.C. Toh, Y. Yuan, X.Y. Zhao, SDPNAL +: A Matlab software for semidefinite programming with bound constraints (version 1.0). Optim. Methods Softw. **35**, 87–115 (2020)
119. G. Tavares, New algorithms for quadratic unconstrained binary optimization problem (QUBO) with applications in engineering and social sciences, Ph.D Thesis, Rutgers University, 2008
120. D. Wanga, R. Kleinberg, Analyzing quadratic unconstrained binary optimization problems via multicommodity flows. Discrete Appl. Math. **157**, 3746–3753 (2009)
121. L.J. Watters, Reduction of integer polynomial programming problems to zero-one linear programming problems. Oper. Res. **15**, 1171–1174 (1967)
122. H. Wolkowicz, R. Saigal, L. Vandeberghe (ed.), *Handbook on Semidefinite Programming* (Kluwer Academic, Boston, 2000)
123. W.I. Zangwill, Media selection by decision programming. J. Advertising Res. **5**, 30–36 (1965)

Chapter 7
The Random QUBO

Karthik Natarajan

Abstract In this chapter, we discuss instances of QUBO where the input data is random. Such random instances are often analyzed within the topic of "probabilistic combinatorial optimization" and the goal is to study the behavior of the distribution of the optimal value and the distribution of the optimal solution. The introduction of randomness makes the problem more challenging from a computational perspective. We discuss a probabilistic model with dependent random variables where it is possible to extend many of the known computational results from deterministic QUBO to random QUBO. We also review an interesting asymptotic characterization of the random optimal value under independent and identically distributed random variables.

7.1 Introduction

Consider the quadratic unconstrained binary optimization (QUBO) problem:

$$\text{(QUBO)}\ Z(\mathbf{Q}, \mathbf{c}) = \max_{\mathbf{x} \in \{0,1\}^n} \mathbf{x}^T \mathbf{Q} \mathbf{x} + \mathbf{c}^T \mathbf{x}, \tag{7.1}$$

where $\mathbf{Q} \in \mathcal{S}_n$ is a $n \times n$ real symmetric matrix with zeros along the diagonal and $\mathbf{c} \in \mathbb{R}^n$. This representation of QUBO is without loss of generality since $x_i^2 = x_i$ for $x_i \in \{0, 1\}$ and thus the diagonal entries of $\mathbf{Q}$ can be absorbed into the vector $\mathbf{c}$. Any matrix $\mathbf{Q}$ not necessarily symmetric can be transformed to a symmetric matrix $(\mathbf{Q} + \mathbf{Q}^T)/2$ without changing the objective function value. In this chapter, we consider the random version of QUBO where possibly some or all of the components of the matrix $\tilde{\mathbf{Q}}$ and vector $\tilde{\mathbf{c}}$ are random. The expected optimal objective value of the QUBO problem (7.1) averaged over the possible realizations of the random terms

K. Natarajan (✉)
Engineering Systems and Design, Singapore University of Technology and Design, Singapore, Singapore
e-mail: karthik_natarajan@sutd.edu.sg

A. P. Punnen (ed.), *The Quadratic Unconstrained Binary Optimization Problem*,
https://doi.org/10.1007/978-3-031-04520-2_7

for a continuous distribution is given as:

$$E\left[Z(\tilde{\mathbf{Q}}, \tilde{\mathbf{c}})\right] = \int_{\Omega} Z(\mathbf{Q}, \mathbf{c}) f(\mathbf{Q}, \mathbf{c}) d\mathbf{Q} d\mathbf{c},$$

where $f(\mathbf{Q}, \mathbf{c})$ is the probability density function of $(\tilde{\mathbf{Q}}, \tilde{\mathbf{c}})$ with support in Ω. For a discrete distribution, the expected optimal value is given as:

$$E\left[Z(\tilde{\mathbf{Q}}, \tilde{\mathbf{c}})\right] = \sum_{\Omega} Z(\mathbf{Q}, \mathbf{c}) p(\mathbf{Q}, \mathbf{c}),$$

where $p(\mathbf{Q}, \mathbf{c}) = \mathbb{P}(\tilde{\mathbf{Q}} = \mathbf{Q}, \tilde{\mathbf{c}} = \mathbf{c})$ is the probability mass function of $(\tilde{\mathbf{Q}}, \tilde{\mathbf{c}})$ with support in Ω. There are two challenges in computing the expected optimal value:

1. Solving QUBO with a deterministic objective where $(\tilde{\mathbf{Q}}, \tilde{\mathbf{c}}) = (\mathbf{Q}, \mathbf{c})$ with probability one is already NP-hard.
2. The randomness in the objective function increases the computational challenge since we now need to average the optimal value across a set of QUBOs, possibly exponentially many QUBOs. For example, if $\mathbf{Q}$ is fixed and the random vector $\tilde{\mathbf{c}}$ has independent entries each taking two possible values, we would need to solve a set of 2^n NP-hard QUBO instances to compute the expected optimal value.

In this chapter, we discuss bounds and numerical approaches to compute the expected optimal value with dependent random variables. We also discuss an asymptotic characterization of the expected optimal value under identically and independently distributed random variables. Before doing so, we discuss three applications of the analysis of random QUBO:

1. When comparing different algorithms to solve QUBO problems, it is common practice in numerical experiments to randomly generate instances of QUBO. The algorithms are applied to these randomly generated instances and then compared across the instances in terms of the running times and the best integer solutions generated in a fixed amount of time. Averaging over the instances corresponds to evaluating the expected optimal value by solving all the QUBO instances.
2. In practical applications, one is often interested in solving multiple instances of QUBO where the objective function gets modified slightly. By better understanding the distribution of the optimal solution and exploiting information of random objective, it is possible to solve the random QUBO instances faster as we shall see later in this chapter.
3. There is a significant stream of research in probabilistic combinatorial optimization which shows that asymptotically the optimal value of many random combinatorial optimization problems converges to non trivial limits when the random terms are independent and identically distributed. Examples include the linear assignment problem (see [1]), the minimum spanning tree problem (see [9]), the traveling salesperson problem (see [2]) and the quadratic assignment problem (see [8]). In the case of QUBO, these results have been analyzed in

the context of the Sherrington-Kirkpatrick model [25] with Gaussian random variables which we review in the last part of the chapter.

7.2 An Upper Bound on the Expected Optimal Value

We start by discussing a bound on the expected optimal value that makes use of only the marginal distributions of the random coefficients. Assume the marginal distribution function of the random term $\tilde{Q}_{ij}$ is $F_{ij}(\cdot)$ for $i < j$ where $Q_{ji} = Q_{ij}$ for every realization where $j > i$ and the marginal distribution function of $\tilde{c}_i$ is $F_i(\cdot)$. Let the joint distribution be denoted by θ which lies in the Fréchet class of distributions Θ with the given marginal distributions. For example, using mutually independent random variables with the given marginal distributions provides one such distribution in the set Θ. Hence $\Theta \neq \emptyset$. The exact probability distribution is incompletely specified. We focus on an upper bound that is generated by the multivariate joint distribution of the random components that maximizes the expected maximum objective value of the QUBO over all distributions in the class Θ. The problem of interest is defined as:

$$\phi = \sup_{\theta\in\Theta} E_\theta\left[Z(\tilde{\mathbf{Q}}, \tilde{\mathbf{c}})\right]. \tag{7.2}$$

For any $\mathbf{x} \in \{0, 1\}^n$, symmetric matrix $\mathbf{R} \in \mathcal{S}_n$ with zeros along the diagonal and vector $\mathbf{d} \in \mathbb{R}^n$, we have:

$$\begin{aligned}
\mathbf{x}^T\mathbf{Q}\mathbf{x} + \mathbf{c}^T\mathbf{x} &= \mathbf{x}^T\mathbb{R}\mathbf{x} + \mathbf{d}^T\mathbf{x} + \mathbf{x}^T(\mathbf{Q} - \mathbb{R})\mathbf{x} + (\mathbf{c} - \mathbf{d})^T\mathbf{x} \\
&\quad \text{[adding and subtracting } \mathbf{R} \text{ from } \mathbf{Q} \text{ and } \mathbf{d} \text{ from } \mathbf{c}] \\
&\leq \max_{x\in\{0,1\}^n}\left(\mathbf{x}^T\mathbb{R}\mathbf{x} + \mathbf{d}^T\mathbf{x}\right) + \sum_{i\neq j}\left[Q_{ij} - R_{ij}\right]^+ + \sum_i \left[c_i - d_i\right]^+ \\
&\quad \text{[taking the maximum over } \{0,1\}^n \text{ for the first term and} \\
&\quad \text{over individual components for the second term]},
\end{aligned}$$

where $z^+ = \max(0, z)$. Since the right hand side of the inequality is a valid upper bound for all $\mathbf{x} \in \{0, 1\}^n$, by taking the maximum on the left hand side over all feasible solutions $\mathbf{x} \in \{0, 1\}^n$ and computing expectations, we obtain:

$$E\left[Z(\tilde{\mathbf{Q}}, \tilde{\mathbf{c}})\right] \leq Z(\mathbb{R}, \mathbf{d}) + \sum_{i\neq j} E\left[\tilde{Q}_{ij} - R_{ij}\right]^+ + \sum_i E\left[\tilde{c}_i - d_i\right]^+.$$

By minimizing over the $\mathbb{R}$ matrix and vector $\mathbf{d}$, an upper bound on the expected optimal value is obtained as follows:

$$\min_{R\in\mathcal{S}_n:diag(R)=\mathbf{0},d\in\mathbb{R}^n}\left(Z(\mathbb{R}, \mathbf{d}) + \sum_{i\neq j} E\left[\tilde{Q}_{ij} - R_{ij}\right]^+ + \sum_i E\left[\tilde{c}_i - d_i\right]^+\right),$$

where diag($\mathbf{R}$) denotes the vector formed with the diagonal entries of the matrix $\mathbb{R}$ and $\mathbf{0}$ is the vector of all zeros.

In fact this upper bound is tight; namely there is a joint distribution of the random $\tilde{\mathbf{Q}}$ and $\tilde{\mathbf{c}}$ such that given any set of marginals with well-defined expected values, the upper bound is attained. Namely:

$$\phi = \min_{R \in S_n : diag(\boldsymbol{R}) = \mathbf{0}, \boldsymbol{d} \in \mathbb{R}^n} \left(Z(\mathbb{R}, \mathbf{d}) + \sum_{i \neq j} E\left[\tilde{Q}_{ij} - R_{ij}\right]^+ + \sum_i E\left[\tilde{c}_i - d_i\right]^+ \right).$$

We refer the reader to [3, 4, 14, 16, 18, 19, 31] for the proof of results along this line arising in probabilistic combinatorial optimization. Let us discuss a few properties of the upper bound ϕ:

1. The reformulation of ϕ above can be interpreted as finding a deterministic estimate $\mathbf{R}$ and $\mathbf{d}$ of the random matrix $\tilde{\mathbf{Q}}$ and random vector $\tilde{\mathbf{c}}$ such that the objective function of the deterministic QUBO problem is balanced with penalty terms for incorrectly estimating the random terms. An alternative maximization formulation for the bound ϕ is given by (see [16, 18]):

$$\begin{aligned} \phi = \max_{\boldsymbol{x}, \boldsymbol{X}} & \sum_{i \neq j} \int_{1 - X_{ij}}^{1} F_{ij}^{-1}(t)dt + \sum_i \int_{1 - x_i}^{1} F_i^{-1}(t)dt \\ \text{s.t. } & (\mathbf{x}, \mathbf{X}) \in \text{conv}\left\{(\mathbf{z}, \mathbf{z}\mathbf{z}^T) \mid \mathbf{z} \in \{0, 1\}^n\right\}, \end{aligned}$$

   where the optimization is over the Boolean quadric polytope. Under the assumption that the random variables are absolutely continuous, the optimal decision variables $\mathbf{x}$ and $\mathbf{X}$ are exactly the expectation of the random optimal solution $\mathbf{x}(\tilde{\mathbf{Q}}, \tilde{\mathbf{c}})$ and the quadratic form $\mathbf{x}(\tilde{\mathbf{Q}}, \tilde{\mathbf{c}})\mathbf{x}(\tilde{\mathbf{Q}}, \tilde{\mathbf{c}})^T$ under the distribution θ that attains the bound ϕ (see [18] for a detailed proof of this result in general integer programs).
2. Clearly when $\tilde{\mathbf{Q}} = \mathbf{Q}$ and $\tilde{\mathbf{c}} = \mathbf{c}$, we have:

$$\begin{aligned} \phi &= \min_{R \in S_n : diag(\boldsymbol{R}) = \mathbf{0}, \boldsymbol{d} \in \mathbb{R}^n} \left(Z(\mathbb{R}, \mathbf{d}) + \sum_{i \neq j} [Q_{ij} - R_{ij}]^+ + \sum_i [c_i - d_i]^+ \right) \\ &= Z(\mathbf{Q}, \mathbf{c}) \\ &\quad [\text{where the optimal solution is } \mathbb{R} = \mathbf{Q} \text{ and } \mathbf{d} = \mathbf{c}]. \end{aligned}$$

   The second equality follows from observing that if $R_{ij} < Q_{ij}$, by increasing R_{ij} to Q_{ij}, the first term $Z(\mathbb{R}, \mathbf{d})$ increases at most at a rate of one while the penalty term $[Q_{ij} - R_{ij}]^+$ decreases at a rate of one. If $R_{ij} > Q_{ij}$, by decreasing R_{ij} to Q_{ij}, the first term $Z(\mathbb{R}, \mathbf{d})$ does not increase and might possibly decrease while the penalty term $[Q_{ij} - R_{ij}]^+$ stays at zero. A similar observation can be made for the c_i and d_i terms. Thus in the deterministic case, the bound reduces to the optimal QUBO value.

3. We can associate a QUBO with an undirected graph as follows: $G = (V, E)$ is a graph on the n nodes where $|V| = n$. For nodes i and j where $\tilde{Q}_{ij} = \tilde{Q}_{ji} = 0$ with probability one, there is no edge connecting i and j while for all other pairs of nodes i and j where $\tilde{Q}_{ij} = \tilde{Q}_{ji}$, we add an undirected edge $\{i, j\}$ to E such that $i < j$. When the graph is acyclic or series-parallel, QUBO is known to be solvable in polynomial time (see [20]). Interestingly, in this case the tightest upper bound on the expected optimal value is also solvable in polynomial time. To see this observe that ϕ is the optimal value to

$$\begin{aligned}\min_{t,\mathbf{R},\mathbf{d}} \; & t + \sum_{\{i,j\}\in E} \left(E\left[\tilde{Q}_{ij} - R_{ij}\right]^+ + E\left[\tilde{Q}_{ji} - R_{ji}\right]^+ \right) + \sum_i E\left[\tilde{c}_i - d_i\right]^+ \\ \text{s.t. } \; & t \geq Z(\mathbb{R}, \mathbf{d}) \\ & \mathbf{R} \in \mathcal{S}_n, diag(\mathbf{R}) = \mathbf{0}, R_{ij} = 0 \; \forall \{i, j\} \notin E, \\ & \mathbf{d} \in \mathbb{R}^n,\end{aligned}$$

The key step in solving the separation problem for this convex optimization problem is:
Separation problem: Given t^* and symmetric matrix $\mathbb{R}^*$ with zeros along the diagonal with $R^*_{ij} = 0$ for $\{i, j\} \notin E$ and vector $\mathbf{d}^*$, decide whether $t^* \geq Z(\mathbb{R}^*, \mathbf{d}^*)$, and if not, find a violated inequality.
For sparse graphs such as acyclic or series-parallel graphs, the optimal value of the QUBO $Z(\mathbb{R}^*, \mathbf{d}^*)$ is computable in polynomial time. Let the optimal solution be $\mathbf{x}^* \in \{0, 1\}^n$. If $t^* < Z(\mathbb{R}^*, \mathbf{d}^*)$, the violated inequality is given by $t \geq \mathbf{x}^{*T}\mathbb{R}\mathbf{x}^* + \mathbf{d}^T\mathbf{x}^*$. The separation problem is hence solvable in polynomial time. From the equivalence of separation and optimization (see [12]), the bound ϕ is computable in polynomial time for sparse graphs where the deterministic QUBO is solvable in polynomial time.

7.3 Quadratic Convex Reformulation (QCR)

We next discuss convex reformulations for QUBO. Such formulations are particularly useful at the root node in a branch and bound algorithm to obtain powerful upper bounds. We consider its application in both deterministic and stochastic versions.

7.3.1 QCR Method for Determinsitic QUBO

One useful preprocessing approach to solve QUBO problems and more generally binary quadratic programs and mixed integer quadratic programs is the Quadratic Convex Reformulation (QCR) method proposed in [5–7]. Their method is inspired from the SDP relaxation for discrete optimization problems developed in [15, 24, 26] among others. We review the key idea of the approach next and then discuss its applications to random QUBO.

For any $\mathbf{u} \in \mathbb{R}^n$, consider the optimization problem:

$$Z(\mathbf{u}; \mathbf{Q}, \mathbf{c}) = \max_{\mathbf{x} \in \{0,1\}^n} \mathbf{x}^T (\mathbf{Q} - \mathrm{Diag}(\mathbf{u}))\mathbf{x} + (\mathbf{c} + \mathbf{u})^T \mathbf{x}, \tag{7.3}$$

where $\mathrm{Diag}(\mathbf{u})$ denotes a diagonal $n \times n$ matrix with the elements of $\mathbf{u}$ along the diagonal. Since $x_i^2 = x_i$ for $x_i \in \{0, 1\}$, we have:

$$Z(\mathbf{Q}, \mathbf{c}) = Z(\mathbf{u}; \mathbf{Q}, \mathbf{c}).$$

The key advantage of reformulation (7.3) is when one chooses a vector $\mathbf{u}$ such that the objective function is concave (this corresponds to $\mathrm{Diag}(\mathbf{u}) - \mathbf{Q} \succeq 0$). This transforms a nonconvex QUBO to an equivalent convex QUBO (maximization of a concave quadratic function). It is then possible to use algorithms that solve convex relaxations at the various steps of a branch and bound method to solve the problem to optimality. Off-the-shelf mixed integer quadratic programming solvers such as CPLEX and Gurobi now also allow the user to automatically convexify such instances. An upper bound on the optimal value of the convex QUBO in (7.3) is obtained by solving the convex relaxation:

$$\bar{Z}(\mathbf{u}; \mathbf{Q}, \mathbf{c}) = \max_{\mathbf{x} \in [0,1]^n} \mathbf{x}^T (\mathbf{Q} - \mathrm{Diag}(\mathbf{u}))\mathbf{x} + (\mathbf{c} + \mathbf{u})^T \mathbf{x}, \tag{7.4}$$

where:

$$Z(\mathbf{Q}, \mathbf{c}) = Z(\mathbf{u}; \mathbf{Q}, \mathbf{c}) \leq \bar{Z}(\mathbf{u}; \mathbf{Q}, \mathbf{c}).$$

An “optimal" choice of the vector $\mathbf{u}$ is one that makes the upper bound (7.4) obtained from the convex relaxation as small as possible:

$$\mathbf{u}_{opt} = \mathrm{argmin} \left\{ \bar{Z}(\mathbf{u}; \mathbf{Q}, \mathbf{c}) \,\middle|\, \mathrm{Diag}(\mathbf{u}) - \mathbf{Q} \succeq 0 \right\}. \tag{7.5}$$

Using standard duality arguments, it can be shown that (see [5, 7]) $\mathbf{u}_{opt}$ is the optimal vector $\mathbf{u}$ of the following SDP:

$$\begin{aligned} \bar{Z}(\mathbf{u}_{opt}; \mathbf{Q}, \mathbf{c}) = & \min_{r \in \mathbb{R}, u \in \mathbb{R}^n} \; r \\ & \text{s.t.} \begin{bmatrix} r & (-\mathbf{c} - \mathbf{u})^T/2 \\ (-\mathbf{c} - \mathbf{u})/2 & \mathrm{Diag}(\mathbf{u}) - \mathbf{Q} \end{bmatrix} \succeq 0, \end{aligned} \tag{7.6}$$

where the positive semidefiniteness constraint is for a matrix of size $(n+1) \times (n+1)$. The semidefinite program (7.6) is the dual to the classic semidefinite programming relaxation of the QUBO problem:

$$\begin{aligned} \bar{Z}(\mathbf{u}_{opt}; \mathbf{Q}, \mathbf{c}) = \max_{\mathbf{x} \in \mathbb{R}^n, X \in \mathcal{S}_n} \quad & \mathbf{Q}.\mathbf{X} + \mathbf{c}^T \mathbf{x} \\ \text{s.t.} \quad & X_{ii} = x_i, \quad i = 1, \ldots, n, \\ & \begin{bmatrix} 1 & \mathbf{x}^T \\ \mathbf{x} & \mathbf{X} \end{bmatrix} \succeq 0. \end{aligned} \tag{7.7}$$

where $\mathbf{u}_{opt}$ is the optimal dual vector to the first set of constraints in (7.7). For QUBO, [5] proposed the use of this semidefinite program as a preprocessing phase to convexify the objective before applying an exact branch and bound method to solve the QUBO. In the numerical experiments in [5], the relative gap between the optimum value of the QUBO and the continuous SDP based relaxation is shown to be about half the relative gap between the optimum value and a simpler relaxation using eigenvalues of the matrix $\mathbf{Q}$ (see [13] and the discussion in the next section of this chapter). Solving the QUBO with the CPLEX solver was also shown to be faster using such a SDP based preprocessing step as compared to the eigenvalue based preprocessing step.

7.3.2 *QCR Methods for Random QUBO*

In this section, we discuss the application of the QCR method to random QUBO where the matrix $\mathbf{Q}$ is fixed but the vector $\tilde{\mathbf{c}}$ is random. Specifically, we are interested in the random QUBO:

$$Z(\mathbf{Q}, \tilde{\mathbf{c}}) = \max_{\mathbf{x} \in \{0,1\}^n} \mathbf{x}^T \mathbf{Q} \mathbf{x} + \tilde{\mathbf{c}}^T \mathbf{x}, \tag{7.8}$$

We discuss a variety of preprocessing approaches to tackle this problem. In each of these approaches, we are interested in finding a preprocessing vector $\mathbf{u}$ such that $\text{Diag}(\mathbf{u}) - \mathbf{Q} \succeq 0$. This helps convexify the objective function and make the optimal value of the convex relaxation as tight as possible. We then solve the problem using convex MIQP solution techniques implemented in solvers such as CPLEX. The convex random QUBO is given as:

$$Z(\mathbf{u}; \mathbf{Q}, \tilde{\mathbf{c}}) = \max_{\mathbf{x} \in \{0,1\}^n} \mathbf{x}^T (\mathbf{Q} - \text{Diag}(\mathbf{u}))\mathbf{x} + (\tilde{\mathbf{c}} + \mathbf{u})^T \mathbf{x}. \tag{7.9}$$

Let us discuss four different preprocessing approaches to solve the random instances of QUBO.

(a) Eigenvalue Based Method The simplest possible choice of the preprocessing vector $\mathbf{u}$ is to use $\mathbf{u}_{eig} = \lambda_{max}(\mathbf{Q})\mathbf{e}$, where $\lambda_{max}(\mathbf{Q})$ is the largest eigenvalue of the matrix $\mathbf{Q}$ and $\mathbf{e}$ is an n dimensional vector with all entries equal to 1. Clearly $\mathbf{Q} - \text{Diag}(\mathbf{u}_{eig})$ is negative semidefinite and the objective function is concave with respect to the decision variable $\mathbf{x}$. Such an eigenvalue based preprocessing method was first proposed in [13].

(b) Joint Distribution Based Method For each realization of the vector $\tilde{\mathbf{c}} = \mathbf{c}$, the "optimal" choice of the vector $\mathbf{u}$ is obtained by solving (7.5) based on the discussion on the QCR method. This is equivalent to solving the SDP problem (7.6) for every $\mathbf{c}$. Namely for any realization $\tilde{\mathbf{c}} = \mathbf{c}$, the SDP is given as:

$$\begin{aligned} \bar{Z}(\mathbf{u}_{opt}(\mathbf{c}); \mathbf{Q}, \mathbf{c}) = & \min_{r \in \mathbb{R}, u \in \mathbb{R}^n} \; r \\ & \text{s.t.} \begin{bmatrix} r & (-\mathbf{c} - \mathbf{u})^T/2 \\ (-\mathbf{c} - \mathbf{u})/2 & \text{Diag}(\mathbf{u}) - \mathbf{Q} \end{bmatrix} \succeq 0, \end{aligned} \tag{7.10}$$

where $\mathbf{u}_{opt}(\mathbf{c})$ is the optimal $\mathbf{u}$ decision vector for the SDP viewed as a function of the realization $\mathbf{c}$. The main challenge in this approach is that for every realization, we need to solve a SDP before solving the QUBO. Since the number of realizations of the joint distribution can be very large, even infinite, this is numerically very challenging to implement in practice.

(c) Mean Based Method In this method, we choose a common preprocessing vector $\mathbf{u}_\mu$ by simply using the mean of the random vector. Namely, we solve a single SDP:

$$\begin{aligned} \bar{Z}(\mathbf{u}_\mu; \mathbf{Q}, \boldsymbol{\mu}) = & \min_{r \in \mathbb{R}, u \in \mathbb{R}^n} \; r \\ & \text{s.t.} \begin{bmatrix} r & (-\boldsymbol{\mu} - \mathbf{u})^T/2 \\ (-\boldsymbol{\mu} - \mathbf{u})/2 & \text{Diag}(\mathbf{u}) - \mathbf{Q} \end{bmatrix} \succeq 0, \end{aligned} \tag{7.11}$$

where $\mathbf{u}_\mu$ is the optimal $\mathbf{u}$ vector in (7.11). Here we need to solve a single SDP to obtain $\mathbf{u}_\mu$.

(d) Marginal Based Method In this method, we use the formulation for the bound ϕ from the earlier section to develop a common preprocessing vector. Consider the simple case where only the mean and standard deviation of each $\tilde{c}_i$ is known and given as μ_i and σ_i. Consider the tight upper bound:

$$\phi = \sup_{\theta \in \Theta} E_\theta \left[Z(\mathbf{Q}, \tilde{\mathbf{c}}) \right], \tag{7.12}$$

where the set of distributions Θ describes all random vectors $\tilde{\mathbf{c}}$ with mean μ_i and standard deviation σ_i for each $\tilde{c}_i$. The tightest upper bound is then given as:

$$\phi = \min_{d \in \mathbb{R}^n} \left\{ Z(\mathbf{Q}, \mathbf{d}) + \sum_i \frac{1}{2}\left(\mu_i - d_i + \sqrt{(\mu_i - d_i)^2 + \sigma_i^2}\right) \right\},$$

where we use the fact that the maximum value of $E[\tilde{c}_i - d_i]^+$ given only the mean and standard deviation of $\tilde{c}_i$ is equal to $\frac{1}{2}(\mu_i - d_i + \sqrt{(\mu_i - d_i)^2 + \sigma_i^2})$ (this univariate bound comes from the Cauchy-Schwarz inequality and is tight; see [19]). Let us now apply the QCR method to the inner QUBO problem. We then have:

$$\phi = \min_{d \in \mathbb{R}^n} \left\{ Z(\mathbf{u}; \mathbf{Q}, \mathbf{d}) + \sum_i \frac{1}{2}\left(\mu_i - d_i + \sqrt{(\mu_i - d_i)^2 + \sigma_i^2}\right) \right\}. \tag{7.13}$$

An upper bound on ϕ is then given by:

$$\phi \leq \min_{d \in \mathbb{R}^n} \left\{ \bar{Z}(\mathbf{u}; \mathbf{Q}, \mathbf{d}) + \sum_i \frac{1}{2}\left(\mu_i - d_i + \sqrt{(\mu_i - d_i)^2 + \sigma_i^2}\right) \right\},$$

where $\bar{Z}(\mathbf{u}; \mathbf{Q}, \mathbf{d})$ is the optimal value using the convex relaxation in (7.4). The "optimal" choice of the vector $\mathbf{u}$ is to then minimize the upper bound obtained from using the convex relaxation satisfying the constraint $\text{Diag}(\mathbf{u}) - \mathbf{Q} \succeq 0$. Thus we need to solve:

$$\mathbf{u}_{\mu,\sigma} = \text{argmin}_{\text{Diag}(u) - \mathbf{Q} \succeq 0} \min_{d \in \mathbb{R}^n} \left\{ \bar{Z}(\mathbf{u}; \mathbf{Q}, \mathbf{d}) + \sum_i \frac{1}{2}\left(\mu_i - d_i + \sqrt{(\mu_i - d_i)^2 + \sigma_i^2}\right) \right\},$$

where $\boldsymbol{\sigma} = (\sigma_1, \ldots, \sigma_n)$. Now again using standard conic duality, it is easy to see that $\mathbf{u}_{\mu,\sigma}$ is obtained by solving the SDP:

$$\begin{aligned} \min_{d,r,u} \; & r + \sum_i \frac{1}{2}\left(\mu_i - d_i + \sqrt{(\mu_i - d_i)^2 + \sigma_i^2}\right) \\ \text{s.t.} \; & \begin{bmatrix} r & -(\mathbf{d} + \mathbf{u})^T/2 \\ -(\mathbf{d} + \mathbf{u})/2 & \text{diag}(\mathbf{u}) - \mathbf{Q} \end{bmatrix} \succeq 0, \end{aligned} \tag{7.14}$$

where $\mathbf{u}_{\mu,\sigma}$ is the optimal $\mathbf{u}$ vector in (7.14). This can be solved as a linear SDP. The key difference in this formulation is the presence of the penalty function in the objective function which affects the choice of the optimal vector $\mathbf{u}$ unlike the deterministic QCR formulation.

7.3.3 Numerical Experiments

In this section, we discuss the application of the different QCR methods in solving a set of instances of random QUBO. The computational studies were implemented in Matlab R2012a on an Intel Core 2 Duo CPU (2.8 GHz) laptop with 4 GB of RAM. The SDP problems were solved with CVX [10, 11] and SDPT3 [29, 30] and the QUBOs were solved with CPLEX 12.6 using the Matlab interface. In our computational experiments, we set $T = 10$ minutes as the maximum time to solve any instance of QUBO. Define the value $\text{obj}(\mathbf{Q}, \mathbf{c})$ as follows:

$$\text{obj}(\mathbf{Q}, \mathbf{c}) = \begin{cases} Z(\mathbf{Q}, \mathbf{c}), & \text{if the QUBO is solvable within } T \text{ minutes,} \\ \text{Best lower bound,} & \text{otherwise,} \end{cases}$$

where the best lower bound is derived from the best feasible solution found in T minutes. We define $\text{gap}(\mathbf{u}; \mathbf{Q}, \mathbf{c})$ as the relative difference between the convex relaxation of the objective function for a given preprocessing vector $\mathbf{u}$ and the value of $\text{obj}(\mathbf{Q}, \mathbf{c})$:

$$\text{gap}(\mathbf{u}; \mathbf{Q}, \mathbf{c}) = \frac{\bar{Z}(\mathbf{u}; \mathbf{c}, \mathbf{Q}) - \text{obj}(\mathbf{Q}, \mathbf{c})}{\text{obj}(\mathbf{Q}, \mathbf{c})}.$$

Since the running time of the branch-and-bound method to solve the QUBO depends on the strength of its convex relaxation, we say that a vector $\mathbf{u}$ is preferable to $\mathbf{u}'$ for the kth instance if $\text{gap}(\mathbf{u}; \mathbf{Q}, \mathbf{c}) < \text{gap}(\mathbf{u}'; \mathbf{Q}, \mathbf{c})$.

7.3.3.1 Random Instances

We use the set of randomly generated instances as in [5] and [22]. The parameters are chosen as follows:

1. The linear coefficients c_i are chosen uniformly and independently in the range $[-100, 100]$.
2. The diagonal entries of $\mathbf{Q} \in \mathcal{S}_n$ are all 0, and the off-diagonal coefficients of the symmetric matrix $\mathbf{Q}$ are in the range $[-50, 50]$.
3. The matrix $\mathbf{Q}$ has density d. The density refers to the probability that a nonzero will occur in any off-diagonal entry.

In the numerical experiments, we use 100 samples from the joint distribution. We use the sample mean and the sample standard deviation to compute the preprocessing parameters $\mathbf{u}_{\mu}$ and $\mathbf{u}_{\mu,\sigma}$ by solving the respective SDPs. The results are listed in Tables 7.1 and 7.2. In the tables, we report the following values for the four different choices of preprocessing vectors $\mathbf{u} = \mathbf{u}_{eig}$, $\mathbf{u} = \mathbf{u}_{opt}(\mathbf{c})$, $\mathbf{u} = \mathbf{u}_{\mu}$ and $\mathbf{u} = \mathbf{u}_{\mu,\sigma}$:

Table 7.1 Gap and CPU time for different parameters $\mathbf{u}$

		$\mathbf{u} = \mathbf{u}_{eig}$				$\mathbf{u} = \mathbf{u}_{opt}(\mathbf{c})$			
n	d	Gap	t_**u**	t_01QP	Solved	Gap	t_**u**	t_01QP	Solved
50	0.4	13.8	0.02	106.0	100	4.9	45.27	23.0	100
50	0.6	15.1	0.01	109.0	100	6.7	45.39	28.2	100
50	1.0	12.5	0.02	85.9	100	6.6	45.84	30.9	100
60	0.2	17.2	0.01	1022.4	100	5.8	53.90	43.5	100
60	0.4	14.0	0.01	1136.3	100	4.4	51.96	50.8	100
70	0.3	18.5	0.02	1249.4	100	8.2	58.50	280.0	100
80	0.2	18.4	0.03	16569.2(218.0)	76	7.8	70.68	450.9	100
90	0.6	18.3	0.03	13764.3(327.7)	42	9.1	83.60	7040.1	100
100	0.1	16.5	0.02	4761.2(297.6)	16	3.5	100.83	178.2	100
120	0.2	21.1	0.05	**	0	10.4	129.52	5770.7(360.7)	16

** denotes the instances were not solvable within the time limit

Table 7.2 Gap and CPU time for different parameters $\mathbf{u}$

		$\mathbf{u} = \mathbf{u}_{\mu}$				$\mathbf{u} = \mathbf{u}_{\mu,\sigma}$			
n	d	Gap	t_**u**	t_01QP	Solved	Gap	t_**u**	t_01QP	Solved
50	0.4	8.9	0.48	20.5	100	7.3	1.52	18.0	100
50	0.6	10.0	0.46	24.3	100	9.0	1.55	19.4	100
50	1.0	8.8	0.45	26.0	100	8.5	1.55	24.1	100
60	0.2	10.7	0.63	53.6	100	8.7	1.82	35.8	100
60	0.4	7.1	0.50	42.9	100	6.1	1.78	37.1	100
70	0.3	12.2	0.69	292.5	100	10.9	2.20	261.5	100
80	0.2	12.4	0.67	545.1	100	10.8	2.40	435.3	100
90	0.6	11.4	0.83	7095.1	100	11.0	3.26	6783.9	100
100	0.1	7.9	1.21	317.1	100	5.9	4.05	210.7	100
120	0.2	13.6	1.35	3610.4(361.0)	10	13.0	4.03	5500.0(343.8)	16

1. The average gap over the 100 instances.
2. The CPU time taken to compute the preprocessing parameter **u** denoted by "t_**u**". For the distribution method using the samples, "t_**u**" is the total CPU time taken to solve the 100 SDPs. For the mean based and marginal based method (using only the mean and standard deviation), "t_**u**" is the CPU time taken to solve a single SDP each. For the eigenvalue based method, "t_**u**" is the CPU time taken to compute the largest eigenvalue of $\mathbf{Q}$ which is almost immediate.
3. The total CPU time taken to compute all the convex QUBOs is denoted by "t_01QP". If we solve every instance within 10 minutes, we report the total CPU time. If there are $m < 100$ instances that are solvable within 10 minutes each, we report the total CPU time to solve these m instances and report the average time for the m solved instances in the parentheses.
4. The number of instances (out of 100) which are solved within 10 minutes is denoted by "solved".

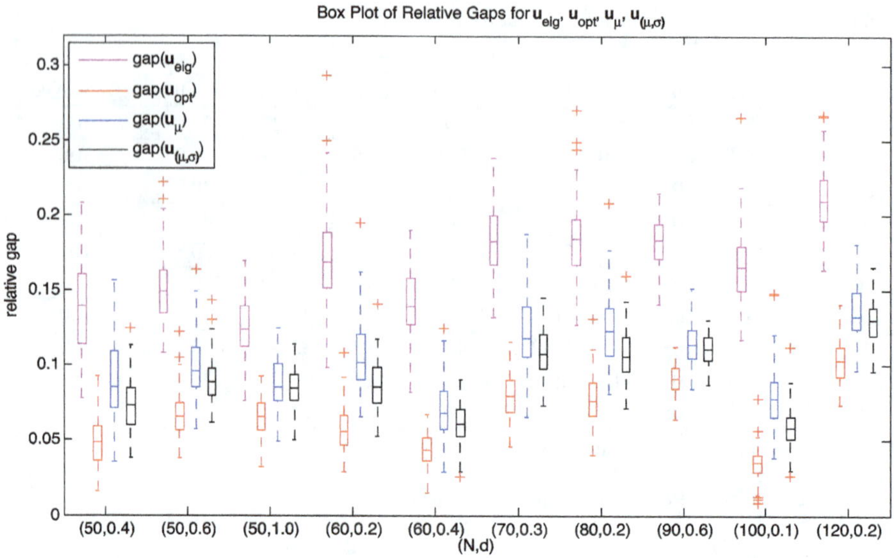

Fig. 7.1 Boxplot of the relative gaps for all the 100 scenarios

In addition to the average gap, we plot the distributions of the relative gaps for the 100 scenarios using the boxplot in Fig. 7.1. From Tables 7.1 and 7.2, we observe that the average relative gap of using $\mathbf{u}_{\mu,\sigma}$ is always smaller than using $\mathbf{u}_{eig}$ and $\mathbf{u}_{\mu}$. In addition to the average value, from Fig. 7.1 we observe that the relative gap of using $\mathbf{u}_{\mu,\sigma}$ has a smaller sample minimum, lower quartile (25th percentile), median, upper quartile (75th percentile), and sample maximum than using $\mathbf{u}_{eig}$ and $\mathbf{u}_{\mu}$. The relative gap using the exact samples from the distribution is the smallest as should be expected. Hence, in terms of the relative gap between the optimal value of the QUBO and its convex relaxation, parameter $\mathbf{u}_{\mu,\sigma}$ is better than $\mathbf{u}_{eig}$ and $\mathbf{u}_{\mu}$ and closest to the distribution based method using samples.

We also plot the CPU time taken to solve the QUBO for every scenario $\mathbf{c}$. In Fig. 7.2, "t_01QP($\mathbf{u}$)", denotes the CPU time taken to solve the convex QUBO with the preprocessing parameter $\mathbf{u}$ for scenario $\mathbf{c}$. Since the CPU time taken to solve the QUBO by using $\mathbf{u}_{eig}$ is much larger than the other three methods, we exclude the eigenvalue based method from consideration. Since $\mathbf{u}_{\mu}$ and $\mathbf{u}_{\mu,\sigma}$ are common preprocessing vectors for all the 100 sampled instances, and we can compute them quickly by solving a single SDP problem, the CPU time of getting $\mathbf{u}_{\mu}$ and $\mathbf{u}_{\mu,\sigma}$ turns out to be negligible in Fig. 7.2. However, to use $\mathbf{u}_{opt}(\mathbf{c})$ we must solve an SDP problem for every instance. Hence in the plot of the CPU time to solve the QUBO, the time $t\_\mathbf{u}_{opt}$ taken to compute $\mathbf{u}_{opt}$ is added.

From Fig. 7.2, we see that for small size instances (subfigure (a)) and medium size instances (subfigure (b)), the marginal based method (using mean and standard deviation) is better than the distribution based method using samples and the mean based method. The CPU time of using $\mathbf{u}_{\mu,\sigma}$ to solve the QUBO have the smallest

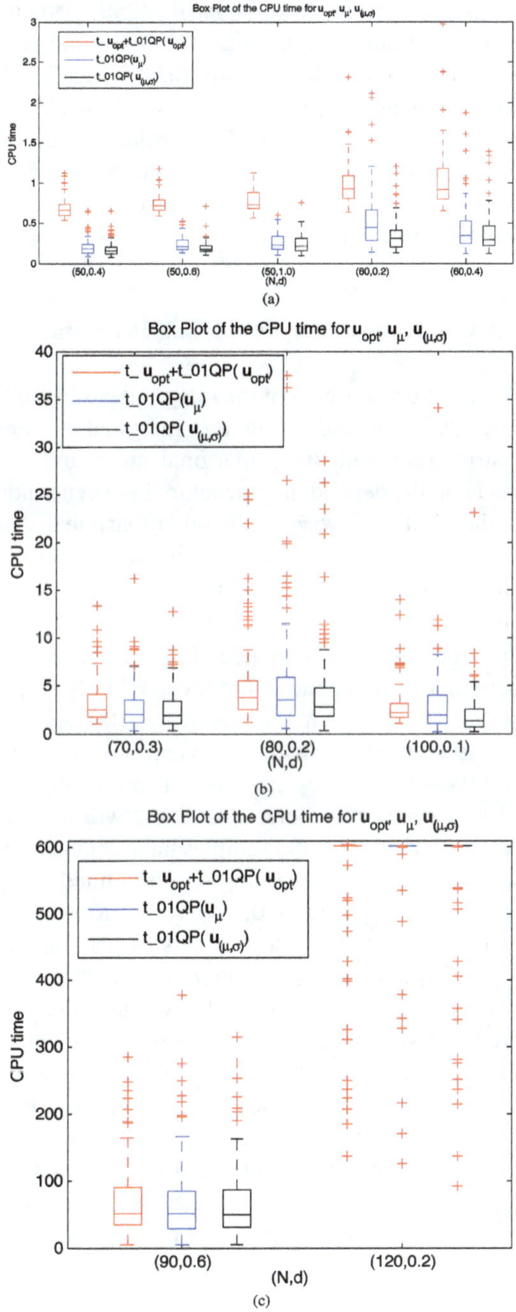

Fig. 7.2 Boxplot of the CPU time: (for the instances which can not be solved in 10 minutes, we just plot its CPU time as 600 seconds in the figure). (**a**) Small size instances. (**b**) Medium size instances. (**c**) Hard to solve instances

sample minimum, lower quartile (25th percentile), median, upper quartile (75th percentile), and sample maximum. For the difficult instances (subfigure (c)), the three methods look more similar in Fig. 7.2. From Tables 7.1 and 7.2, we can see that using $\mathbf{u}_{\mu,\sigma}$, we need the smallest CPU time to solve all the 100 instances when $n = 90, d = 0.6$. For the largest and most difficult set of instances with $n = 120, d = 0.2$, very few instances can be solved to optimality in 10 minutes. By using $\mathbf{u}_{\mu,\sigma}$, we solve 16 instances to optimality which is the same as using the sample information.

7.3.3.2 Robustness Tests Using Permutations

We test the robustness of the marginal based method using permutation experiments. The QCR method using the marginals is developed for the Fréchet class of distributions with fixed marginal mean and variance. However no assumption is made on the dependency structure between random variables. To test the robustness of the solutions, we generate other feasible distributions in this set by permuting the individual components of the randomly generated samples in the following manner. Given the sample data vectors $\mathbf{c}^{(1)}, \ldots, \mathbf{c}^{(100)}$, we compute the sample mean $\boldsymbol{\mu}$ and the sample standard deviation $\boldsymbol{\sigma}$. For $i = 1, \ldots, n$, we randomly permute the ith component sequence of the vectors $c_i^{(1)}, c_i^{(2)}, \ldots, c_i^{(100)}$. By performing this permutation independently for each $i = 1, \ldots, n$, we generate a new set of samples $\{\mathbf{c}^{(k)}, k = 1, \ldots, K\}$ where $K = 100$; see [18] for a similar set of experiments in the context of stochastic knapsack problems. Note that the sample mean $\boldsymbol{\mu}$ and the standard deviation $\boldsymbol{\sigma}$ will not change after these permutations. Hence the preprocessing parameter $\mathbf{u}_{\mu}$ and $\mathbf{u}_{\mu,\sigma}$ will not change. However $\mathbf{u}_{opt}(\mathbf{c})$ will almost surely change since the samples have changed. As a control, we also use the average sample based preprocessing vector defined as $\mathbf{u}_{ave} := \sum_{k=1}^{100} \mathbf{u}_{opt}(\mathbf{c}^{(k)})/100$. Since $\mathbf{Q} - diag(\mathbf{u}_{opt}(\mathbf{c}^{(k)}) \succeq 0, k = 1, \ldots, K$, we have $\mathbf{Q} - diag(\mathbf{u}_{ave}) \succeq 0$.

For the tests, we use two sets of parameters ($(n, d) = (50, 0.6)$ and $(n, d) = (70, 0.3)$) and perform numerical tests for the samples after the random permutations. For each set of data, we test the results across 15 permutations. The preprocessing vectors $\mathbf{u}_{ave}$, $\mathbf{u}_{\mu}$ and $\mathbf{u}_{\mu,\sigma}$ are computed only once and hence the CPU time of computing these preprocessing parameters can be ignored. The numerical results are shown in Tables 7.3 and 7.4. From Tables 7.3 and 7.4, we see that by using $\mathbf{u}_{\mu,\sigma}$ the average gap is smaller than using $\mathbf{u}_{\mu}$ and $\mathbf{u}_{ave}$. Moreover the total CPU time taken to solve the QUBO is always the smallest for all the permutations by using $\mathbf{u}_{\mu,\sigma}$. This seems to indicate that exploiting the marginal distribution information with the QCR method provides robustness.

Table 7.3 Gap and CPU time with 15 permutations: $n = 50$, $d = 0.6$

	$\mathbf{u}_{opt}(\mathbf{c}^{(k)}), k = 1, \ldots, K$				$\mathbf{u}_{ave}$			$\mathbf{u}_{\mu}$			$\mathbf{u}_{\mu,\sigma}$		
No.	Gap	t_**u**	t_01QP	Solved	Gap	t_01QP	Solved	*Gap*	t_01QP	Solved	Gap	t_01QP	Solved
1	6.6	47.78	22.6	100	10.5	22.1	100	10.0	19.7	100	8.9	15.2	100
2	6.5	55.79	27.9	100	10.4	25.3	100	9.9	23.9	100	8.8	18.7	100
3	6.7	56.25	29.6	100	10.6	27.7	100	10.1	25.8	100	9.0	19.8	100
4	6.7	58.73	27.7	100	10.7	26.5	100	10.1	24.5	100	9.0	19.1	100
5	6.5	53.95	27.8	100	10.4	26.3	100	9.9	24.2	100	8.8	19.0	100
6	6.5	54.24	27.3	100	10.4	25.6	100	9.9	23.8	100	8.8	18.8	100
7	6.7	56.78	28.4	100	10.9	27.8	100	10.4	25.7	100	9.1	19.5	100
8	6.6	54.65	27.2	100	10.5	26.0	100	9.9	23.8	100	8.9	18.7	100
9	6.5	55.60	26.7	100	10.4	24.7	100	9.9	23.4	100	8.8	18.3	100
10	6.5	45.54	26.7	100	10.4	25.0	100	9.9	23.1	100	8.8	18.0	100
11	6.6	52.42	27.4	100	10.5	26.6	100	10.0	24.7	100	8.9	18.9	100
12	6.5	52.29	27.4	100	10.4	26.2	100	9.8	23.5	100	8.8	18.8	100
13	6.7	54.48	28.3	100	10.6	26.8	100	10.1	24.9	100	9.0	19.4	100
14	6.6	53.38	26.9	100	10.5	25.0	100	10.0	22.8	100	8.9	18.3	100
15	6.4	56.66	24.5	100	10.3	22.8	100	9.8	21.5	100	8.7	17.0	100

Table 7.4 Gap and CPU time with 15 permutations: $n = 70$, $d = 0.3$

	$\mathbf{u}_{opt}(\mathbf{c}^{(k)}), k = 1, \ldots, K$				$\mathbf{u}_{ave}$			$\mathbf{u}_{\mu}$			$\mathbf{u}_{\mu,\sigma}$		
No.	Gap	t_**u**	t_01QP	Solved	Gap	t_01QP	Solved	Gap	t_01QP	Solved	Gap	t_01QP	Solved
1	8.3	62.42	303.7	100	13.1	500.2	100	12.3	322.9	100	11.0	278.3	100
2	8.3	63.63	275.1	100	13.1	498.1	100	12.2	300.0	100	11.0	267.7	100
3	8.5	62.99	340.7	100	13.4	580.9	100	12.6	368.2	100	11.3	328.8	100
4	8.3	62.75	275.3	100	13.1	481.2	100	12.2	293.1	100	11.0	267.7	100
5	8.2	62.87	291.5	100	13.0	489.0	100	12.2	301.8	100	10.9	267.4	100
6	8.4	63.07	339.4	100	13.1	600.4	100	12.2	350.8	100	11.0	321.6	100
7	8.2	62.71	278.8	100	13.1	453.9	100	12.3	288.4	100	11.0	265.8	100
8	8.2	62.96	271.5	100	13.0	471.8	100	12.2	286.9	100	10.9	258.2	100
9	8.2	62.95	274.4	100	13.1	493.4	100	12.2	307.3	100	10.9	264.0	100
10	8.1	62.76	246.7	100	12.9	423.5	100	12.0	264.0	100	10.8	237.4	100
11	8.3	63.12	276.3	100	13.2	471.6	100	12.4	297.0	100	11.0	261.6	100
12	8.2	63.09	279.6	100	13.1	501.2	100	12.2	302.1	100	10.9	266.1	100
13	8.2	63.15	282.8	100	13.0	485.3	100	12.2	302.2	100	10.9	266.5	100
14	8.5	63.37	330.5	100	13.4	579.2	100	12.6	350.0	100	11.2	311.7	100
15	8.2	66.04	293.7	100	13.0	497.6	100	12.2	311.1	100	10.9	274.0	100

7.4 The Sherrington-Kirkpatrick Model

In this section, we discuss the Sherrington-Kirkpatrick model [25], the study of which was initiated in the statistical physics community for the study of spin glasses and has recently found applications in the machine learning and computer science community. Consider an optimization problem of the form:

$$\text{(SK)}\ Z_n^{SK}(\mathbf{Q}) = \max_{\mathbf{x}\in\{-1,1\}^n} \mathbf{x}^T\mathbf{Q}\mathbf{x}, \tag{7.15}$$

where $\mathbf{Q} \in \mathbb{R}^{n\times n}$ is a $n \times n$ real matrix, not necessarily symmetric. In the Sherrington-Kirkpatrick model [25], the entries $\tilde{Q}_{ij}$ for all i, j of the random matrix $\tilde{Q}$ are standard normal random variables with mean 0 and variance 1, independently distributed. A characterization of the asymptotic expected optimal value for this model was first conjectured in [23] and later proved formally in [27]; see the books [21, 28] for a detailed exposition on the result, its extensions and proof. This characterization is given by:

$$\lim_{n\to\infty} \frac{1}{\sqrt{2}n^{3/2}} E\left[Z_n^{SK}(\tilde{\mathbf{Q}})\right] = \text{constant},$$

where the constant is approximately 0.763166 and obtained through a complicated expression known as the Parisi formula. The proof of this result requires detailed mathematical analysis and is beyond the scope of this chapter. We only provide some brief ideas below.

For any fixed solution $\mathbf{x} \in \{-1, 1\}^n$, $\mathbf{x}^T\tilde{\mathbf{Q}}\mathbf{x}$ is a normal random variable with the mean and variance given by:

$$E\left[\mathbf{x}^T\tilde{\mathbf{Q}}\mathbf{x}\right] = 0 \text{ and Variance}\left[\mathbf{x}^T\tilde{\mathbf{Q}}\mathbf{x}\right] = n^2,$$

where for any two solutions $\mathbf{x}, \mathbf{y} \in \{-1, 1\}^n$, we have:

$$E\left[(\mathbf{x}^T\tilde{\mathbf{Q}}\mathbf{x})(\mathbf{y}^T\tilde{\mathbf{Q}}\mathbf{y})\right] = \sum_{i=1}^n\sum_{j=1}^n x_ix_jy_iy_j = \left(\sum_{i=1}^n x_iy_i\right)^2.$$

Intuitively, we would expect many of the pairs $\mathbf{x}, \mathbf{y} \in \{-1, 1\}^n$ in high dimension to be close to orthogonal and hence the objective value of the solutions will be almost uncorrelated in many cases with $\sum_i x_iy_i \approx 0$. Given a collection of k independent and identically distributed Gaussian random variables $\tilde{c}_i$ with mean 0 and variance σ^2, from extreme value theory, the expected maximum is known to satisfy:

$$\lim_{k\to\infty} \frac{E\left[\max_{i=1,\ldots,k} \tilde{c}_i\right]}{\sqrt{2\log(k)}} = \sigma,$$

Since $Z_n^{SK}(\tilde{\mathbf{Q}})$ is the maximum of 2^n Gaussian random variables, many of which will be close to uncorrelated, plugging in $k = 2^n$ and $\sigma^2 = n^2$, we would expect $E\left[Z_n^{SK}(\tilde{\mathbf{Q}})\right]$ to roughly scale as $n^{3/2}$. The precise behavior is much more challenging to characterize. To derive the actual expression of the expected optimal value, it is useful to study the related partition function which is defined for any $\alpha > 0$ as follows:

$$P_n^{SK}(\alpha, \mathbf{Q}) = \sum_{\mathbf{x}\in\{-1,1\}^n} \exp\left(\alpha \mathbf{x}^T \mathbf{Q}\mathbf{x}\right).$$

It is straightforward to see that:

$$\exp(\alpha Z_n^{SK}(\mathbf{Q})) \le P_n^{SK}(\alpha, \mathbf{Q}) \le 2^n \exp\left(\alpha Z_n^{SK}(\mathbf{Q})\right),$$

where the first inequality is from choosing only the maximum value in the sum defining the partition function and the second inequality is from using the maximum value as an upper bound for each of the 2^n terms in the sum defining the partition function. The expected optimal value and the expected partition function are then related as follows:

$$\begin{aligned}
\lim_{n\to\infty} \frac{1}{\sqrt{2}n^{3/2}} E\left[Z_n^{SK}(\tilde{\mathbf{Q}})\right] &\le \lim_{\alpha\to\infty, n\to\infty} \frac{1}{\sqrt{2}\alpha n^{3/2}} E\left[\log\left(P_n^{SK}(\alpha, \tilde{\mathbf{Q}})\right)\right] \\
&\le \lim_{\alpha\to\infty, n\to\infty} \frac{1}{\sqrt{2}\alpha n^{1/2}} + \frac{1}{\sqrt{2}n^{3/2}} E\left[Z_n^{SK}(\tilde{\mathbf{Q}})\right] \\
&= \lim_{n\to\infty} \frac{1}{\sqrt{2}n^{3/2}} E\left[Z_n^{SK}(\tilde{\mathbf{Q}})\right],
\end{aligned}$$

which implies:

$$\lim_{n\to\infty} \frac{1}{\sqrt{2}n^{3/2}} E\left[Z_n^{SK}(\tilde{\mathbf{Q}})\right] = \lim_{\alpha\to\infty, n\to\infty} \frac{1}{\sqrt{2}\alpha n^{3/2}} E\left[\log\left(P_n^{SK}(\alpha, \tilde{\mathbf{Q}})\right)\right].$$

Much of the analysis is then to prove that the limit on the right hand side is well defined and given by Parisi formula [23]. Recent work by Montanari [17] builds on this observation to show that it is possible to obtain an algorithm that for any $\epsilon > 0$, gives a vector $\mathbf{x}^* \in \{-1, 1\}^n$ such that $\mathbf{x}^{*T}\tilde{\mathbf{Q}}\mathbf{x}^*$ is at least $(1 - \epsilon)$ of the optimal value with probability converging to one as $n \to \infty$. Such a results indicates that on average with appropriate assumptions on the randomness, the optimization problem might be easier to solve in comparison to the worst-case.

Acknowledgments We would like to thanks Dongjian Shi and Toh Kim Chuan for their significant contributions to the work discussed in this chapter. This work was done as part of the PhD thesis titled "Regret models and preprocessing techniques for combinatorial optimization under uncertainty" by Dongjian Shi at the National University of Singapore. The author of this chapter and Toh Kim Chuan were his doctoral advisors.

References

1. D. Aldous, The $\zeta(2)$ limit in the random assignment problem. Random Struct. Algorithms **18**, 381–418 (2001)
2. J.H. Beardwood, J. Halton, J.M. Hammersley, The shortest path through many points. Math. Proc. Cambridge Philos. Soc. **55**, 299–327 (1959)
3. D. Bertsimas, K. Natarajan, C.P. Teo, Probabilistic combinatorial optimization: moments, semidefinite programming, and asymptotic bounds. SIAM J. Optim. **15**, 185–209 (2004)
4. D. Bertsimas, K. Natarajan, C.P. Teo, Persistence in discrete optimization under data uncertainty. Math. Program. **108**, 251–274 (2006)
5. A. Billionnet, S. Elloumi, Using mixed integer quadratic programming solver for the unconstrained quadratic 0-1 problem. Math. Program. A **109**, 55–68 (2007)
6. A. Billionnet, S. Elloumi, A. Lambert, Extending the QCR method to general mixed-integer programs. Math. Program. **131**, 381–401 (2002)
7. A. Billionnet, S. Elloumi, M.-C. Plateau, Improving the performance of standard solvers for quadratic 0–1 programs by a tight convex reformulation: the QCR method. Discrete Appl. Math. **157**, 1185–1197 (2009)
8. R.E. Burkard, U. Fincke, Probabilistic analysis of some combinatorial optimization problems. Discrete Appl. Math. **12**, 21–29 (1985)
9. A.M. Frieze, On the value of a random minimum spanning tree problem. Discrete Appl. Math. **10**, 47–56 (1985)
10. M. Grant, S. Boyd, Graph implementations for nonsmooth convex programs, in *Recent Advances in Learning and Control (A Tribute to M. Vidyasagar)*, ed. by V. Blondel, S. Boyd, H. Kimura. Lecture Notes in Control and Information Sciences, pp. 95–110. Springer, Berlin (2008)
11. M. Grant, S. Boyd, CVX: matlab software for disciplined convex programming, version 2.0. beta (2013)
12. M. Grötschel, L. Lovász, A.J. Schrijver, *Geometric Algorithms and Combinatorial Optimization* (Wiley, New York, 1988)
13. P.L. Hammer, A.A. Rubin, Some remarks on quadratic programming with 0-1 variables. RAIRO-Oper. Research **3**, 67–79 (1970)
14. W.K.K. Haneveld, Robustness against dependence in PERT: an application of duality and distributions with known marginals. Math. Program. Study **27**, 153–182 (1986)
15. F. Körner, A tight bound for the Boolean quadratic optimization problem and its use in a branch and bound algorithm. Optimization **19**, 711–721 (1988)
16. I. Meilijson, A. Nadas, Convex majorization with an application to the length of critical path. J. Appl. Probab. **16**, 671–677 (1979)
17. A. Montanari, Optimization of the Sherrington-Kirkpatrick Hamiltonian. SIAM J. Comput. (2021, to appear)
18. K. Natarajan, M. Song, C.P. Teo, Persistency model and its applications in choice modeling. Manag. Sci. **55**, 453–469 (2009)
19. K. Natarajan, D. Shi, K.C. Toh, Bounds for random binary quadratic programs. SIAM J. Optim. **28**, 671–692 (2018)
20. M. Padberg, The Boolean quadric polytope: some characteristics, facets and relatives. Math. Program. **45**, 134–172 (1989)
21. D. Panchenko, *The Sherrington-Kirkpatrick Model*. Springer Monographs in Mathematics (Springer, Berlin, 2013)
22. P.M. Pardalos, G.P. Rodgers, Computational aspects of a branch and bound algorithm for quadratic zero-one programming. Computing **45**, 131–144 (1990)
23. G. Parisi, Infinite number of order parameters for spin-glasses. Phys. Rev. Lett. **43**, 1754–1756 (1979)
24. S. Poljak, F. Rendl, H. Wolkowicz, A recipe for semidefinite relaxation for (0,1)-quadratic programming. J. Global Optim. **7**, 51–73 (1995)

25. D. Sherrington, S. Kirkpatrick, Solvable model of a spin-glass. Phys. Rev. Lett. **35**, 1792–1796 (1975)
26. N.Z. Shor, Class of global minimum bounds of polynomial functions. Cybern. Syst. Anal. **23**, 731–734 (1987)
27. M. Talagrand, The Parisi formula. Ann. Math. **163**, 221–263 (2006)
28. M. Talagrand, *Mean Field Models for Spin Glasses*. A Series of Modern Surveys in Mathematics, vol. 55 (Springer, Berlin, 2011)
29. K.C. Toh, M.J. Todd, R.H. Tutuncu, SDPT3 — a Matlab software package for semidefinite programming. Optim. Methods Softw. **11**, 545–581 (1999)
30. K.C. Toh, M.J. Todd, R.H. Tutuncu, Solving semidefinite-quadratic-linear programs using SDPT3. Math. Program. **95**, 189–217 (2003)
31. G. Weiss, Stochastic bounds on distributions of optimal value functions with applications to PERT, network flows and reliability. Oper. Res. **34**, 595–605 (1986)

Chapter 8
Fast Heuristics and Approximation Algorithms

Abraham P. Punnen

Abstract This chapter discusses theoretical analysis of approximation algorithms for QUBO and the Ising QUBO. We point out that the standard performance measure of relative performance ratio is not suitable for analysing the quality of approximation algorithms for the general versions of QUBO and the Ising QUBO. Thus, polynomial time approximation algorithms are considered and analyzed for the general problems using normalized relative error and domination analysis. More specifically, we discuss no worse than average performance bounds and how to improve them. Also, we present polynomial time approximation algorithms for QUBO and the Ising QUBO with a constant domination ratio. Then, the standard performance ratio of polynomial time approximation algorithms is considered for special cases of the Ising QUBO and QUBO. Relationships between ϵ-approximation algorithms for QUBO and the Ising QUBO are established when the associated data is specially structured. Further, a polynomial time $\frac{1}{2\Delta}$-approximation algorithm and a polynomial time $\frac{1}{\log n}$-approximation algorithm, in expectation, are presented for the Ising QUBO when the cost matrix has all diagonal entries zero and Δ is the maximum degree of a node in the associated support graph. Performance ratio of approximation algorithms for other special cases are briefly discussed.

8.1 Introduction

Recall that the quadratic unconstrained binary optimization problem (QUBO) is the mathematical programming problem

$$\begin{aligned} \text{QP:} \quad & \text{Maximize} \quad \phi(\mathbf{x}) = \mathbf{x}^T \mathbf{Q}\mathbf{x} + \mathbf{c}^T \mathbf{x} \\ & \text{Subject to:} \quad \mathbf{x} \in \{0, 1\}^n, \end{aligned}$$

A. P. Punnen (✉)
Department of Mathematics, Simon Fraser University, Surrey, BC, Canada
e-mail: apunnen@sfu.ca

A. P. Punnen (ed.), *The Quadratic Unconstrained Binary Optimization Problem*,
https://doi.org/10.1007/978-3-031-04520-2_8

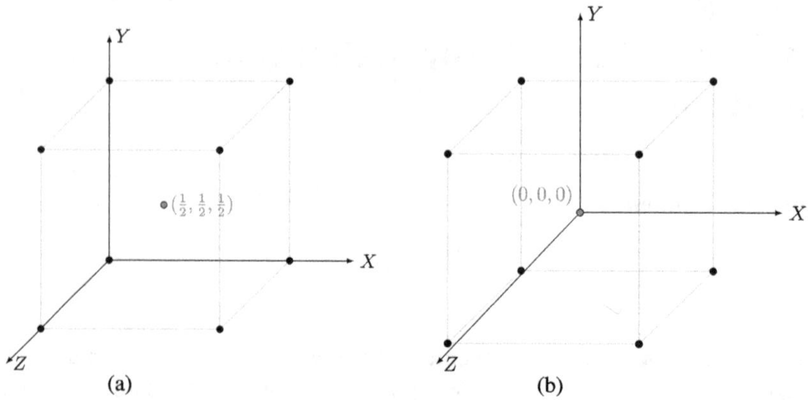

Fig. 8.1 QUBO vs Ising QUBO: shifted centre. (**a**) QUBO: centre (1/2,1/2,1/2). (**b**) Ising QUBO: centre (0,0,0)

where $\mathbf{Q} = (q_{ij})$ is an $n \times n$ symmetric matrix and $\mathbf{c}^T = (c_1, c_2, \ldots, c_n)$ is a row vector in $\mathbb{R}^n$. Without loss of generality, the diagonal elements of $\mathbf{Q}$ are assumed to be zero and $\mathbf{Q}$ is not the zero matrix. Another closely related problem, equivalent to QUBO (up to optimality), is the Ising QUBO which is to

$$\text{IQ:} \quad \text{Maximize} \quad \varphi(\mathbf{x}) = \mathbf{x}^{\mathbf{T}}\mathbf{A}\mathbf{x} + \mathbf{b}^T\mathbf{x}$$
$$\text{Subject to:} \quad \mathbf{x} \in \{-1, 1\}^n,$$

where $\mathbf{A} = (a_{ij})$ is a symmetric $n \times n$ matrix and $\mathbf{b}^T = (b_1, b_2, \ldots, b_n)$ is a row vector in $\mathbb{R}^n$. An instance of the Ising QUBO is completely defined by the matrix $\mathbf{A}$ and the vector $\mathbf{b}$ and hence we sometimes use the ordered pair $(\mathbf{A}, \mathbf{b})$ to represent an instance of the Ising QUBO. Similarly, an instance of QUBO is represented by the ordered pair $(\mathbf{Q}, \mathbf{c})$. Note that the Ising QUBO can be obtained from QUBO by shifting the centre $(1/2, 1/2, \ldots, 1/2)$ of the hypercube $[0, 1]^n$ to the centre $\mathbf{0} = (0, 0, ...0)$ of the hypercube $[-1, 1]^n$ (Fig. 8.1).

Both QUBO and the Ising QUBO can be represented in various forms. For a detailed discussion on this, we refer to Chapter 1. We will sometimes use the graph theoretic version of QUBO which can be stated as follows. Let $V = \{1, 2, \ldots, n\}$ and $E = \{(i, j), (j, i) : q_{ij} \neq 0\}$. The directed graph $G = (V, E)$ is called the *directed support graph of* $\mathbf{Q}$ and the weight of the edge $(i, j) \in E$ is q_{ij}. Note that $q_{ij} = q_{ji}$ and G has no loops. The weight of the node $i \in V$ is c_i. Then, the QUBO is to find a subset S of V such that

$$\sum_{(i,j)\in G[S]} q_{ij} + \sum_{i \in S} c_i$$

is maximized, where $G[S]$ is the subgraph of G induced by S. The graph G can also be viewed as an undirected graph, where edges (i, j) and (j, i) are replaced by one undirected edge (i, j) of weight $2q_{ij}$. In this case, we call G the *support graph* of $\mathbf{Q}$ and the edges of G can be labelled as an ordered pair (i, j) with $i < j$.

In the directed graph G discussed above when vertex weights are chosen as b_i for $i \in V$ and edge weights are chosen as a_{ij} for $(i, j) \in E$ the Ising QUBO is to assign positive or negative signs ($x_i = 1$ or -1) to nodes of G such that

$$\sum_{(i,j)\in E} a_{ij}x_i x_j + \sum_{i\in V} b_i x_i$$

is maximized. We also have the analogous undirected version where the weight of the edge (i, j) is $2a_{ij}$ and $i < j$.

The problems QUBO and Ising QUBO are closely related. The Ising QUBO $(\mathbf{A}, \mathbf{b})$ can be converted to an equivalent QUBO $(\mathbf{Q}^a, \mathbf{c}^a)$ (here the equivalence is only up to optimality) using the linear transformation

$$\mathbf{x} = 2\tilde{\mathbf{x}} - \mathbf{e}, \tag{8.1}$$

where $\mathbf{e}$ is the all-one vector. This transformation also generates an additive constant K^a in the objective function. It can be verified that $\mathbf{Q}^a = 4\mathbf{A}$, $(\mathbf{c}^a)^T = 2(\mathbf{b}^T - \mathbf{e}^T\mathbf{A} - \mathbf{e}^T\mathbf{A}^T)$ and $K^a = \mathbf{e}^T\mathbf{A}\mathbf{e} - \mathbf{b}^T\mathbf{e}$. Thus, for each $\mathbf{x} \in \{-1, 1\}^n$ there is a corresponding $\tilde{\mathbf{x}} = \frac{1}{2}(\mathbf{x} + \mathbf{e}) \in \{0, 1\}^n$ such that

$$\varphi(\mathbf{x}) = \mathbf{x}^T\mathbf{A}\mathbf{x} + \mathbf{b}^T\mathbf{x} = \tilde{\mathbf{x}}^T\mathbf{Q}^a\tilde{\mathbf{x}} + (\mathbf{c}^a)^T\tilde{\mathbf{x}} + K^a = \phi(\tilde{\mathbf{x}}) + K^a \tag{8.2}$$

Similarly, the instance $(\mathbf{Q}, \mathbf{c})$ of a QUBO can be converted into an equivalent (up to optimality) instance $(\mathbf{A}^q, \mathbf{b}^q)$ of the Ising QUBO using the linear transformation

$$\mathbf{x} = \frac{1}{2}(\hat{\mathbf{x}} + \mathbf{e}), \tag{8.3}$$

where $\mathbf{A}^q = \frac{1}{4}\mathbf{Q}$, $(\mathbf{b}^q)^T = \frac{1}{4}(\mathbf{e}^T\mathbf{Q} + \mathbf{e}^T\mathbf{Q}^T + 2\mathbf{c}^T)$ along with an additive constant $K^q = \frac{1}{4}\mathbf{e}^T\mathbf{Q}\mathbf{e} + \frac{1}{2}\mathbf{c}^T\mathbf{e}$. Thus, for any $\mathbf{x} \in \{0, 1\}^n$, there is a corresponding $\hat{\mathbf{x}} = 2\mathbf{x} - \mathbf{e} \in \{-1, 1\}^n$ such that

$$\phi(\mathbf{x}) = \mathbf{x}^T\mathbf{Q}\mathbf{x} + \mathbf{c}^T\mathbf{x} = \hat{\mathbf{x}}^T\mathbf{A}^q\hat{\mathbf{x}} + (\mathbf{b}^q)^T\hat{\mathbf{x}} + K^q = \varphi(\hat{\mathbf{x}}) + K^q \tag{8.4}$$

Both QUBO and Ising QUBO are strongly NP-hard. In Chap. 3 we have seen various polynomially solvable special cases of these problems. In this chapter we focus primarily on approximation algorithms.

Before getting into the technical details, let us discuss briefly some of the notations used in this chapter. The continuous relaxation of an optimization problem P is denoted by P* and the optimal objective function value of P (P*) by OPT(P) (OPT(P*)). The terminology QUBO refers to the quadratic unconstrained binary optimization problem and QP refers to the specific mathematical programming representation given at the beginning of this chapter. Similarly, the terminology

Ising QUBO refers to the Ising version of the quadratic unconstrained binary optimization problem and IQ refers to the specific mathematical programming representation given at the beginning of this chapter for the Ising QUBO. However, we sometimes use these notations interchangeably. Although $\phi(\mathbf{x})$ is defined for $\mathbf{x} \in \{0, 1\}^n$, we extend the notation to $\mathbf{x} \in [0, 1]^n$ as well. Likewise, the notation $\varphi(\mathbf{x})$ is extended to $\mathbf{x} \in [-1, 1]^n$. A vector with all components equal to 1 is denoted by $\mathbf{e}$, all components equal to $\frac{1}{2}$ is denoted by $\mathbf{h}$, and with all the components equal to zero is denoted by $\mathbf{0}$. The expected value of a random variable x is denoted by $\mathbf{E}(x)$. For all other notations, we follow the convention introduced in Chap. 1.

An algorithm for an optimization problem is said to be a *heuristic* if the solution produced by it is not guaranteed to be optimal. Heuristics are also referred to as approximation algorithms, but sometimes, when talking about approximation algorithms, a polynomial time complexity is assumed. However, we make no distinction between heuristics and approximation algorithms and whenever polynomial complexity is required, we explicitly state 'polynomial time approximation algorithms'. Zemel [41] analyzed various performance measures of heuristic algorithms for combinatorial optimization problems and suggested desirable properties of good performance measures. Let us now look at some of the popular theoretical performance measures of a heuristic algorithm.

Let $\mathcal{H}$ be a heuristic algorithm for an optimization problem OP (in maximization form) which produces a solution with objective function value $H(P)$ for an instance P. Then, $\mathcal{H}$ is said to be a $\frac{1}{\epsilon}$-approximation algorithm if

$$H(P) \geq \frac{1}{\epsilon}\text{OPT}(P) \text{ for any instance P of OP}$$

where $0 < \frac{1}{\epsilon} \leq 1$. A solution produced by a $\frac{1}{\epsilon}$-approximation algorithm is called a $\frac{1}{\epsilon}$-optimal solution. The quantity $\frac{1}{\epsilon}$ is called the *relative performance ratio* of $\mathcal{H}$. In this definition, we assume that $\text{OPT}(P) \geq 0$. For QUBO, since $\phi(\mathbf{0}) = 0$, $\text{OPT}(P) \geq 0$ and hence a heuristic $\mathcal{H}$ with $H(P) < 0$ is not a $\frac{1}{\epsilon}$-approximation algorithm for any $0 \leq \frac{1}{\epsilon} \leq 1$. As we will see later, when P is an instance of the Ising QUBO, $\text{OPT}(P) \geq 0$ when the diagonal elements of $\mathbf{A}$ are non-negative and hence any heuristic algorithm H with $H(P) < 0$ is not a $\frac{1}{\epsilon}$-approximation algorithm for such a problem.

Although QUBO is equivalent to the maximum weight cut problem (see Chap. 1 for the definition), this equivalence is established only at optimality. A $\frac{1}{\epsilon}$-approximation algorithm for the maximum weight cut problem does not necessarily give a $\frac{1}{\epsilon}$-approximation algorithm for QUBO. In the same way, existence of a $\frac{1}{\epsilon}$-approximation algorithm for the Ising QUBO need not imply the existence of a $\frac{1}{\epsilon}$-approximation algorithm for QUBO, although there is a one-to-one linear transformation which maps the solution space of QUBO to the solution space of the Ising QUBO and vice versa. Thus, the QUBO and the Ising QUBO needs to be analyzed independently when the relative performance ratio is considered as the measure of quality. However, as we will see later, there are relative performance ratio preserving reductions between QUBO and Ising QUBO when additional restrictions are imposed on the problem data. Moreover, unless P=NP, no polynomial time $\frac{1}{\epsilon}$-approximation algorithm exists for the general Ising QUBO, for any $\epsilon > 0$ [4]. In fact, testing if the general Ising QUBO has a solution with a non-negative objective function value is itself NP-hard [4]. However, polynomial time $\frac{1}{\epsilon}$-approximation algorithms are possible for the Ising QUBO when the diagonal elements of $\mathbf{A}$ are zeros [4, 8, 11, 19, 24, 30] for appropriate values of ϵ.

Another performance measure of a heuristic algorithm is the *normalized relative error* [34, 40] which is defined as

$$\kappa = \inf_{P \in I} \frac{\text{OPT}(P) - H(P)}{\text{OPT}(P) - \beta}$$

where β is the average of the objective function values of all solutions of P and I is the collection of all 'non-trivial' instances of P. When $\kappa \leq 1$ for a heuristic algorithm H, the solution produced by H will have a value of at least β (i.e. the solution is no worse than average). When $\kappa = 0$, H guarantees an optimal solution. For QUBO, $\text{OPT}(P) \neq \beta$ for all non-trivial instances, where trivial instances of QUBO are those with $\mathbf{Q}$ is the zero matrix and $\mathbf{c}$ is the zero vector. Palubeckis [34] considered theoretical analysis of algorithms with normalized relative error while Tavares [40] examined the measure using experimental analysis.

Yet another performance measure, closely related to κ, is the *differential approximation ratio* δ [9] which is defined as

$$\delta = \inf_{P \in I} \frac{H(P) - \min(P)}{\text{OPT}(P) - \min(P)},$$

where $\min(P)$ is the worst objective function value. For details on the analysis of approximation algorithms using the differential approximation ratio for various combinatorial optimization problems, we refer to [9]. Nesterov [31] used the differential approximation ratio as the performance measure for a semidefinite programming based polynomial time approximation algorithm for a special case of the Ising QUBO.

Domination analysis of algorithms [12, 15] is another way to measure the performance of a heuristic algorithm. A solution $\mathbf{x}$ of a maximization problem P is dominated by another solution $\mathbf{x}'$ if $\text{OBJ}(\mathbf{x}) \leq \text{OBJ}(\mathbf{x}')$, where $\text{OBJ}(\mathbf{x})$ is the objective function value of x. Let $\mathbf{x}^*$ be the solution produced by the heuristic algorithm H for the maximization problem P which is defined over the finite family $\mathcal{F}$ of the feasible solutions and $D(H) = \{\mathbf{x} \in \mathcal{F} : \text{OBJ}(\mathbf{x}) \leq \text{OBJ}(\mathbf{x}^*)\}$. Then, $D(H)$ is the collection of all solutions of P dominated by $\mathbf{x}^*$. The heuristic H is said to have *domination number* δ if $|D(H)| \geq \delta$ for any instance of the problem P [12, 41]. Further, H is said to have *domination ratio* ρ [12] if

$$\rho = \inf_{p \in I} \frac{|D(H)|}{|\mathcal{F}|}$$

Note that $0 \leq \rho \leq 1$ for all instances of P and when $\rho = 1$, the algorithm H guarantees an optimal solution. For QUBO and the Ising QUBO, $|\mathcal{F}| = 2^n$, a fixed number that depends only on the problem dimension and is independent of the other parts of the problem data. The domination ratio has been used in the analysis of heuristic algorithms for the travelling salesman problem [14, 37]. For a survey on domination analysis of algorithms for various combinatorial optimization problems, we refer to [15]. Each of the performance measures discussed above has its own merits and drawbacks. However, each of these measures also provides distinct insights into the quality of a heuristic algorithm.

In this chapter, we consider polynomial time heuristic algorithms for QUBO and the Ising QUBO. Whenever applicable, we discuss appropriate performance measures in terms of relative performance ratio, differential approximation ratio, normalized relative error, or domination ratio.

8.2 Rounding Algorithms

Let us first discuss a very simple class of algorithms for QUBO which 'rounds' a solution $\mathbf{x} \in [0, 1]^n$ to obtain a solution $\mathbf{x}' \in \{0, 1\}^n$ such that $\phi(\mathbf{x}') \geq \phi(\mathbf{x})$. These types of algorithms for QUBO were considered by many authors [28, 40] in different contexts.

For any $k \in \{1, 2, \ldots, n\}$, $\phi(\mathbf{x})$ can be written as

$$\phi(\mathbf{x}) = \left(\sum_{\substack{j=1 \\ j \neq k}}^{n} c_j x_j + \sum_{\substack{i=1 \\ i \neq k}}^{n} \sum_{\substack{j=1 \\ j \neq k}}^{n} q_{ij} x_i x_j \right) + \left(x_k \Big(c_k + \sum_{\substack{j=1 \\ j \neq k}}^{n} 2 q_{kj} x_j \Big) \right) \tag{8.5}$$

Since we assumed that the diagonal elements of $\mathbf{Q}$ are zeros, the condition $j \neq k$ in the second bracket of Eq. (8.5) is redundant, but we are keeping it to emphasise that the associated summation is independent of x_k. In Eq. (8.5), the first bracket is independent of x_k. Let

$$q(k) = c_k + \sum_{\substack{j=1 \\ j \neq k}}^{n} 2 q_{ij} x_j.$$

Now, consider the solution $\hat{\mathbf{x}}$ defined by

$$\hat{x}_j = \begin{cases} x_j \text{ if } j \neq k, \\ 1 \text{ if } j = k \text{ and } q(k) \geq 0 \\ 0 \text{ if } j = k \text{ and } q(k) < 0 \end{cases} \tag{8.6}$$

Then, it is easy to verify that $\phi(\hat{\mathbf{x}}) \geq \phi(\mathbf{x})$ and the number of fractional components of $\hat{\mathbf{x}}$ is one less than that of $\mathbf{x}$, whenever x_k is fractional. Repeating this rounding operation by using $\hat{\mathbf{x}}$ in place of $\mathbf{x}$ and continuing the process, after at most n iterations we get a 0–1 solution $\mathbf{x}^0$ with $\phi(\mathbf{x}^0) \geq \phi(\mathbf{x})$. These steps are summarized in the [0, 1]-*rounding algorithm* below.

Algorithm 1: The [0, 1]-rounding algorithm

Let $\mathbf{x} \in [0, 1]^n$ *be a given fractional solution and set* $\mathbf{x}^0 = \mathbf{x}$;
$S \leftarrow \{i : x_i \notin \{0, 1\}\}$;
for $i \in S$ **do** $q(i) = c_i + 2\sum_{\substack{j=1 \\ j \neq i}}^{n} q_{ij}x_j$;
while $S \neq \emptyset$ **do**
  choose $k \in S$ and update $S = S \setminus \{k\}$;
  if $q(k) \geq 0$ **then** set $x_k^0 = 1$;
  else set $x_k^0 = 0$;
  for $i \in S$ **do** $q(i) = q(i) + 2q_{ik}(x_k^0 - x_k)$// `updating` $q(i)$ `for` $i \in S$;
end
output $\mathbf{x}^0$

Theorem 8.1 *The* [0, 1]*-rounding algorithm constructs a solution* $\mathbf{x}^0 \in \{0, 1\}^n$ *from a solution* $\mathbf{x} \in [0, 1]^n$ *such that* $\phi(\mathbf{x}^0) \geq \phi(\mathbf{x})$ *in* $O(n^2)$ *time.*

The proof of this theorem follows from repetend application of Eq. (8.5) and the rounding scheme (8.6). Formula (8.5) and the validity of the [0, 1]-rounding algorithm as presented above depends on the fact that the diagonal elements of $\mathbf{Q}$ are zeros.

Since $\mathbf{Q}$ is a symmetric, non-zero matrix with diagonal entries zero, $tr(\mathbf{Q}) = 0$ and hence the sum of its eigenvalues is zero. Then, Q has at least one negative eigenvalue and one positive eigenvalue. (Recall that $\mathbf{Q}$ is not a zero matrix.) Therefore $\mathbf{Q}$ is neither positive semidefinite nor negative semidefinite. When $\mathbf{Q}$ is positive semidefinite, $\phi(\mathbf{x})$ is convex over $[0, 1]^n$ and hence there exists an optimal solution to QP* which is at an extreme point of the hypercube $[0, 1]^n$. The [0, 1]-rounding algorithm establishes the extreme point optimality property when the diagonal elements of $\mathbf{Q}$ are zeros and offers a construction scheme to produce an extreme point optimal solution from an arbitrary optimal solution. Note that convexity is not a necessary condition for the extreme point optimality property. When the diagonal elements of Q are non-negative $\phi(\mathbf{x})$ is coordinate-wise convex (convex in each variable) and this is sufficient to establish extreme point optimality and the [0, 1]-rounding algorithm can easily be amended to handle this case.

Theorem 8.2 *The optimal objective function value of QP and QP* are the same when the diagonal elements of* $\mathbf{Q}$ *are non-negative. Further, there exists an optimal solution to QP* which is at an extreme point of the hypercube* $[0, 1]^n$

Proof Let $\mathbf{x}^*$ be an optimal solution to QP*. If $\mathbf{x}^* \notin \{0, 1\}^n$, apply the [0, 1]-rounding algorithm (with appropriate changes to handle non-negative diagonal elements of $\mathbf{Q}$, if applicable) to compute a solution $\mathbf{x}^0 \in \{0, 1\}^n$. Then, $\phi(\mathbf{x}^0) \geq \phi(\mathbf{x}^*)$. Since $\mathbf{x}^*$ is an optimal solution of QP*, $\phi(\mathbf{x}^*) \geq \phi(\mathbf{x}^0)$. Thus, $\phi(\mathbf{x}^*) = \phi(\mathbf{x}^0)$ and the result follows.

The [0, 1]-rounding algorithm can be implemented using an incremental construction of a solution to a $p \times p$ QP for $p = 1, 2, \ldots, n$ converging to a solution to QP, which is the solution for the $n \times n$ QP obtained. The worst case complexity of this implementation is still $O(n^2)$ but could be slightly faster in computational experiments compared to the implementation discussed in the [0, 1]-rounding algorithm.

Algorithm 2: The incremental [0, 1]-rounding algorithm

Input: $\mathbf{x}^* \in [0, 1]^n$;
$\bar{S} \leftarrow \{j : 0 < x_j^* < 1\}$ and $S \leftarrow \{j : x_j^* \in \{0, 1\}\}$;
$x_j^0 \leftarrow x_j^*$ for $j \in S$;
while $\bar{S} \neq \emptyset$ **do**
  Choose $k \in \bar{S}$;
  if $c_k + 2\sum_{j \in S} q_{kj} x_j^* > 0$ **then** $x_k^0 = 1$;
  else $x_k^0 = 0$;
  $\bar{S} \leftarrow \bar{S} \setminus \{k\}$, $S \leftarrow S \cup \{k\}$;
end
output $\mathbf{x}^0$

Lemma 8.1 *The incremental* [0, 1]*-rounding algorithm terminates in* $O(n^2)$ *time with a solution* $\mathbf{x}^0 \in \{0, 1\}^n$ *such that* $\phi(\mathbf{x}^0) \geq \phi(\mathbf{x}^*)$.

Let us now discuss our rounding algorithm for the Ising QUBO [8, 11, 24]. It works almost the same way as the [0, 1]-rounding algorithm above. The algorithm takes the input solution $\mathbf{x}^* \in [-1, 1]^n$ and produce the output solution $\mathbf{x}^0 \in \{-1, 1\}^n$ such that $\varphi(\mathbf{x}^0) \geq \varphi(\mathbf{x}^*)$.

Algorithm 3: The [−1, 1]-rounding algorithm

Input: $\mathbf{x}^* \in [-1, 1]^n$;
$S \leftarrow \{j : -1 < x_j^* < 1\}$;
$\mathbf{x}^0 \leftarrow \mathbf{x}^*$;
while $S \neq \emptyset$ **do**
  choose $k \in S$;
  if $b_k + \sum_{j=1, j \neq k}^{n} (a_{kj} + a_{jk}) x_j^0 \geq 0$ **then** set $x_k^0 = 1$;
  else $x_k^0 = -1$;
  $S \leftarrow S \setminus \{k\}$;
end
output $\mathbf{x}^0$

Theorem 8.3 *The* $[-1, 1]$*-rounding algorithm produces a solution* $\mathbf{x}^0$ *of IQ satisfying* $\varphi(\mathbf{x}^0) \geq \varphi(\mathbf{x}^*)$ *in* $O(n^2)$ *time.*

Proof Note that the diagonal elements of A are zeros. Then, for any $k \in \{1, 2, \ldots, n\}$

$$\varphi(\mathbf{x}) = \sum_{\substack{j=1 \\ j\neq k}}^{n} b_j x_j + \sum_{\substack{i=1 \\ i\neq k}}^{n} \sum_{\substack{j=1 \\ j\neq k}}^{n} a_{ij} x_i x_j + x_k \left(b_k + \sum_{\substack{j=1 \\ j\neq k}}^{n} (a_{kj} + a_{jk}) x_j \right)$$

If $b_k + \sum_{\substack{j=1 \\ j\neq k}}^{n} (a_{kj} + a_{jk}) x_j^* > 0$, we can increase x_k^* to 1 to obtain a better solution. Likewise, if $b_k + \sum_{\substack{j=1 \\ j\neq k}}^{n} (a_{kj} + a_{jk}) x_j^* < 0$ we can decrease the value of x_k^* to -1 to obtain a better solution. This observation establishes the validity of the theorem. The complexity calculation is straightforward.

By choosing $\mathbf{x}^* = 0$ and using the [-1,1]-rounding algorithm, we can construct a solution $\mathbf{x}^0$ of IQ such that $\varphi(\mathbf{x}^0) \geq 0$. This shows that the optimal objective function value of the Ising QUBO is non-negative. Theorem 8.3 has another important consequence. As in the case of QUBO, for the Ising QUBO, the $[-1, 1]$-rounding algorithm can be modified easily to handle the case when the diagonal elements of **A** are non-negative.

Corollary 8.1 ([8, 26]) *When the diagonal elements of A are non-negative, there exists an optimal solution* $\mathbf{x}^0 \in \{-1, 1\}^n$ *to the continuous relaxation* IQ^* *of IQ.*

The $[-1, 1]$-rounding algorithm and the $[0, 1]$-rounding algorithm can be repeated multiple times to obtain improved solutions, in practice. The solution quality of the rounding algorithms depends on two factors:

1. the starting solution **x** and
2. the order in which variables are selected to round.

By choosing these elements randomly and repeating the algorithm multiple times, followed by changing the starting solution vector, good solutions can be obtained which can be further improved using local search or metaheuristic algorithms (see Chap. 9). In this sense, the rounding algorithms play a major role in search path diversification of metaheuristic algorithms.

The empirical performance of the rounding algorithms depend on the criteria applied to select the index k to 'round'. As mentioned above, making this a random choice, the algorithm can be repeated multiple times and then we could choose the overall best solution obtained. We can also select the index k using $|q(k)| = \max_{i \in S}\{|q(i)|\}$ for the [-1,1]-rounding algorithm or can develop other specific selection rules, depending on the nature of the rounding algorithms. Palubeckis [34] considered the special cases of the $[0, 1]$-rounding algorithm when the starting solution is the all-half vector **h**. The algorithm can also be repeated by selecting different fractional vectors carefully, to achieve improved performance in practice. For example, later, we will discuss how to generate a 'good' $\mathbf{x} \in [-1, 1]^n$ from an optimal solution to a semidefinite programming (SDP) relaxation of IQ. This construction depends on choosing a random n-dimensional Gaussian and hence can be repeated multiple times to produce improved solutions. For experimental analysis on the rounding algorithms and other simple construction heuristics, and for further theoretical analysis of these types of algorithms, we refer to [36].

Although our rounding algorithms produced some performance guarantee, we will see later that by choosing carefully the starting solution, different (and better) performance guarantees can be obtained for both QUBO and the Ising QUBO.

8.3 The Normalized Relative Error

Let us now consider algorithms for QUBO and the Ising QUBO that produce solutions with objective function values guaranteed to be no worse than average. For such algorithms, the normalized relative error $\kappa \leq 1$. Consequently, we mostly talk about the performance guarantee in terms of the average value of solutions, rather than κ, but the value of κ will be implicit. A solution is said to be *no-worse than average* if its objective function value is greater than or equal to the average of the objective function values of all solutions.

We first establish a closed form formula to compute the average of the objective function values of all the solutions of QUBO. Note that the centre of the hypercube $[0, 1]^n$ is the all-half vector $\mathbf{h}$. Considering the symmetry of the hypercube, it is reasonable to expect that the average of the objective function values of all the solutions of QUBO is attained at $\mathbf{h}$. Our next theorem shows that this is indeed true. Let $\mathbb{Q} = \sum_{i=1}^{n} \sum_{\substack{j=1 \\ j \neq i}}^{n} q_{ij}$ and $\mathbb{C} = \sum_{i=1}^{n} c_i$.

Theorem 8.4 ([34]) *The average of the objective function values of all 2^n solutions of the QUBO* $(\mathbf{Q}, \mathbf{c})$ *is* $\mathcal{A}(\mathbf{Q}, \mathbf{c}) = \frac{1}{4}\mathbb{Q} + \frac{1}{2}\mathbb{C}$.

Proof Consider n independent random variables $x_1, x_2, \ldots, x_n$ taking values 0 or 1, each with probability $\frac{1}{2}$. Then, the expected value $\mathbf{E}(x_i)$ of x_i is $\frac{1}{2}$. For $i \neq j$, x_i and x_j are independent and hence $\mathbf{E}(x_i x_j) = \mathbf{E}(x_i)\mathbf{E}(x_j) = \frac{1}{4}$. Now, consider the random variable

$$X = \sum_{i=1}^{n} \sum_{j=1, j \neq i}^{n} q_{ij} x_i x_j + \sum_{j=1}^{n} c_j x_j.$$

By the linearity property of expectation and using the fact that $q_{ii} = 0$ for all i, the expected value of X is

$$\mathbf{E}(X) = \frac{1}{4}\sum_{i=1}^{n}\sum_{j=1}^{n} q_{ij} + \frac{1}{2}\sum_{i=1}^{n} c_i$$

and the result follows.

Remark 8.1 When $\mathbf{Q}$ is allowed to have non-zero diagonal elements, the average of the objective function values of all 2^n solutions of QUBO is $\mathcal{A}(\mathbf{Q}, \mathbf{c}) = \frac{1}{4}\mathbb{Q} + \frac{1}{2}\mathbb{C} + \frac{1}{2}\text{tr}(\mathbf{Q})$, where $\text{tr}(\mathbf{Q})$ is the trace of the matrix $\mathbf{Q}$. Note that in the definition of $\mathbb{Q}$, the diagonal elements are excluded.

Corollary 8.2 *A solution to the QUBO* $(\mathbf{Q}, \mathbf{c})$ *with objective function value no worse than* $\mathcal{A}(\mathbf{Q}, \mathbf{c})$ *can be identified in* $O(n^2)$ *time.*

Proof Note that the objective function value of the fractional solution $\mathbf{x} = \mathbf{h}$, the all-$\frac{1}{2}$ vector, is $\mathcal{A}(\mathbf{Q}, \mathbf{c})$. Then, by Theorem 8.1, the solution $\mathbf{x}^*$ produced by the [0, 1]-rounding algorithm, starting with $\mathbf{h}$, satisfies $\phi(\mathbf{x}^*) \geq \phi(\mathbf{h}) = \mathcal{A}(\mathbf{Q}, \mathbf{c})$ and the result follows.

In fact, the [0, 1]-rounding algorithm can be used to guarantee a performance bound that is better than $\mathcal{A}(\mathbf{Q}, \mathbf{c})$.

Theorem 8.5 $\mathcal{A}(\mathbf{Q}, \mathbf{c}) \leq \zeta(\mathbf{Q}, \mathbf{c}) = \max\left\{0, \mathbb{Q} + \mathbb{C}, -\frac{3\mathbb{C}^2}{4\mathbb{Q}}\right\}$. *Further, a solution with objective function value no worse than* $\zeta(\mathbf{Q}, \mathbf{c})$ *can be identified in* $O(n^2)$ *time.*

Proof Consider the single variable quadratic function $f(y) = \mathbb{Q}y^2 + \mathbb{C}y$ for $y \in [0, 1]$. Note that $\mathcal{A}(\mathbf{Q}, \mathbf{c}) = f(1/2)$ and hence $\mathcal{A}(\mathbf{Q}, \mathbf{c}) \leq \max\{f(y) : y \in [0, 1]\}$. If $\mathbb{Q} \geq 0$ then $f(y)$ is a convex function and hence its maximum is attained either at 0 or at 1. If $\mathbb{Q} < 0$, then $f(y)$ is concave and its maximum is attained either at 0 or at 1 if $\frac{-\mathbb{C}}{2\mathbb{Q}} \notin [0, 1]$ and at $y^* = \frac{-\mathbb{C}}{2\mathbb{Q}}$ if $0 \leq y^* \leq 1$. Let $\mathbf{x}^0$ be the all-zero solution and $\mathbf{x}^1$ be the all-one solution. If $0 < \frac{-\mathbb{C}}{2\mathbb{Q}} < 1$ then choose the fractional solution $\mathbf{x}$ with $x_i = -\frac{\mathbb{C}}{2\mathbb{Q}}$ for all i and then $\phi(\mathbf{x}) = -\frac{3\mathbb{C}^2}{4\mathbb{Q}}$. Now apply the [0, 1]-rounding algorithm, starting with this $\mathbf{x}$, and let the resulting solution be $\mathbf{x}^*$. Theorem 8.1 guarantees that $\phi(\mathbf{x}^*) \geq -\frac{3\mathbb{C}^2}{4\mathbb{Q}}$. Then, the best of $\mathbf{x}^0, \mathbf{x}^1, \mathbf{x}^*$ is guaranteed to have an objective function value no worse than $\zeta(\mathbf{Q}, \mathbf{c})$ and the result follows.

From Theorem 8.5, it follows that when $\mathbb{Q} \geq 0$, one of the trivial solutions, the all-one or all-zero solutions, is no worse than average. This probably raises questions on the merit of using the performance guarantee $\mathcal{A}(\mathbf{Q}, \mathbf{c})$. To highlight its relevance, it may be noted that even powerful local search algorithms could get trapped at a solution with value less than $\mathcal{A}(\mathbf{Q}, \mathbf{c})$ [34].

Consider the instance of QUBO where the matrix $\mathbf{Q}$ has all elements equal to one, except the diagonal elements which are zeros. Let $c_i = -\frac{1}{2}(k - 1 + \epsilon)$ where $\epsilon > 0$ is a very small number. Then, $\mathcal{A}(\mathbf{Q}, \mathbf{c}) = \frac{n}{4}(n - k - \epsilon) > 0$ for all $k < n$

for an appropriately small $\epsilon > 0$. Now consider the neighborhood consisting of all r-flip moves where $r \leq k$ [33, 34]. Then the all-zero solution is locally optimal with respect to this neighborhood and this local optimum has value zero which is less than $\mathcal{A}(\mathbf{Q}, \mathbf{c})$. These types of neighborhoods are very powerful, even for small values of k and is the basis for many successful implementations of complex metaheuristics for QUBO. See Chap. 9 for a detailed discussion on metaheuristic algorithms for QUBO.

Let us now look at another performance measure related to $\mathcal{A}(\mathbf{Q}, \mathbf{c})$ introduced by Palubeckis [34]. Let QP(r) be the optimization problem obtained from QP with the additional restriction that $\sum_{j=1}^{n} x_j = r$ and let $F(r)$ be the family of all feasible solutions of QP(r). Then, $|F(r)| = \binom{n}{r}$ and $F(r) \cap F(s) = \emptyset$ for $r \neq s$. The average of the objective function values of all solutions of QP(r) is denoted by $\mathcal{A}_r(\mathbf{Q}, \mathbf{c})$ and $F^* = \max\limits_{0 \leq r \leq n} \mathcal{A}_r(\mathbf{Q}, \mathbf{c})$.

Theorem 8.6 $F^* \geq \mathcal{A}(\mathbf{Q}, \mathbf{c})$

Proof Since

$$\mathcal{A}(\mathbf{Q}, \mathbf{c}) = \frac{\sum_{r=0}^{n} |F(r)| \mathcal{A}_r(\mathbf{Q}, \mathbf{c})}{\sum_{r=0}^{n} |F(r)|},$$

we have $\mathcal{A}(\mathbf{Q}, \mathbf{c}) \leq \max_{0 \leq r \leq n} \mathcal{A}_r(\mathbf{Q}, \mathbf{c}) = F^*$.

It is not difficult to construct instances $(\mathbf{Q}, \mathbf{c})$ of QUBO such that $F^* > \mathcal{A}(\mathbf{Q}, \mathbf{c})$. Thus, it will be interesting to develop heuristic algorithms for QUBO that guarantee solutions with objective function values no worse than F^*. We now discuss one such algorithm based on the works of Palubeckis [34]. First, let us derive a closed form formula for $\mathcal{A}_r(\mathbf{Q}, \mathbf{c})$.

Theorem 8.7 $\mathcal{A}_r(\mathbf{Q}, \mathbf{c}) = \dfrac{r(r-1)}{n(n-1)} \displaystyle\sum_{i=1}^{n} \sum_{j=1}^{n} q_{ij} + \frac{r}{n} \sum_{i=1}^{n} c_i$

Proof Let $\mathbf{x}^1, \mathbf{x}^2 \ldots, \mathbf{x}^t$ be the collection of all solutions of QP(r), where $t = \binom{n}{r}$. Then,

$$\mathcal{A}_r(\mathbf{Q}, \mathbf{c}) = \frac{1}{t} \sum_{k=1}^{t} \left(\sum_{i=1}^{n} \sum_{j=1}^{n} q_{ij} x_i^k x_j^k + \sum_{i=1}^{n} c_i x_i^k \right)$$

$$= \frac{1}{t} \left(\sum_{i=1}^{n} \sum_{j=1}^{n} q_{ij} \sum_{k=1}^{t} x_i^k x_j^k + \sum_{i=1}^{n} c_i \sum_{k=1}^{t} x_i^k \right)$$

For simplicity, we assume that $r \geq 2$ but the formula works for all $r = 0, 1, \ldots, n$. Note that $\sum_{k=1}^{t} x_i^k$ counts the total number of solutions of QP(r) containing $x_i = 1$ which is $\binom{n-1}{r-1}$. Likewise, $\sum_{k=1}^{t} x_i^k x_j^k$ counts the total number of solutions of QP(r)

containing both $x_i = x_j = 1$ and this is $\binom{n-2}{r-2}$. Thus,

$$\mathcal{A}_r(\mathbf{Q}, \mathbf{c}) = \frac{1}{t}\left(\binom{n-2}{r-2}\sum_{i=1}^{n}\sum_{j=1}^{n} q_{ij} + \binom{n-1}{r-1}\sum_{i=1}^{n} c_i\right)$$

Substituting $t = \binom{n}{r}$ and simplifying, we get the result.

In view of Theorem 8.6, to efficiently construct a solution of QP with objective function value no worse than F^*, it is sufficient to find a solution for QP(r) with objective function value no worse than $\mathcal{A}_r(\mathbf{Q}, \mathbf{c})$ for $r = 0, 1, \ldots, n$ and choose the best overall solution. So, let us explore this. Besides the connection with generating a heuristic solution for QUBO with some level of performance guarantee, this problem is interesting by itself since QP(r) is the cardinality constrained QUBO [27] which also has various applications. Let $\mathcal{A}_r(\mathbf{Q}, \mathbf{c}/x_k = 1)$ be the average value of all solutions of QP(r) given that $x_k = 1$ for a specified index k, for $1 \le k \le n$.

Lemma 8.2 $\mathcal{A}_r(\mathbf{Q}, \mathbf{c}/x_k = 1) = \frac{r-1}{n-1}\left(\frac{r-2}{n-2}\mathbb{Q} + \mathbb{C}\right) + \frac{n-r}{n-1}c_k + 2\frac{n-r}{n-1}\frac{r-1}{n-2}\sum_{i=1}^{n} q_{ik}$, *where* $r \neq 0$

Proof When x_k is fixed at value 1, we can reduce the cost matrix of QP(r) by deleting row k and column k and adding the sum of row k and column k to the cost vector $\mathbf{c}$. This also releases a constant term c_k. Then, $\mathcal{A}_r(\mathbf{Q}, \mathbf{c}/x_k = 1)$ is $c_k + \mathcal{A}_{r-1}(\mathbf{Q}', \mathbf{c}')$ where $\mathbf{Q}'$ and $\mathbf{c}'$, respectively, are the cost matrix and cost vector obtained for the reduced problem. Note that $\mathbf{Q}$ is symmetric and $q_{ii} = 0$ for all i. Using the formula for $\mathcal{A}_r(\mathbf{Q}, \mathbf{c})$, we get,

$$\begin{aligned}
\mathcal{A}_r(\mathbf{Q}, \mathbf{c}/x_k = 1) &= c_k + \mathcal{A}_{r-1}(\mathbf{Q}', \mathbf{c}') \\
&= c_k + \frac{r-1}{n-1}\left(\frac{r-2}{n-2}\left[\mathbb{Q} - 2\sum_{i=1}^{n} q_{ik}\right] + \mathbb{C} - c_k + 2\sum_{i=1}^{n} q_{ik}\right) \\
&= \frac{r-1}{n-1}\left(\frac{r-2}{n-2}\mathbb{Q} + \mathbb{C}\right) + \frac{n-r}{n-1}c_k + 2\frac{n-r}{n-1}\frac{r-1}{n-2}\sum_{i=1}^{n} q_{ik},
\end{aligned}$$

and this completes the proof.

Now, for $r \neq 0$ choose p such that

$$\mathcal{A}_r(\mathbf{Q}, \mathbf{c}/x_p = 1) = \max_{1 \le k \le n} \mathcal{A}_r(\mathbf{Q}, \mathbf{c}/x_k = 1) \ge \mathcal{A}_r(\mathbf{Q}, \mathbf{c}) \tag{8.7}$$

In the expression of $\mathcal{A}_r(\mathbf{Q}, \mathbf{c}/x_k = 1)$ in Lemma 8.2, $\frac{r-1}{n-1}\left(\frac{r-2}{n-2}\mathbb{Q} + \mathbb{C}\right)$ is a constant independent of k. Thus, to identify p in Eq. (8.7), it is sufficient to compute

$$g(p) = \max_{1 \le k \le n} g(k), \tag{8.8}$$

where

$$g(k) = \frac{n-r}{n-1}c_k + 2\frac{n-r}{n-1}\frac{r-1}{n-2}\sum_{i=1}^{n} q_{ik}.$$

Now, fix $x_p = 1$ and we get a reduced problem of the type QP($r-1$) with data ($\mathbf{Q}'$, $\mathbf{c}'$) for appropriate $\mathbf{Q}'$ and $\mathbf{c}'$. The process of fixing a selected variable at value 1 as discussed above can be continued recursively, until r variables have been fixed. Now, set the remaining variables at value zero. By the specific construction employed and using Eq. (8.7), it can be verified that the solution, say $\mathbf{x}^r$, obtained will have objective function value at least $\mathcal{A}_r(\mathbf{Q}, \mathbf{c})$. We call this construction scheme *the averaged variable selection algorithm*.

Theorem 8.8 *The averaged variable selection algorithm produces a solution* $\mathbf{x}^r$ *to QP(r) with* $\phi(\mathbf{x}^r) \geq \mathcal{A}_r(\mathbf{Q}, \mathbf{c})$ *in* $O(n^2)$ *time.*

The proof of the theorem follows from the preceding discussions. Let $F^* = \mathcal{A}_p(\mathbf{Q}, \mathbf{c}) = \max\{\mathcal{A}_r(\mathbf{Q}, \mathbf{c}) : r = 0, 1, \ldots, n\}$. Then $\phi(\mathbf{x}^p) \geq F^*$.

Theorem 8.9 (Palubeckis [34]) $\phi(\mathbf{x}^p) \geq F^* \geq \mathcal{A}(\mathbf{Q}, \mathbf{c})$. *A solution* $\mathbf{x}^*$ *of QUBO with* $\phi(\mathbf{x}^*) \geq F^*$ *can be identified in* $O(n^3)$ *time.*

Proof By construction $\phi(\mathbf{x}^p) \geq \max\{\phi(\mathbf{x}^r) : r = 0, 1, 2, \ldots, n\}$. The result follows from Theorems 8.6 and 8.8.

The $O(n^3)$ complexity indicated in Theorem 8.8 can be reduced to $O(n^2)$ by identifying and eliminating all but two values of r from consideration and then solving QP(r) only for these two values of r. Note that for $r = 0, 1, \ldots, n$,

$$\mathcal{A}_r(\mathbf{Q}, \mathbf{c}) = \frac{r(r-1)}{n(n-1}\mathbb{Q} + \frac{p}{n}\mathbb{C} \tag{8.9}$$

$$= r^2\frac{\mathbb{Q}}{n(n-1)} + r\frac{(n-1)\mathbb{C} - \mathbb{Q}}{n(n-1)} \tag{8.10}$$

Now, consider the single variable quadratic function

$$\psi(r) = r^2\frac{\mathbb{Q}}{n(n-1)} + r\frac{(n-1)\mathbb{C} - \mathbb{Q}}{n(n-1)} \text{ for } r \in [0, n] \tag{8.11}$$

If $\mathbb{Q} \geq 0$ then $\psi(r)$ is a convex function and its maximum is achieved at $r = 0$ or $r = n$. In this case $F^* = \max\{\mathcal{A}_0(\mathbf{Q}, \mathbf{c}), \mathcal{A}_n(\mathbf{Q}, \mathbf{c})\}$. If $\mathbb{Q} < 0$ then $\psi(r)$ is a concave function. Thus its maximum is achieved at $r = r^0$ where $r^0 = -\frac{(n-1)\mathbb{C} - \mathbb{Q}}{2\mathbb{Q}}$. If $r^0 \notin (0, n)$ then the maximum of $\psi(r)$ for $r \in [0, n]$ is achieved at $r = 0$ or $r = n$. If $r^0 \in (0, n)$ then maximum of $\psi(r)$ for $r \in [0, n]$ is achieved at r^0. But r^0 need not be an integer. In this case, maximum of $\psi(r)$ for $r \in \{0, 1, 2, \ldots, n\}$ is achieved at $r^1 = \lfloor r^0 \rfloor$ or at $r^2 = \lceil r^0 \rceil$. Summarizing the foregoing discussion,

$$F^* = \begin{cases} \max\{\mathcal{A}_0(\mathbf{Q},\mathbf{c}), \mathcal{A}_n(\mathbf{Q},\mathbf{c})\} \text{ if } \mathbb{Q} \geq 0, \textit{or } \mathbb{Q} < 0 \text{ and } r^0 \notin (0,n) \\ \max\{\mathcal{A}_{r^1}(\mathbf{Q},\mathbf{c}), \mathcal{A}_{r^2}(\mathbf{Q},\mathbf{c})\} \text{ if } \mathbb{Q} < 0 \text{ and } r^0 \in (0,n). \end{cases}$$

Thus, to identify F^* and the corresponding solution $\mathbf{x}^p$ with $\phi(\mathbf{x}^p) \geq F^*$ we need to consider only two values of r. i.e. $r = 0, 1$ if $\mathbb{Q} \geq 0$ or $\mathbb{Q} < 0$ and $r^0 \notin (0,n)$, and $r = r^1, r^2$ if $\mathbb{Q} < 0$ and $r^0 \in (0,n)$. Since $\mathcal{A}_r(\mathbf{Q},\mathbf{c})$ and an associated solution $\mathbf{x}^r$ with $\phi(\mathbf{x}^r) \geq \mathcal{A}(\mathbf{Q},\mathbf{c})$ can be identified in $O(n^2)$ time for any given r, F^* and the associated solution $\mathbf{x}^p$ with $\phi(\mathbf{x}^p) \geq F^*$ can be identified in $O(n^2)$ time.

When $\mathbb{Q}$ is non-negative, we can in fact construct a solution $\mathbf{x}$ with $\phi(\mathbf{x}) > F^*$ using our [0, 1]-rounding algorithm. Note that, for non-negative integers r and n with $r \leq n$, $\frac{r-1}{n-1} \leq \frac{r}{n}$. Then,

$$\mathcal{A}_r(\mathbf{Q},\mathbf{c}) = \frac{r(r-1)}{n(n-1)}\mathbb{Q} + \frac{r}{n}\mathbb{C} < \left(\frac{r}{n}\right)^2 \mathbb{Q} + \frac{r}{n}\mathbb{C}.$$

Consider the fractional solution $\bar{\mathbf{x}}^r$ where $\bar{x}_i^r = \frac{r}{n}$ for all $i = 1, 2, \ldots, n$. Then, $\phi(\bar{\mathbf{x}}^r) = \left(\frac{r}{n}\right)^2 \mathbb{Q} + \frac{r}{n}\mathbb{C}$. Now, apply the [0, 1]-rounding algorithm discussed earlier to obtain a binary solution $\mathbf{x}^r$ with objective function value $\phi(\mathbf{x}^r) \geq \left(\frac{r}{n}\right)^2 \mathbb{Q} + \frac{r}{n}\mathbb{C}$. It may be noted that the number of components of $\mathbf{x}^r$ with value 1 however need not be equal to r. Now, compute $\mathbf{x}^r$ for $r = 0, 1, 2 \ldots, n$ and choose the overall best solution, say $\mathbf{x}^*$. Then, $\phi(\mathbf{x}^*) > F^*$. When $\mathbb{Q} < 0$ and $r^0 \in (0,n)$, and $r^0 \neq r^1, r^2$, then also we can find a solution $\mathbf{x}^r$ with $\phi(\mathbf{x}^r) > F^*$ by using the [0, 1]-rounding algorithm and the fractional solution $\bar{x}_i^r = \frac{r^0}{n}$ for all $i = 1, 2, \ldots, n$.

If $\mathbf{c}$ is the zero vector and the row sum and column sum of $\mathbf{Q}$ are zeros, $F^* = \mathcal{A}(\mathbf{Q},\mathbf{c}) = \mathcal{A}_r(\mathbf{Q},\mathbf{c}) = 0$ for $r = 0, 1, \ldots, n$. However, in this case $\phi(\mathbf{x}) = \phi(\bar{\mathbf{x}})$ where $\bar{\mathbf{x}}$ is the complement of $\mathbf{x}$ and hence one of the components of $\mathbf{x}$ can be fixed at 0 (see Chap. 1). Now apply the algorithms discussed earlier on the reduced problem to obtain a better performance guarantee than zero on the objective function value of the resulting solution.

8.3.1 Average Value of Solutions of the Ising QUBO

Let us now consider some algorithms for the Ising QUBO that produce no worse than average solutions. In other words, algorithms with the normalized relative error $\kappa \leq 1$. We first present a closed form formula to compute the average of the objective function values of all 2^n solutions of the Ising QUBO. This can be derived from the formula for $\mathcal{A}(\mathbf{Q},\mathbf{c})$ obtained for QUBO and the transformation between QUBO and Ising QUBO. However, a direct proof is given below. Let $\mathcal{V}(\mathbf{A},\mathbf{b})$ be the average of the objective function values of all solutions of the Ising QUBO instance $(\mathbf{A}, \mathbf{b})$.

Theorem 8.10 $\mathcal{V}(\mathbf{A}, \mathbf{b}) = \text{tr}(\mathbf{A})$

Proof Consider n independent random variables $x_1, x_2, \ldots, x_n$ which take values -1 or 1 with probability $\frac{1}{2}$ each. Then, the expected value of $x_i = 0$. For $i \neq j$, since x_i and x_j are independent, $\mathbf{E}(x_i x_j) = \mathbf{E}(x_i)\mathbf{E}(x_j) = 0$. Also, $\mathbf{E}(x_i^2) = 1$. Now, consider the random variable $X = \sum_{i=1}^n \sum_{j=1}^n a_{ij} x_i x_j + \sum_{i=1}^n b_i x_i$. By the linearity property of expectation, the expected value $\mathbf{E}(X)$ of X is $\text{tr}(\mathbf{A})$ and hence $\mathcal{V}(\mathbf{A}, \mathbf{b}) = \text{tr}(\mathbf{A})$.

Unless otherwise stated, hereafter, we assume that the diagonal elements of $\mathbf{A}$ are zeros and hence $\mathcal{V}(\mathbf{A}, \mathbf{b}) = 0$. For QUBO, computing a solution with a non-negative objective function value is trivial since $\mathbf{x} = \mathbf{0}$ satisfies this. Further, the problem of extending any partial solution, to a full solution without worsening the objective function value is also trivial since we can force the remaining variables to have value zero. However, both these problems are not trivial (though simple) for the case of the Ising QUBO. Since $\mathcal{V}(\mathbf{A}, \mathbf{b}) = 0$, we can convert the instance $(\mathbf{A}, \mathbf{b})$ into an equivalent instance of QUBO and compute a solution that is no worse than average. Now, convert the solution obtained back to that of the Ising QUBO and we have a solution with a non-negative objective function value.

Let us now discuss a very simple and direct algorithm to compute a solution to the Ising QUBO with a non-negative objective function value (no worse than average solution). In fact, this algorithm can be used to compute a no-worse than average solution for QUBO as well because of the equivalent transformation between QUBO and the Ising QUBO discussed earlier. The algorithm starts with a partial solution and extends it to full solution by incrementally fixing a free variable at value 1 or -1. Let $\mathbf{x}^S$ be a partial solution to the Ising QUBO defined by the set $S \subseteq \{1, 2, \ldots, n\}$. That is, x_j for $j \in S$ already has an assigned value of -1 or 1. The objective function value of this partial solution is

$$\varphi^S(\mathbf{x}^S) = \sum_{i \in S} \sum_{j \in S} a_{ij} x_i x_j + \sum_{j \in S} b_j x_j \tag{8.12}$$

Algorithm 4: The partial solution extension algorithm

Let $\{x_j^* : j \in S\}$ *be a given partial solution*;
$\bar{S} \leftarrow \{1, 2, \ldots, n\} \setminus S$;
while $\bar{S} \neq \emptyset$ **do**
  Choose $k \in \bar{S}$ and set $x_k^* = 1$;
  if $b_k + \sum_{j \in S}(a_{kj} + a_{jk})x_j^* < 0$ **then** $x_k^* \leftarrow -1$;
  $S \leftarrow S \cup \{k\}$, $\bar{S} \leftarrow \bar{S} \setminus \{k\}$;
end
output $\mathbf{x}^*$

Theorem 8.11 *The partial solution extension algorithm, starting with a partial solution* $\mathbf{x}^s$ *of the Ising QUBO corresponding to* $S \subseteq \{1, 2, \ldots, n\}$*, produces a solution* $\mathbf{x}^*$ *of the Ising QUBO such that* $\varphi(\mathbf{x}^*) \geq \varphi^s(\mathbf{x}^s)$ *in* $O(n(n - |S|))$ *time. Further, when* $S = \emptyset$*,* $\varphi(\mathbf{x}^*) \geq 0$*.*

Proof Without loss of generality, assume that $S = \{1, 2, \ldots, r\}$. For any solution $\mathbf{x}$ and $i = 1, 2, \ldots, n$, let $\alpha_i(\mathbf{x}) = b_i + \sum_{j=1}^{i-1}(a_{ij} + a_{ji})x_j$. Then $\varphi(\mathbf{x}) = \sum_{i=1}^{n} x_i \alpha_i(\mathbf{x})$. Let $\mathbf{x}^k$ and $\mathbf{x}^{k+1}$ be the partial solutions produced in the iterations k and $k+1$ respectively of the algorithm and S_k and S_{k+1} be the associated values of S in these iterations. Then, by construction, $\varphi^{s_{k+1}}(\mathbf{x}^{k+1}) \geq \varphi^{s_k}(\mathbf{x}^k)$ and this establishes that $\varphi(\mathbf{x}^*) \geq \varphi^s(\mathbf{x}^s)$. Also, the computation in each iteration can be carried out in $O(n)$ time and this proves the first part. For the second part, since $S = \emptyset$ we can consider that the starting partial solution has its objective function value zero and from the arguments for the first part of the theorem, the result follows.

The partial solution extension algorithm is described in [19] and it guarantees a solution to the Ising QUBO with objective function value no worse than $\mathcal{V}(\mathbf{A}, \mathbf{0})$ (which is zero) when the diagonal elements of $\mathbf{A}$ are zeros. The same solution will have value at least $\mathrm{tr}(\mathbf{A})$ if diagonal elements are arbitrary. Since $\mathbf{b}$ can be assumed to be zero without loss of generality, we have

Corollary 8.3 *A solution to the general Ising QUBO* $(\mathbf{A}, \mathbf{b})$ *with objective function value no worse than* $\mathcal{V}(\mathbf{A}, \mathbf{b})$ *can be identified in* $O(n^2)$ *time.*

8.4 Domination Analysis of Algorithms

Let us now consider the performance measure of domination ratio introduced by Glover and Punnen [12]. We first consider the Ising QUBO and then discuss how to extend the results obtained for the Ising QUBO to QUBO.

Let $F(0) = \{\mathbf{x} \in \{-1, 1\}^n : \varphi(\mathbf{x}) \leq 0\}$. To carry out domination analysis of algorithms that produce no worse than average solutions, we first want to estimate $|F(0)|$. Note that $|\{-1, 1\}^n| = 2^n$ and we will show that $\frac{|F(0)|}{2^n} \geq \frac{1}{38}$. To establish this, we closely follow the analysis provided by Alon et al. [3] for the equivalent maximum cut problem. We first assume that $\mathbf{A}$ is an upper triangular non-zero matrix and the diagonal elements of $\mathbf{A}$ are zeros. Further, for simplicity, we first assume that $\mathbf{b} = \mathbf{0}$. For such an instance $(\mathbf{A}, \mathbf{0})$ of the Ising QUBO, $\mathcal{V}(\mathbf{A}, \mathbf{0}) = 0$.

Let $x_1, x_2, \ldots, x_n$ be n independent random variables, each distributed uniformly on $\{-1, 1\}$ and let X be a random variable defined by

$$X = \sum_{i=1}^{n-1} \sum_{j=i+1}^{n} a_{ij} x_i x_j. \tag{8.13}$$

Since X is defined using an upper triangular matrix with zeros on the diagonal, it is convenient to associate with X a complete undirected graph K_n having edge set $(e_1, e_2, \ldots, e_m)$ where $m = n(n-1)/2$ such that each e_k maps uniquely to an ordered pair $(i, j), i < j$ and the weight $a(e_k)$ of the edge e_k is a_{ij}.

Lemma 8.3 ([3]) *The random variable X in Eq.* (8.13) *satisfies the following properties:*

1. $\mathbf{E}(X) = 0$ *and* $\mathbf{E}(X^2) = \sum_{k=1}^{m} a^2(e_k) > 0.$
2. $\left(\mathbf{E}(X^2)\right)^2 = \sum_{k=1}^{m} a^4(e_k) + 2 \sum_{1 \le k < \ell \le m} a^2(e_k)a^2(e_\ell)$
3. $\mathbf{E}\left(X^4\right) \le \sum_{k=1}^{m} a^4(e_k) + 30 \left(\sum_{1 \le k < \ell \le m} a^2(e_k)a^2(e_\ell) \right).$

Proof Since the random variable x_i takes values 1 or -1, each with probability $\frac{1}{2}$, $\mathbf{E}(x_i^k) = 0$ if k is an odd number and 1 if k is an even number. Now, (1) and (2) follow from the linearity property of expectation. Let us now prove (3). In the multinomial expansion of $X^4 = \left(\sum_{i=1}^{n-1} \sum_{j=i+1}^{n} a_{ij}x_ix_j\right)^4$, any term having an odd power of x_i as a factor contributes nothing to $\mathbf{E}(X^4)$ since $\mathbf{E}(x_i^k) = 0$ for odd k. Thus, we need to consider only three types of terms.

1. Terms of the form $(a_{ij}x_ix_j)^4$. These terms contribute to the expectation a_{ij}^4 and the corresponding multinomial coefficient is 1. So, the overall contribution of such terms to $\mathbf{E}(X^4)$ is $\sum_{i=1}^{n-1} \sum_{j=i+1}^{n} a_{ij}^4$.
2. Terms of the form $(a_{ij}x_ix_j)^2(a_{rs}x_rx_s)^2$, where (i, j) and (r, s) are distinct edges of K_n. The multinomial coefficient in this case is 6 and the overall contribution of the terms of this type to $\mathbf{E}(X^4)$ is $\sum_{1 \le k < \ell \le m} a(e_k)a(e_\ell)$.
3. Terms of the form $(a_{ij}x_ix_j)(a_{k\ell}x_kx_\ell)(a_{rs}x_rx_s)(a_{uv}x_ux_v)$. This makes a non-zero contribution to the expectation when $(i, j), (k, \ell), (r, s), (u, v)$ are distinct edges that form a 4-cycle in K_n. Let C be the collection of all such 4-cycles. Then, the resulting contribution of any $C \in C$ to $\mathbf{E}(X^4)$ is $\prod_{(i,j) \in C} a_{ij}$. The multinomial coefficient in this case is 24. Thus the overall contribution of terms of this type to $\mathbf{E}(X^4)$ is $24 \sum_{C \in C} \prod_{e \in C} a_e$.

Thus,

$$\mathbf{E}(X^4) = \sum_{k=1}^{m} a(e_k)^4 + 6 \sum_{1 \le k < \ell \le m} a^2(e_k)a^2(e_\ell) + 24 \sum_{C \in C} \prod_{e \in C} a(e) \tag{8.14}$$

Now, for any 4-cycle with edges e_i, e_j, e_k, e_ℓ appearing in this order,

$$a(e_i)a(e_j)a(e_k)a(e_\ell) \le \frac{1}{2}\left(a^2(e_i)a^2(e_k) + a^2(e_j)a^2(e_\ell)\right).$$

Thus, from (8.14),

$$\mathbf{E}(X^4) \le \sum_{k=1}^{m} a(e_k)^4 + 30\left(\sum_{1\le k<\ell\le m} a^2(e_k)a^2(e_\ell)\right)$$

and this completes the proof.

Lemma 8.4 ([3]) $\mathbf{E}(X^4) \le 15\left(\mathbf{E}(X^2)\right)^2$.

Proof Let $\alpha = \sum_{k=1}^{m} a(e_k)^4$ and $\beta = \sum_{1\le k<\ell\le m} a^2(e_k)a^2(e_\ell)$. Then,

$$\frac{\mathbf{E}(X^4)}{\left(\mathbf{E}(X^2)\right)^2} \le \frac{\alpha+30\beta}{\alpha+2\beta} \le 15 - \frac{14\alpha}{\alpha+2\beta} \le 15$$

and the result follows.

It may be noted that Lemma 8.4 is a strengthened version of Bonami's Lemma [32], which was proved in a more general context, but yielding a weaker constant for our problem. Let us now state a bound, proved in [3], on the probability of a random variable with bounded fourth moment being non-negative (non-positive).

Lemma 8.5 ([3]) *Let X be a real valued random variable satisfying $\mathbf{E}(X) = 0$, $\mathbf{E}(X^2) = \sigma^2 > 0$ and $\mathbf{E}(X^4) \le b\sigma^4$. Then, $Prob\,(X > 0) \ge \frac{1}{2^{4/3}b}$ and $Prob\,(X < 0) \ge \frac{1}{2^{4/3}b}$.*

Combining Lemmas 8.4 and 8.5, the probability of X being non-negative (non-positive) is at least $\frac{1}{2^{4/3}15} > \frac{1}{38}$. Thus we

Lemma 8.6 *For the Ising QUBO $(\mathbf{A},\mathbf{0})$ where $\mathbf{A}$ is an upper triangular matrix with zeros on the diagonal, $|F(0)| \ge \frac{2^n}{38}$.*

In other words, for the Ising QUBO $(\mathbf{A},\mathbf{0})$, where $\mathbf{A}$ is an upper triangular matrix with zeros on the diagonal, any algorithm that guarantees to produce a non-negative solution has domination ratio of at least $\frac{1}{38}$. In particular, the domination ratio of the partial solution extension algorithm for the Ising QUBO $(\mathbf{A},\mathbf{0})$ is at least $\frac{1}{38}$.

Theorem 8.12 *For the general Ising QUBO $(\mathbf{A},\mathbf{b})$ any algorithm that produces a solution with objective function value no worse than $\mathrm{tr}(\mathbf{A})$ has domination ratio of at least $\frac{1}{38}$.*

Proof First, assume that $\mathbf{A}$ is upper triangular with zeros along the diagonal but $\mathbf{b}$ is arbitrary. Construct the $(n+1)\times(n+1)$ matrix $\mathbf{A}'$ with its $(n+1)$th column as $(\mathbf{b},0)^T$ and the $(n+1)$th row as the zero vector $\mathbf{0}$ in $\mathbb{R}^{n+1}$. Consider the Ising QUBO $(\mathbf{A}',\mathbf{0})$ with variables $\mathbf{x}' = (x_1', x_2', \ldots, x_n', x_{n+1}') \in \{-1,1\}^{n+1}$. Let $\varphi'(\mathbf{x})$ denote the objective function of the Ising QUBO $(\mathbf{A}',\mathbf{0})$. Then $\varphi'(\mathbf{x}') = \varphi'(-\mathbf{x})$. Take any solution $\mathbf{x}'$ of the Ising QUBO $(\mathbf{A}',\mathbf{0})$ with $x_{n+1} = 1$ and let $\mathbf{x} \in \{-1,1\}^n$ be the solution obtained from $\mathbf{x}'$ by truncating the $(n+1)$th component. Then, $\varphi'(\mathbf{x}') = \varphi(\mathbf{x})$. Let $F'(0) = \{\mathbf{x}' \in \{-1, 1\}^{n+1} : \varphi(\mathbf{x}') \leq 0\}$. By Lemma 8.6, $\frac{|F'(0)|}{2^{n+1}} \geq \frac{1}{38}$. Let $F(0) = \{\mathbf{x} \in \{-1, 1\}^n : (\mathbf{x}, 1) \in F'(0)\}$. Since $\varphi(\mathbf{x}') = \varphi(-\mathbf{x}')$ for all $\mathbf{x}' \in F'(0)$, $|F'(0)| = 2|F(0)|$. Further, $F(0)$ is precisely $\{\mathbf{x} \in \{-1, 1\}^n : \varphi(\mathbf{x}) \geq 0\}$. Thus,

$$\frac{|F(0)|}{2^n} = \frac{2|F(0)|}{2^{n+1}} = \frac{|F'(0)|}{2^{n+1}} \geq \frac{1}{38}.$$

The last inequality follows from Lemma 8.6. This proves the theorem for the Ising QUBO instance $(\mathbf{A}, \mathbf{b})$ when $\mathbf{A}$ is upper triangular with zeros on diagonal, but arbitrary $\mathbf{b}$ vector. Including arbitrary elements on the diagonal simply add tr$(\mathbf{A})$ to the objective function value of all solutions and hence the result is true for this case as well. Finally, replacing $\mathbf{A}$ by $\frac{1}{2}(\mathbf{A} + \mathbf{A}^T)$ will not alter the objective function values and hence the result is true for all symmetric matrices and this concludes the proof.

Corollary 8.4 *The domination ratio of any algorithm for the Ising QUBO that produces a solution with objective function value no worse than $\mathcal{V}(\mathbf{A}, \mathbf{b})$ is at least $\frac{1}{38}$.*

In the previous section, we discussed heuristic algorithms for QUBO that guarantee solutions with objective function value no worse than $\mathcal{A}(\mathbf{Q}, \mathbf{c})$, but deferred the discussion of domination analysis. To identify the estimates for the domination ratio of such algorithms, we want to determine the number of solutions with objective function value no worse than $\mathcal{A}(\mathbf{Q}, \mathbf{c})$.

Theorem 8.13 *For $\mathbf{x} \in \{0, 1\}^n$ and $\hat{\mathbf{x}} = 2\mathbf{x} - \mathbf{e} \in \{-1, 1\}^n$, if $\phi(\mathbf{x}) \leq \mathcal{A}(\mathbf{Q}, \mathbf{c})$ then $\varphi(\hat{\mathbf{x}}) \leq \mathcal{V}(\mathbf{A}^q, \mathbf{b}^q) + K^q$. Likewise, for $\mathbf{x} \in \{-1, 1\}^n$ and $\tilde{\mathbf{x}} = \frac{1}{2}(\mathbf{x} + \mathbf{e}) \in \{0, 1\}^n$ if $\varphi(\mathbf{x}) \leq \mathcal{V}(\mathbf{A}, \mathbf{b})$ then $\phi(\tilde{\mathbf{x}}) \leq \mathcal{A}(\mathbf{Q}^a, \mathbf{c}^a) + K^a$.*

Proof Note that $\mathcal{A}(\mathbf{Q}, \mathbf{c}) = \mathcal{V}(\mathbf{A}^q, \mathbf{b}^q) + K^q$ and $\phi(\mathbf{x}) = \varphi(\hat{\mathbf{x}}) + K^q$. Since $\phi(\mathbf{x}) \leq \mathcal{A}(\mathbf{Q}, \mathbf{c})$, we have $\varphi(\hat{\mathbf{x}}) + K^q \leq \mathcal{A}(\mathbf{Q}, \mathbf{c}) = \mathcal{V}(\mathbf{A}^q, \mathbf{b}^q) + K^q$ and hence $\varphi(\mathbf{x}) \leq \mathcal{V}(\mathbf{A}^q, \mathbf{b}^q)$. The proof of the second part follows analogously.

Let $\Omega(\mathbf{Q}, \mathbf{c}) = \{\mathbf{x} \in \{0, 1\}^n : \phi(\mathbf{x}) \leq \mathcal{A}(\mathbf{Q}, \mathbf{c})\}$ and $\Gamma(\mathbf{A}, \mathbf{b}) = \{\mathbf{x} \in \{-1, 1\}^n : \varphi(\mathbf{x}) \leq \mathcal{V}(\mathbf{A}, \mathbf{b})\}$.

Corollary 8.5 $\Omega(\mathbf{Q}, \mathbf{c}) = \Gamma(\mathbf{A}^q, \mathbf{b}^q)$ *and* $\Gamma(\mathbf{A}, \mathbf{b}) = \Omega(\mathbf{Q}^a, \mathbf{c}^a)$.

As a consequence of Theorem 8.13 and Corollaries 8.4 and 8.5, we have

Theorem 8.14 *Any algorithm that produces a solution with value no worse than $\mathcal{A}(\mathbf{Q}, \mathbf{c})$ for the instance $(\mathbf{Q}, \mathbf{c})$ of a QUBO have domination ratio of at least $\frac{1}{38}$*

A proof of Theorem 8.14 is explored in [35] using a different random variable than X and the expressions there are too cumbersome. The domination ratio of any algorithm for QUBO or the Ising QUBO that guarantees no worse than average solutions (and no other guarantees) cannot be more than $\frac{1}{4}$. Consider the instance $(\mathbf{Q}, \mathbf{c})$ of QUBO where $\mathbf{c} = (-1, 0, 0, \ldots, 0)$ and $q_{12} = 1$ with all other elements of $\mathbf{Q}$ are zeros. Here $\mathcal{A}(\mathbf{Q}, \mathbf{c}) = -\frac{1}{4}$. The set of solutions with objective function value less than or equal to $-\frac{1}{4}$ is precisely those solutions with $x_1 = 1$ and $x_2 = 0$. There are 2^{n-2} such solutions and so the domination ratio cannot be more than $\frac{1}{4}$. When G is bipartite, this lower bound is the precise value of the domination ratio of algorithms that guarantee no worse than average solutions. For a nice proof of this, see Chap. 10.

Let us now state a non-approximability result for QUBO in terms of domination ratio.

Theorem 8.15 *Let α and β be relatively prime natural numbers bounded above by a polynomial function of the input length of the associated QUBO such that $\alpha > \beta$ and let $\delta = \frac{\alpha}{\beta}$. Then, unless P=NP, no polynomial time heuristic algorithm for QUBO can have domination ratio more than $1 - 2^{\lceil \frac{n}{\delta} \rceil - n}$.*

For an analogues result, along with a validity proof for the case of the bipartite QUBO, we refer to Chap. 10.

8.5 Relative Performance Ratio and Approximation Algorithms

As observed earlier in this chapter, the Ising QUBO can be transformed into an equivalent QUBO and viceversa in such a way that the corresponding optimal solutions sets are preserved. However, because of the resulting constants K^a and K^q in the objective function, the relative performance ratio is not preserved under these transformations. Thus, existence of an ϵ-approximation algorithm for the Ising QUBO need not imply directly the existence of an ϵ-approximation for QUBO and viceversa. However, for a fairly large class of problem instances, the existence of a polynomial time ϵ-approximation algorithm for QUBO implies the existence of a polynomial time ϵ-approximation algorithm for the Ising QUBO and viceversa.

We start with a negative result that, the Ising QUBO $(\mathbf{A}, \mathbf{b})$, when the diagonal elements of $\mathbf{A}$ are permitted to take at least one negative value, cannot be approximated by any factor in polynomial time [4]. We now show that

Theorem 8.16 *Unless P=NP, no polynomial time algorithm exists to determine if the optimal objective function value of the Ising QUBO is non-negative or not.*

Proof Consider the Ising QUBO $(\mathbf{A}, \mathbf{0})$ where the diagonal elements of $\mathbf{A}$ are zeros and $\mathbf{A}$ has integer entries. We call such an Ising QUBO as the *zero diagonal integer Ising* QUBO. Suppose that there is a polynomial time algorithm $\mathcal{L}$ which determines if the optimal objective function value of an instance of the Ising QUBO (without any restrictions on $\mathbf{A}$) is non-negative or not. We will show that using this algorithm, we can compute precisely the optimal objective function value of the zero diagonal integer Ising QUBO in polynomial time. Note that the optimal objective function value of the zero diagonal integer Ising QUBO $(\mathbf{A}, \mathbf{0})$ is non-negative (Lemma 8.10) and belongs to $[0, n^2 a_{\max}]$ where $a_{\max} = \max_{1 \le i,j \le n} |a_{ij}|$. Let $\mathbf{A}'$ be the matrix obtained from $\mathbf{A}$ by assigning $a_{11} = -K$ for some constant K. Now by applying $\mathcal{L}$ on the Ising QUBO $(\mathbf{A}', \mathbf{0})$ by choosing $K \in [0, n^2 a_{\max}]$ in a binary search fashion, we can determine the precise optimal objective function value of the zero diagonal integer Ising QUBO $(\mathbf{A}, \mathbf{0})$ in $O(f(\mathcal{L}) \log(n^2 a_{\max}))$ time where $O(f(\mathcal{L}))$ is the complexity of $\mathcal{L}$. Since the evaluation version of the zero diagonal integer Ising QUBO is NP-hard, the result follows.

The result of Theorem 8.16 was mentioned in[4] and suggested a reduction from the maximum cut problem. Despite the negative result established in Theorem 8.16, when the diagonal elements of $\mathbf{A}$ are restricted to be zero, as we will observe later in this subsection, interesting ϵ-approximation algorithms can be obtained for the Ising QUBO. For QUBO, restricting diagonal elements of $\mathbf{Q}$ to be zero is not much of an advantage. Our next theorem shows that obtaining meaningful bounds on the relative performance ratio for a polynomial time heuristic algorithm for QUBO is difficult, even in some very restricted form.

Theorem 8.17 *Unless NP = ZPP, there does not exist any polynomial time $n^{1-\epsilon}$-approximation algorithm for QUBO for any $\epsilon > 0$, even if the diagonal elements of $\mathbf{Q}$ are zeros, $\mathbf{Q}$ contains only two distinct non-zero entries, and all elements of $\mathbf{c}$ are unity.*

Proof Let $G = (V, E)$ be a graph on n nodes with $c_i = 1$ for all $i \in V$ and $q_{ij} = q_{ji} = 0$ if $(i, j) \notin E$. Choose $q_{ij} = q_{ji} = -(n + 1)$ for all $(i, j) \in E$. Now, suppose that there is a $n^{1-\epsilon}$-approximation algorithm for QP and let S be the solution produced by such an algorithm for the instance $(\mathbf{Q}, \mathbf{c})$ of QP. (Here we use the graph theoretic form of QP.) Note that $\text{OPT}(QP)$ is the size of the maximum stable set in G and it is at least one. Further, the subgraph $G[S]$ of G induced by S must be a stable set and the heuristic solution value is precisely the size of this stable set. Thus, the existence of an $n^{1-\epsilon}$-approximation algorithm for QP implies the existence of such an algorithm for the maximum stable set problem. The result now follows from the fact that no polynomial time $n^{1-\epsilon}$-approximation algorithm for the maximum stable set problem can exist unless NP=ZPP [10, 17, 18].

It may be noted that ZPP (zero-error probabilistic polynomial time) is the complexity class of decision problems where a probabilistic Turing machine exists which returns a correct YES or NO answer in time polynomial in expectation for all possible input [6]. For more approximation hardness results for the maximum stable set problem (maximum clique problem) that translate to approximation hardness results for QP, we refer to [10, 16, 38, 39]. Despite the negative result established in Theorem 8.17, it is possible to identify special cases of QUBO that admit ϵ-approximation algorithms and even fully polynomial approximation schemes (FPTAS) [7, 20]. We will come to this later. From Theorem 8.17 we have the corollary

Corollary 8.6 *No polynomial time $n^{1-\epsilon}$-approximation algorithm exists for the QUBO $(\mathbf{Q}, \mathbf{0})$ for any $\epsilon > 0$, unless NP = ZPP, if $\mathbf{Q}$ is an arbitrary matrix.*

Let us now explore the question: under what conditions an ϵ-approximation algorithm for the QUBO guarantees an ϵ-approximation algorithm for the Ising QUBO and vice versa.

A matrix M is said to be a *zero sum matrix* if the sum of its elements is zero and a *strong zero sum matrix* if the sum of its elements in each row is zero and the sum of the elements of each column is also zero. There are many special cases of this class of matrices. For example, the Laplacian matrix of a graph is a strong zero sum matrix and so is the matrix $I - D$, where D is a doubly stochastic matrix and I is an identity matrix. A vector $\mathbf{c}$ is said to be a *zero sum vector* if the sum of its elements is zero.

Theorem 8.18 *If the Ising QUBO* $(\mathbf{A}, \mathbf{0})$, *where* $\mathbf{A}$ *is a zero sum matrix with diagonal entries zero, can be solved by a polynomial time* ϵ*-approximation algorithm then the QUBO* $(\mathbf{Q}, \mathbf{c})$ *where* $\mathbf{Q}$ *is a zero sum matrix with diagonal elements zero and* $\mathbf{c}$ *is a zero sum vector, can be solved by a polynomial time* ϵ*-approximation algorithm.*

Proof Consider the instance $(\mathbf{Q}, \mathbf{c})$ of the QUBO where $\mathbf{Q}$ is a zero sum matrix and $\mathbf{c}$ is a zero sum vector. Construct the equivalent Ising QUBO instance $(\mathbf{A}^q, \mathbf{b}^q)$ using the transformation $\mathbf{x} = \frac{1}{2}(\hat{\mathbf{x}}+\mathbf{e})$ (See Eqs. (8.3) and (8.4) for the definition of $\mathbf{A}^q$ and $\mathbf{b}^q$). Since $\mathbf{Q}$ is a zero sum matrix with zeros on the diagonal, $\mathbf{A}^q$ is also a zero sum matrix with zeros on the diagonal. Further, $\mathbf{c}$ is a zero sum vector and therefore the resulting constant value K^q is also zero. Thus, $\mathbf{x}^T\mathbf{Q}\mathbf{x} + \mathbf{c}^T\mathbf{x} = \hat{\mathbf{x}}^T\mathbf{A}^q\hat{\mathbf{x}} + \mathbf{b}^q\hat{\mathbf{x}}$ where $\hat{\mathbf{x}} = 2\mathbf{x} - \mathbf{e} \in \{-1, 1\}$. Now, if $\hat{\mathbf{x}}^*$ is an ϵ-optimal solution for the Ising QUBO $(\mathbf{A}^q, \mathbf{b}^q)$ then its pre-image, $\mathbf{x} = \frac{1}{2}(\hat{\mathbf{x}} + \mathbf{e})$ is an ϵ-optimal solution of $(\mathbf{Q}, \mathbf{c})$. Now consider the matrix

$$\mathbf{A}' = \left[\begin{array}{c|c} \mathbf{A}^q & \frac{1}{2}\mathbf{b}^T \\ \hline \frac{1}{2}\mathbf{b} & 0 \end{array}\right]$$

and the $n+1$ dimensional vector $\tilde{\mathbf{x}} \in \{-1, 1\}^{n+1}$. For the Ising QUBO $(\mathbf{A}', \mathbf{0})$, if $\tilde{\mathbf{x}}$ is an ϵ-optimal solution, then $-\tilde{\mathbf{x}}$ is also an ϵ-optimal solution. Thus, for any ϵ-optimal solution to $(\mathbf{A}', \mathbf{0})$, there exists an ϵ-optimal solution with the $(n + 1)$th component equal to 1 and the first n components of such a solution will be an ϵ-optimal solution to the Ising QUBO $(\mathbf{A}^q, \mathbf{b}^q)$, and the result follows.

We have an analogous result for the Ising QUBO as well. The proof is similar to that of the above theorem.

Theorem 8.19 *If the QUBO* $(\mathbf{Q}, \mathbf{0})$, *where* $\mathbf{Q}$ *is a strong zero sum matrix with diagonal entries zero can be solved by a polynomial time* ϵ*-approximation algorithm then the Ising QUBO* $(\mathbf{A}, \mathbf{0})$ *where* $\mathbf{A}$ *is a strong zero sum matrix with diagonal elements zero can be solved by a polynomial time* ϵ*-approximation algorithm.*

Theorem 8.19 is also true if $\mathbf{A}$ is a zero sum matrix (not necessarily strong zero sum matrix) provided the QUBO $(\mathbf{Q}, \mathbf{c})$ with $\mathbf{Q}$ is a zero sum matrix and $\mathbf{c}$ is a zero sum vector can be solved by a polynomial time ϵ-approximation algorithm.

Theorems 8.18 and 8.19 are valid even with the assumption that diagonal elements of $\mathbf{A}$ and $\mathbf{Q}$ are arbitrary and with this assumption, one can always force the strong zero sum property (see Chap. 1) to $\mathbf{A}$ and $\mathbf{Q}$ leading to a much more general result, but with less applicability because of the resulting stronger hypothesis. Note that, for most usable ϵ-approximation algorithms for the Ising QUBO and the QUBO, zero-diagonal assumption is made and the condition is added to the theorem to highlight this and also to emphasize that the zero-digonality is preserved under the transformation we used.

8.5.1 ϵ-Approximation Algorithms for the Ising QUBO

Hereafter, without loss of generality we assume that $\mathbf{b} = \mathbf{0}$. In addition, we assume that the diagonal elements of $\mathbf{A}$ are zeros, which is indeed a restriction (See Theorem 8.16). Let G' be the support graph of $\mathbf{A}$ with the weight of the edge (i, j) in G' being $\omega_{ij} = 2|a_{ij}|$.

The value $\omega(M)$ of any matching $M = \{(u_1, v_1), (u_2, v_2), \ldots, (u_t, v_t)\}$ in G' is given by $\sum_{i=1}^{t} \omega_{ij} = 2\sum_{i=1}^{t} |a_{u_i v_i}|$. We now present a theorem linking the objective function value of a particular solution to the Ising QUBO and a matching M in G' as proved in [19].

Theorem 8.20 *If M is a matching in G' then a solution $\mathbf{x}^*$ of IQ with $\varphi(\mathbf{x}^*) \geq \omega(M)$ can be identified in $O(n^2)$ time.*

Proof Let $M = \{(u_1, v_1), (u_2, v_2), \ldots, (u_t, v_t)\}$ and for $1 \leq k \leq t$, let $M_k = \{(u_1, v_1), (u_2, v_2), \ldots, (u_k, v_k)\}$. Then, obviously $M_t = M$. We prove the result using mathematical induction on k. When $k = 1$, we have $M_1 = \{(u_1, v_1)\}$ and let $P = \{u_1, v_1\} \subseteq \{1, 2, \ldots, n\}$. Define the partial solution $\mathbf{x}^p$ corresponding to the set P as follows: If $a_{u_1 v_1} \geq 0$ assign $x^p_{u_1} = x^p_{v_1} = 1$ else assign $x^p_{u_1} = -x^p_{v_1} = 1$. Note that $\varphi^p(\mathbf{x}^p) \geq \omega(M_1)$ (in fact, equality holds here and recall the definition of the value $\varphi^p(.)$ of a partial solution from Eq. (8.12)). Now, extend the partial solution $\mathbf{x}^p$ corresponding to M_1 to a solution $\mathbf{x}^*$ with $\varphi(\mathbf{x}^*) \geq \varphi^p(\mathbf{x}^p) = \omega(M_1)$ using the partial solution extension algorithm discussed earlier. Thus, the result is true for $k = 1$. Now, suppose that there is a partial solution $\mathbf{x}^p$ defined by the set $P = \{u_1, u_2, \ldots, u_k, v_1, v_2, \ldots, v_k\}$ such that $\varphi^p(\mathbf{x}^p) \geq \omega(M_k)$ for some $k < t$. Let $S = \{u_{k+1}, v_{k+1}\}$. Note that $P \cap S = \emptyset$. Let $\mathbf{x}^s$ be the partial solution defined by $x^s_{u_{k+1}} = x^s_{v_{k+1}} = 1$ if $a_{u_{k+1} v_{k+1}} \geq 0$ and $x^s_{u_{k+1}} = -x^s_{v_{k+1}} = 1$ if $a_{u_{k+1} v_{k+1}} < 0$. Now, consider the partial solutions $\bar{x} = \mathbf{x}^p \cup \mathbf{x}^s$ and $\hat{\mathbf{x}} = -\mathbf{x}^p \cup \mathbf{x}^s$ defined by the set $D = P \cup S$. Note that $\varphi^p(\mathbf{x}^p) + \varphi^s(\mathbf{x}^s) \geq \omega(M_{k+1})$ and $\varphi^s(\mathbf{x}^s) \geq 0, \varphi^p(\mathbf{x}^p) \geq 0$. If $\varphi^d(\bar{\mathbf{x}}) \geq \varphi(\mathbf{x}^p) + \varphi(\mathbf{x}^s)$ then extend the partial solution $\bar{\mathbf{x}}$ to a solution $\mathbf{x}^*$ of IQ using the partial solution extension algorithm and we have $\varphi(\mathbf{x}^*) \geq \varphi(\mathbf{x}^p) + \varphi(\mathbf{x}^s)$, which implies $\varphi(\mathbf{x}^*) \geq \omega(M_{k+1})$. If $\varphi^d(\bar{\mathbf{x}}) < \varphi(\mathbf{x}^p) + \varphi(\mathbf{x}^s)$, then the contribution $\delta(P, S)$ of all elements in the cut set (P, S) on the subgraph of G' induced by $P \cup S$

to $\varphi^d(\bar{\mathbf{x}})$ is negative. i.e., $\delta(P, S) < 0$. This implies that

$$\varphi^d(\hat{\mathbf{x}}) = \varphi^d(\mathbf{x}^p) + \varphi(\mathbf{x}^s) - \delta(P, S) > \varphi^d(\mathbf{x}^p) + \varphi(\mathbf{x}^s).$$

Thus, as discussed earlier, we can extend the partial solution $\hat{\mathbf{x}}$ to a solution $\mathbf{x}^*$ of IQ such that $\varphi(\mathbf{x}^*) \geq \omega(M_{k+1})$. The first part of the theorem now follows from mathematical induction.

To prove the $O(n^2)$ complexity of constructing $\mathbf{x}^*$, we incrementally build partial solutions corresponding to $M_1, M_2, \ldots, M_t$ and if necessary, extend the final partial solution to a solution of IQ without worsening the objective function value. By careful updating, from a partial solution corresponding to M_k, a partial solution corresponding to M_{k+1} can be obtained in $O(n)$ time and hence the overall complexity of constructing $\mathbf{x}^*$ is $O(n^2)$.

Following [19], we now show that if the matching M is selected using a greedy algorithm, a $\frac{1}{2\Delta}$-optimal solution can be obtained for the Ising QUBO $(\mathbf{A}, \mathbf{0})$ whenever the diagonal elements of $\mathbf{A}$ are zeros. Consider the following greedy matching algorithm on G'.

Algorithm 5: The greedy matching algorithm

Input: *The graph* $G' = (V, E)$ *and edge-weight vector* ω. $G^* \leftarrow G'$ and $E^* \leftarrow E$;
$k \leftarrow 0$;
while $E^* \neq \emptyset$ **do**
  $k \leftarrow k + 1$;
  choose an $e_k = (u_k, v_k) \in G^*$ with maximum weight;
  $E_k = \{(i, j) : (i, j) \in E^*$ and either $i \in \{u_k, v_k\}$ or $j \in \{u_k, v_k\}$;
  $E^* \leftarrow E^* - E_k$;
  $G^* \leftarrow G^* - \{u_k, v_k\}$ // remove nodes u_k and v_k from G^*;
end
output the matching $M^* = \{e_1, e_2 \ldots e_k\}$ in G'

Lemma 8.7 ([19]) *The matching* M^* *produced by the greedy matching algorithm satisfies* $\omega(M^*) \geq \frac{1}{2\Delta}\omega(E)$, *where* $\omega(E) = \sum_{(i,j)\in E} |a_{ij}|$.

Proof By construction $\omega(e_1) \geq \omega(e_2) \geq \cdots \geq \omega(e_k)$. Note that $E_i = \{e : e$ has a common end point with $e_i\} - \cup_{j=1}^{i-1} E_j$ and $|E_i| \leq deg(u_i) + deg(v_i) \leq 2\Delta$. Since e_i is an edge with maximum weight in E_i,

$$\omega(e_i) \geq \frac{\omega(E_i)}{|E_i|} \geq \frac{\omega(E_i)}{2\Delta}$$

Since $E_1, E_2, \ldots, E_k$ is a partition of E, we have

$$\omega(M^*) = \sum_{i=1}^{k} \omega(e_i) \geq \sum_{i=1}^{k} \frac{\omega(E_i)}{2\Delta} = \frac{\omega(E)}{2\Delta}.$$

Theorem 8.21 ([19]) *A $\frac{1}{2\Delta}$-optimal solution to IQ can be identified in $O(n^2 \log n)$ time, where $\Delta > 1$ is the maximum degree of nodes in the support graph G of $\mathbf{A}$.*

Proof Let M^* be the matching produced by the greedy matching algorithm. Note that the greedy matching algorithm can be implemented in $O(n^2 \log n)$ time by sorting the edge-weight vector ω. From Theorem 8.20, we can compute a solution $\mathbf{x}^*$ of IQ such that $\phi(\mathbf{x}^*) \geq \omega(M^*)$ in $O(n^2)$ time. Let $|A| = \sum_{i=1}^{n}\sum_{j=1}^{n} |a_{ij}|$. Since $\text{OPT}(QP) \leq |A|$, from Lemma 8.7

$$\phi(\mathbf{x}^*) \geq \frac{1}{2\Delta}\omega(E) = \frac{1}{2\Delta}|A| \geq \frac{1}{2\Delta}\text{OPT}(IQ)$$

and the result follows.

When the support graph of $\mathbf{A}$ is a complete graph K_n, the performance ratio of the above algorithm becomes $\frac{1}{2(n-1)}$. We now discuss a polynomial time approximation algorithm for IQ with better performance ratio, in expectation.

8.5.2 Computing ϵ-Optimal Solutions from SDP Relaxation

Recall that the diagonal elements of $\mathbf{A}$ are assumed to be zeros and we are considering the Ising QUBO $(\mathbf{A}, \mathbf{0})$. Throughout this subsection, we follow the graph theoretic representation of the Ising QUBO. We are going to use the $[-1, 1]$-rounding algorithm again, but the solution $\mathbf{x}^* \in [-1, 1]^n$ used for rounding is constructed from an optimal solution of a semidefinite programming (SDP) relaxation of IQ. Such rounding algorithms are discussed by various authors [4, 8, 11, 24, 26, 30] who established performance guarantees similar to that we discuss here. Our presentation here follows closely the book [11] which builds on the discussions given in [24]. Consider the SDP relaxation of the Ising QUBO (see Chap. 6 for a detailed discussion of SDP relaxation of QUBO and the Ising QUBO):

$$\begin{aligned} \text{SDP(max):} \quad & \text{Maximize} \quad \psi(\mathbf{v}) = \sum_{(i,j)\in E} a_{ij}\mathbf{v}_i^T\mathbf{v}_j \\ & \text{Subject to:} \quad \mathbf{v} \in S^{n-1}, \end{aligned}$$

where $S^{n-1} = \{\mathbf{v} \in \mathbb{R}^n : ||\mathbf{v}|| \leq 1\}$.

Let $\mathbf{v}_1, \mathbf{v}_2, \ldots, \mathbf{v}_n$ be an optimal solution of the SDP(max) with optimal objective function value $P_{\max}$. Also, let $P_{\min}$ be the optimal objective function value of SDP(min), which is similar to SDP(max) except that we are minimizing the objective function.

Lemma 8.8 *Unless $\mathbf{A}$ is the zero matrix, $P_{\max} > 0$ and $P_{\min} < 0$.*

The proof of this lemma follows from the fact that $\mathcal{V}(\mathbf{A}, \mathbf{0}) = 0$. Let $\boldsymbol{\rho} = (\rho_1, \rho_2, \dots \rho_n) \in \mathbb{R}^n$ be a random vector such that its components ρ_i, $i = 1, 2, \dots, n$ are identically and independently distributed (i.i.d) random variables following the standard normal distribution $N(0, 1)$. (i.e. $\boldsymbol{\rho}$ is a random n-dimensional Gaussian vector.) Let $y_i' = \boldsymbol{\rho}^T \mathbf{v}_i, i = 1, 2, \dots, n$.

Lemma 8.9 ([11]) $\mathbf{E}[y_i' y_j'] = \mathbf{v}_i^T \mathbf{v}_j$. *Further,* $\mathbf{E}\Big[\sum_{(i,j)\in E} a_{ij} y_i' y_j'\Big] = P_{\max}$, *where* $\mathbf{E}(.)$ *denotes the expected value.*

Proof Let $\{\mathbf{e}_1, \mathbf{e}_2, \dots, \mathbf{e}_n\}$ be the standard basis of $\mathbb{R}^n$. Then, by definition of $\boldsymbol{\rho}$, $\boldsymbol{\rho}^T \mathbf{e}_1, \boldsymbol{\rho}^T \mathbf{e}_2, \dots, \boldsymbol{\rho}^T \mathbf{e}_n$ are independent standard normal variates with mean zero and standard deviation one. Thus,

$$\mathbf{E}\left[(\boldsymbol{\rho}^T \mathbf{e}_i)(\boldsymbol{\rho}^T \mathbf{e}_j)\right] = 0 \text{ for } i \neq j \tag{8.15}$$

and

$$\mathbf{E}\left[(\boldsymbol{\rho}^T \mathbf{e}_i)^2\right] = 1 \text{ for } i = 1, 2, \dots, n \tag{8.16}$$

Now, let $\mathbf{v}_i = (v_{i1}, v_{i2}, \dots, v_{in})$ and $\mathbf{v}_j = (v_{j1}, v_{j2}, \dots, v_{jn})$. Then,

$$\mathbf{v}_i = \sum_{k=1}^{n} v_{ik} \mathbf{e}_k \text{ and } \mathbf{v}_j = \sum_{k=1}^{n} v_{jk} \mathbf{e}_k.$$

Therefore,

$$y_i' = \boldsymbol{\rho}^T \mathbf{v}_i = \sum_{k=1}^{n} v_{ik} \boldsymbol{\rho}^T \mathbf{e}_k \text{ and } y_j' = \boldsymbol{\rho}^T \mathbf{v}_j = \sum_{k=1}^{n} v_{jk} \boldsymbol{\rho}^T \mathbf{e}_k.$$

Then, from (8.15) and (8.16),

$$\mathbf{E}[y_i' y_j'] = \mathbf{E}\left((\boldsymbol{\rho}^T \mathbf{v}_i)(\boldsymbol{\rho}^T \mathbf{v}_j)\right) = \mathbf{E}\left[\left(\sum_{k=1}^{n} v_{ik} \boldsymbol{\rho}^T \mathbf{e}_k\right)\left(\sum_{k=1}^{n} v_{jk} \boldsymbol{\rho}^T \mathbf{e}_k\right)\right] = \mathbf{v}_i^T \mathbf{v}_j.$$

Now, the last part of the lemma follows from the linearity property of expectation and the optimality of $\mathbf{v}_1, \mathbf{v}_2, \dots, \mathbf{v}_n$ to SDP(max).

Thus, it is easy to construct scalars $y_1', y_2', \dots, y_n'$ from an optimal solution to SDP(max) with expected value equal to the optimal objective function value of SDP(max) leading us to a potential rounding algorithm. Let $y_{\max}' = \max_{1\le i\le n} |y_i|$. Then, $\frac{y_i'}{y_{\max}'} \in [-1, 1]$ and we can apply our $[-1, 1]$-rounding algorithm on these values to obtain a solution $\mathbf{x} \in \{-1, 1\}^n$ with expected value no worse than $\frac{P_{\max}}{(y_{\max}')^2}$.

We don't have much control over $y'_{\max}$ and hence have difficulty in establishing a data independent bound on the solution quality. Since $\boldsymbol{\rho}$ is an n-dimensional Gaussian, the number of y'_i values falls outside $[-T, T]$ may not be too many for an appropriate choice of T. Let

$$y_i = \begin{cases} y'_i \text{ if } y'_i \in [-T, T], \\ 0 \text{ if } y'_i \notin [-T, T]. \end{cases}$$

We will now show that, for $T = 3\sqrt{1 + \ln\left(\frac{R}{P_{\max}}\right)}$, $\mathbf{E}\left[\sum_{(i,j)\in E} a_{ij} y_i y_j\right] \geq \frac{1}{2} P_{\max}$. It may be noted that there is nothing special about the constant 3 used in the definition of T and it is primarily for convenience in handling the related constants later in our analysis and also to adhere to the calculations from [11] which we are following. The solution produced by the $[-1, 1]$-rounding algorithm, starting with $x_i = \frac{y_i}{T}$ guarantees a solution to the Ising QUBO $(\mathbf{A}, \mathbf{0})$ with expected value no worse than $\frac{P_{\max}}{2T^2}$. Thus, our main task is to establish that $\mathbf{E}\left[\sum_{(i,j)\in E} a_{ij} y_i y_j\right] \geq \frac{1}{2} P_{\max}$.

Let $z_i = y'_i - y_i$ be the error generated while truncating y'_i. Then, $y_i = y'_i - z_i$ and

$$\begin{aligned}
\mathbf{E}\Big[\sum_{(i,j)\in E} a_{ij} y_i y_j\Big] &= \mathbf{E}\Big[\sum_{(i,j)\in E} a_{ij}(y'_i - z_i)(y'_j - z_j)\Big] \\
&= \sum_{(i,j)\in E} a_{ij}\left(\mathbf{E}[y'_i y'_j] - \mathbf{E}[z_i y'_j] - \mathbf{E}[y'_i z_j] + \mathbf{E}[z_i z_j]\right) \\
&= P_{\max} - \mathbf{E}\Big[\sum_{(i,j)\in E} a_{ij}(z_i y'_j + y'_i z_j)\Big] + \mathbf{E}\Big[\sum_{(i,j)\in E} a_{ij} z_i z_j\Big] \qquad (8.17)
\end{aligned}$$

To deal with the expected values on the right hand side of Eq. (8.17), we make use of the following general probabilistic result proved in detail in [11, 24].

Lemma 8.10 ([11, 24]) *Let $\alpha_1, \alpha_2, \ldots, \alpha_n$ and $\beta_1, \beta_2, \ldots, \beta_n$ be real valued random variables satisfying $\mathbf{E}(\alpha_i^2) \leq a$ and $\mathbf{E}(\beta_i^2) \leq b$, then*

$$\mathbf{E}\left[\sum_{(i,j)\in E} a_{ij}(\alpha_i \beta_j + \alpha_j \beta_i)\right] \leq 2R\sqrt{ab},$$

where $R = P_{\max} - P_{\min}$.

Note that, in the above lemma, no assumption is made on the independence of the associated random variables. In view of Lemma 8.8, $R \geq P_{\max}$. To make use of Lemma 8.10 for estimating the expected values given on the RHS of Eq. (8.17), we need to find upper bounds on $\mathbf{E}[(y'_i)^2]$ and $\mathbf{E}[z_i^2]$. Since y'_i follows $N(0, 1)$, $\mathbf{E}[(y'_i)^2]$ is the variance, which is 1. Now, let us find a good upper bound on $\mathbf{E}[z_i^2]$. The bound given in the following lemma and its validity proof is from [11]. A somewhat similar result is available in [8], but in a different form.

Lemma 8.11 $\mathbf{E}[z_i^2] \leq \frac{1}{10}\left(\frac{P_{\max}}{R}\right)^2$

Proof Note that $\mathbf{E}[z_i^2] = \frac{2}{\sqrt{2\pi}}\int_T^\infty x^2 e^{-x^2/2}dx$. For $x \in [T, \infty)$, the integrand $x^2e^{-x^2/2} \leq (x^2 + \frac{1}{x^2})e^{-x^2/2}$. The antiderivative of $(x^2 + \frac{1}{x^2})e^{-x^2/2}$ is $-(x + \frac{1}{x})e^{-x^2/2}$. Therefore,

$$\mathbf{E}[z_i^2] \leq \frac{2}{\sqrt{2\pi}}\int_T^\infty (x^2 + \frac{1}{x^2})e^{-x^2/2}dx = \frac{2}{\sqrt{2\pi}}\left(T + \frac{1}{T}\right)e^{-T^2/2} = \sqrt{\frac{2}{\pi}}\left(T + \frac{1}{T}\right)e^{-T^2/2}.$$

Since $T = 3\sqrt{1 + \ln\left(\frac{R}{P_{\max}}\right)} \geq 3$, $\sqrt{\frac{2}{\pi}}\left(T + \frac{1}{T}\right) = \sqrt{\frac{2}{\pi}}\left(\frac{T^2+1}{T^2}\right)T \leq \sqrt{\frac{2}{\pi}}\frac{10}{9}T \leq T$, and we have,

$$\mathbf{E}[z_i^2] \leq Te^{-T^2/2} \tag{8.18}$$

Since $\ln(x) \leq x - 1$, we have $T \leq 3\sqrt{\frac{R}{P_{\max}}}$. Then, from (8.18),

$$\begin{aligned}\mathbf{E}[z_i^2] &\leq 3\sqrt{\frac{R}{P_{\max}}}e^{-\frac{9}{2}(1+\ln(\frac{R}{P_{\max}}))} \leq 3e^{-9/2}\sqrt{\frac{R}{P_{\max}}}\left(\frac{P_{\max}}{R}\right)^{9/2} \\ &< \frac{1}{10}\left(\frac{P_{\max}}{R}\right)^4 \leq \frac{1}{10}\left(\frac{P_{\max}}{R}\right)^2\end{aligned}$$

and this completes the proof.

Now, in Lemma 8.10, choose $\alpha_i = z_i$ and $\beta_i = y_i'$, $a = 1$, $b = \frac{1}{10}\left(\frac{P_{\max}}{R}\right)^2$ we get

$$\mathbf{E}\Big[\sum_{(i,j)\in E} a_{ij}(z_i y_j' + z_j y_i')\Big] \leq 2R\sqrt{ab} = 2R\frac{1}{\sqrt{10}}\left(\frac{P_{\max}}{R}\right) = \frac{2}{\sqrt{10}}P_{\max}. \tag{8.19}$$

Also, in Lemma 8.10, choose $\alpha_i = z_i$, $\beta_i = -z_i$ $a = b = \frac{1}{10}\left(\frac{P_{\max}}{R}\right)^2$ and we have

$$\begin{aligned}\mathbf{E}\Big[\sum_{(i,j)\in E} a_{ij}(z_i(-z_j) + z_j(-z_i))\Big] &= -2\mathbf{E}\Big[\sum_{(i,j)\in E} a_{ij}z_iz_j\Big] \\ &\leq 2R\sqrt{ab} = 2R\frac{1}{10}\left(\frac{P_{\max}}{R}\right)^2 \leq \frac{1}{5}P_{\max}\frac{P_{\max}}{R}\end{aligned} \tag{8.20}$$

Note that $\frac{P_{\max}}{R} \leq 1$ and hence from inequality (8.20),

$$\mathbf{E}\Big[\sum_{(i,j)\in E} a_{ij} z_i z_j\Big] \geq -\frac{1}{10} P_{\max} \tag{8.21}$$

Using the inequalities (8.19) and (8.21) in (8.17), we have,

$$\mathbf{E}\Big[\sum_{(i,j)\in E} a_{ij} y_i y_j\Big] \geq P_{\max} - \frac{2}{\sqrt{10}} P_{\max} - \frac{1}{10} P_{\max} \geq \frac{1}{4} P_{\max}. \tag{8.22}$$

Let us now present our SDP rounding algorithm for the Ising QUBO.

Algorithm 6: The SDP rounding algorithm

1 **Input:** *An optimal solution* $\mathbf{v}_1, \mathbf{v}_2, \ldots, \mathbf{v}_n$ *of SDP.*;
2 $P_{\max}$ is the maximum value of SDP(max) and $P_{\min}$ is the minimum value of SDP(min);
3 Generate a random n-dimensional Gaussian $\boldsymbol{\rho}$;
4 $y_i' \leftarrow \boldsymbol{\rho}^T \mathbf{v}_i, i = 1, 2 \ldots, n$;
5 $R \leftarrow P_{\max} - P_{\min}$ and $T \leftarrow 3\sqrt{1 + \ln\left(\frac{R}{P_{\max}}\right)}$;
6 **for** $i = 1, 2, \ldots, n$ **do**
7 | $y_i \leftarrow y_i'$ if $|y_i'| \leq T$ else $y_i = 0$
8 **end**
9 $x_i \leftarrow y_i/T$;
10 Apply the $[-1, 1]$-rounding algorithm on $\mathbf{x} = (x_1, x_2, \ldots, x_n)$ and output the solution;

In view of inequality (8.22), the expected value of $\varphi(\mathbf{x})$ for $\mathbf{x} = (x_1, x_2, \ldots x_n) \in [-1, 1]^n$ constructed in step 9 of the SDP rounding algorithm, i.e. $\mathbf{E}\Big[\sum_{(i,ju)\in E} a_{ij} x_i x_j\Big]$, is at least $\frac{1}{4}\frac{P_{\max}}{T^2} \geq P_{\max} K\left(\frac{1}{1+\ln\frac{R}{P_{\max}}}\right)$, for some constant K.

Theorem 8.22 ([11]) *The SDP rounding algorithm produces a solution to IQ with objective function value at least* $\text{OPT}(IQ)K\left(\frac{1}{1+\ln\frac{R}{P_{\max}}}\right)$ *in expectation, for some constant* K.

The proof of the theorem follows from the discussion above, Theorem 8.3, and the fact that $P_{\max} \geq \text{OPT}(IQ)$.

Lemma 8.12 $\frac{1}{1+\ln\frac{R}{P_{\max}}} \leq \frac{K'}{\log n}$ *for some constant* K'

Proof Note that $P_{\min} < 0$ (unless A is the zero matrix) and hence $\frac{R}{P_{\max}} = 1 + \frac{|P_{\min}|}{P_{\max}}$. But, $|P_{\min}| \leq \sum_{(i,j)\in E} |a_{ij}|$ and from the proof of Theorem 8.21, $P_{\max} \geq \frac{\sum_{(i,j)\in E} |a_{ij}|}{2n}$. Thus, $\frac{R}{P_{\max}} \leq 3n$ and the result follows.

Thus, Lemma 8.12 provides an $O\left(\frac{1}{\log n}\right)$ bound on the performance ratio of the SDP rounding algorithm, in expectation which was proved by many authors [8, 26, 30]. In fact, a stronger bound is possible since $\frac{R}{P_{\max}} \le \theta(\bar{G})$ [4, 11, 24] where $\theta(.)$ is the Lovász theta function. We restrict our discussion strictly to approximation aspects of QUBO and Ising QUBO only. For very interesting connections of the SDP rounding algorithm with Grothendieck constant of a graph, we refer to [4].

8.5.3 Other Special Cases

Let us now consider briefly, without details, some other special cases of QUBO and the Ising QUBO where polynomial time approximation algorithms exist with guaranteed relative performance ratio.

Let us start with a special case of QUBO. As discussed in Chap. 1, $\mathbf{Q}$ is called a *half product matrix* if it is upper triangular with diagonal elements zero and there exist vectors $\mathbf{a} = (a_1, a_2, \ldots, a_n)$ and $\mathbf{b} = (b_1, b_2, \ldots, b_n)$ such that $q_{ij} = a_i b_j$ for $i = 1, 2, \ldots, n$ and $j > i$. A *half-product QUBO* is an instance of QUBO when $\mathbf{Q}$ is a half product matrix and this special case is also NP-hard [7]. The single-machine variance minimization problem is a special case of the half-product QUBO [7, 23] along with various other machine scheduling problems such as scheduling two machines to minimize total weighted completion time [21], scheduling two machines to minimize makespan [21], scheduling a single machine to minimize total weighted earliness and tardiness[21] and scheduling with controllable processing times [20]. The Ising version of the half-product QUBO was considered by Kubiak [23], Amit [5], and Mattis [25].

Interestingly, the half product QUBO (with minimization objective) can be solved in pseudo-polynomial time. Also, the problem can be solved by a fully polynomial approximation scheme (FPTAS) when $\mathbf{a}, \mathbf{b}$ are non-negative [7]. Janiak et al. [20] considered a scheduling problem with controllable processing times and showed that it is equivalent to the half product QUBO. However, this formulation brings in an additional constant to the objective function, which can be ignored for optimization. Note that, the performance ratio depends on the constant and hence cannot be discarded when establishing performance ratio guarantee. The algorithm of [7] does not handle this constant, but Janiak et al. [20] developed an FPTAS to solve the resulting half-product QUBO which also takes into consideration this additional constant.

Following Hermelin et al. [19], we proved in Theorem 8.21 that for the Ising QUBO $(\mathbf{A}, \mathbf{0})$ with $\mathbf{A}$ having diagonal entries zero, a $\frac{1}{2\Delta}$-optimal solution can be identified in $O(n^2 \log n)$ time, where Δ is the maximum degree of a node in G. Hermelin et al. [19] also considered several other special cases of the Ising QUBO, and proved guaranteed performance ratios for low complexity polynomial time algorithms. This includes the special cases when (1) $\mathbf{A}$ has diagonal entries zero and all other entries belongs to $\{-1, 0, 1\}$, (2) the support graph G has bounded tree-width, and (3) when the support graph is H-minor free. For details of these algorithms, we refer to [19].

When **A** is positive semidefinite, Nesterov [31] established an expected approximation ratio of $\frac{2}{\pi}$ for the Ising QUBO along with a performance guarantee in terms of the differential approximation ratio. When the support graph G of **A** is bipartite, Alon and Naor [1, 2] proposed a polynomial time algorithm with constant relative performance ratio, in expectation. Other relevant works on QUBO related to a rank restricted **Q** matrix and additional assumptions on its elements, we refer to [13, 22, 29].

8.6 Conclusion

In this chapter we discussed several fast heuristic algorithms for QUBO and the Ising QUBO and presented theoretical analysis on their performance guarantees in various forms using different performance measures. These include relative performance ratio, average based analysis, domination analysis, among others. We also pointed out the connections between Grothendieck constant and the approximability of the Ising QUBO. Further refinements on performance guarantees are possible by exploiting the structure of the support graph or using conditions on elements of **Q** and **c** (**A** and **b**) and it would be interesting to explore further in this direction.

References

1. N. Alon, A. Naor, Approximating the cut-norm via Grothendieck's inequality, in *Proceedings of the Thirty-sixth Annual ACM Symposium on Theory of Computing*, (2004), pp. 72–80
2. N. Alon, A. Naor, Approximating the cut-norm via Grothendieck's inequality. SIAM J. Comput. **35**, 787–803 (2006)
3. N. Alon, G. Gutin, M. Krivelevich, Algorithms with large domination ratio. J. Algor. **50**, 118–131 (2004)
4. N. Alon, K. Makarychev, Y. Makarychev3, A. Naor, Quadratic forms on graphs. Invent. Math. **163**, 499–522 (2006)
5. D.J. Amit, *Modeling Brain Function: TheWorld of Attractor Neural Networks* (Cambridge University Press, Cambridge, 1989)
6. S. Arora, B. Barak, *Computational Complexity: A Modern Approach* (Cambridge University Press, New York, 2009)
7. T. Badics, E. Boros, Minimization of half-products. Math. Oper. Res. **23**, 649–660 (1998)
8. M. Charikar, A. Wirth, Maximizing quadratic programs: Extending Grothendieck's Inequality, in *45th Annual IEEE Symposium on Foundations of Computer Science FOCS* (2004), pp. 54–60
9. M. Demange, P. Grisoni, V.Th. Paschos, Differential approximation algorithms for some combinatorial optimization problems. Theoret. Comput. Sci. **209**, 107–122 (1998)
10. L. Engebretsen, J. Holmerin, Clique is hard to approximate within $n^{1-o(1)}$, in *Proceedings of International Colloquium on Automata, Languages and Programming*, Geneva (2000), pp. 2–12

11. B. Gartner, J. Matoušek, *Approximation Algorithms and Semidefinite Programming* (Springer, New York, 2010)
12. F. Glover, A.P. Punnen, The traveling salesman problem: new solvable cases and linkages with the development of approximation algorithms. J. Oper. Res. Soc. **48**, 502–510 (1997)
13. V. Goyal, L. Kaya, R. Ravi, An FPTAS for minimizing the product of two non-negative linear cost functions. Math. Programm. **126**, 401–405 (2011)
14. G. Gutin, A. Yeo, A. Zverovitch, Exponential neighborhoods and domination analysis for TSP, in ed. by G. Gutin, A.P. Punne, *The Travelling Salesman Problem and Its Variations* (Kluwer Academic Publishers, Boston, 2002)
15. G. Gutin, A. Yeo, Domination analysis of combinatorial optimization algorithms and problems, in ed. by M.C. Golumbic, I.B.A. Hartman, *Graph Theory, Combinatorics and Algorithms*. Operations Research/Computer Science Interfaces Series, vol. 34 (Springer, Boston, 2005)
16. M.M. Halldórsson, Approximations of independent sets in graphs, in *Proceedings of APPROX '98 Conference*. Springer-Verlag Lecture Notes in Computer Science, vol. 1444 (Springer, Berlin, 1998), pp. 1–13
17. J. Håstad, Clique is hard to approximate within $n^{(1-\epsilon)}$, in *Acta Mathematica* (1996), pp. 627–636
18. J. Håstad, Some optimal inapproximability results. J. ACM **48**, 798–859 (2001)
19. D. Hermelin, L. Kellerhals, R. Niedermeier, R. Pugatch, Approximating sparse quadratic program (2020). arXiv:2007.01252v4 [cs.DS]
20. A. Janiak, M.Y. Kovalyov, W. Kubiak, F. Werner, Positive half-products and scheduling with controllable processing times. Eur. J. Oper. Res. **165**, 416–422 (2005)
21. B. Jurisch, W. Kubiak, J. Józefowska, Algorithms for minclique scheduling problems. Discret. Appl. Math. **72**, 115–139 (1997)
22. W. Kern, G.J. Woeginger, Quadratic programming and combinatorial minimum weight product problems. Math. Programm. **110**, 641–649 (2007)
23. W. Kubiak, New results on the completion time variance minimization. Discret. Appl. Math. **58**, 157–168 (1995)
24. K. Makarychev, Quadratic forms on graphs and their applications, Ph.D. Thesis, Princeton University, 2008
25. D.C. Mattis, Solvable spin systems with random interaction. Phys. Lett. **6A**, 412 (1976)
26. A. Megretski, Relaxation of quadratic programs in operator theory and system analysis, in *Systems, Approximation, Singular Integral Operators, and Related Topics* (Bordeaux, 2000) (Birkhäuser, Basel, 2001), pp. 365–392
27. A. Mehrotra, Cardinality constrained Boolean quadratic polytope. Discret. Appl. Math. **79**, 137–154 (1997)
28. P. Merz, B. Freislebeng, Greedy and local search heuristics for unconstrained binary quadratic programming. J. Heuristics **8**, 197–213 (2002)
29. S. Mittal, A. Schulz, An FPTAS for optimizing a class of low-rank functions over a polytope. Math. Programm. **141**, 103–120 (2013)
30. A. Nemirovski, C. Roos, T. Terlaky, On Maximization of quadratic form over intersection of ellipsoids with common center. Math. Programm. **86**, 463–473 (1999)
31. Y. Nesterov, Semidefinite relaxation and nonconvex quadratic optimization. Optim. Method Softw. **9**, 141–160 (1998)
32. R. O'Donnell, *Analysis of Boolean Functions* (Cambridge University Press, Cambridge, 2014)
33. G. Palubeckis, Quadratic 0-1 optimization. Informatica **1**, 89–106 (1990)
34. G. Palubeckis, Heuristics with a worst-case bound for unconstrained quadratic 0-1 programming. Informatica **3**, 225–240 (1992)
35. P. Pandey, Topics in quadratic binary optimization problems, Ph.D. Thesis, Simon Fraser University, 2018
36. A.P. Punnen, N. Kaur, Fast heuristics for the quadratic unconstrained binary optimization problem. Research Report, Department of Mathematics, Simon Fraser University, (2021)
37. A.P. Punnen, F.S. Margot, S.N. Kabadi, TSP heuristics: domination analysis and complexity. Algorithmica **35**, 111–127 (2003)

38. A. Samorodnitsky, L. Trevisan, A PCP characterization of NP with optimal amortized query complexity, in *Proceedings of ACM Symposium on Theory of Computing*, Portland, OR (2000), pp. 191–199
39. A. Srinivasan, On the approximability of clique and related maximization problems. J. Comput. Syst. Sci. **67**, 633–651 (2003)
40. G. Tavares, New algorithms for quadratic unconstrained binary optimization problem (QUBO) with applications in engineering and social sciences, Ph.D Thesis, Rutgers University, 2008
41. E. Zemel, Measuring the quality of approximate solutions to zero-one programming problems. Math. Oper. Res. **6**, 319–332 (1981)

Chapter 9
Metaheuristic Algorithms

Yang Wang and Jin-Kao Hao

Abstract Metaheuristic algorithms are practically used to produce approximate solutions to large QUBO instances that cannot be solved exactly due to the high computational complexity. This chapter is dedicated to a review on the general metaheuristic approach for solving the QUBO. First, we present some basic components of local search that are widely used in the design of state-of-the-art metaheuristic algorithms for the problem. Then we overview the metaheuristic algorithms in the literature by groups of fast solving heuristics, local search based methods and population based search methods. Finally, we review some of the most popular and effective metaheuristic algorithms and present experimental results on different sets of instances.

9.1 Basic Ingredients of Local Search

Most of the heuristic and metaheuristic algorithms proposed for solving the QUBO employ a local search procedure to improve the solution quality. Given its importance, we first summarize basic ingredients used in these local search procedures, including the solution representation of a QUBO instance, the move operator along with the fast calculation of the move gain, and the neighbor solution selection strategy.

For a given QUBO instance, its solution $x = \{x_1, x_2, \ldots, x_n\}$ is a boolean vector of length n. To perform neighborhood search, the most widely used move operator is 1-flip, which flips a chosen variable x_i to be the complementary value $1 - x_i$. A total of n neighbor solutions are generated by applying the 1-flip move. Some algorithms introduce the k-flip move that flips k ($2 \le k \le n$) variables of a solution

Y. Wang (✉)
School of Management, Northwestern Polytechnical University, Xi'an, China
e-mail: yangw@nwpu.edu.cn

J.-K. Hao
LERIA, Universitë d'Angers, Angers, France
e-mail: jin-kao.hao@univ-angers.fr

A. P. Punnen (ed.), *The Quadratic Unconstrained Binary Optimization Problem*,
https://doi.org/10.1007/978-3-031-04520-2_9

simultaneously to the corresponding complementary values, respectively. Compared to the 1-flip move, applying the k-flip move yields a much larger number of neighbor solutions up to C_n^k.

The efficiency of local search mostly depends on the evaluation of move gains to determine the next solution on the search trajectory. The move gain of a neighbor solution is the objective variation between a solution and the neighbor solution. Let x and x' represent two binary solutions where x' is obtained from x by a 1-flip move applied to a single variable x_i. The move gain Δ_i of moving from solution x to the neighbor solution x', recorded as $\Delta_i = f(x') - f(x)$, is calculated as:

$$\Delta_i = (1 - 2x_i)(q_{ii} + 2 \sum_{j=1, j \neq i}^{n} q_{ij} x_j) \tag{9.1}$$

$$\Delta_j = \begin{cases} -\Delta_i & if \;\; i = j \\ \Delta_j + 2q_{ij}(1 - 2x_i)(1 - 2x_j) & otherwise \end{cases} \tag{9.2}$$

Since the generalized k-flip move can be considered as a sequence of 1-flip moves, it is straightforward to infer the following equation for the k-flip move. Formally, let $i_1, \ldots, i_k$ are the k variables to flip, then the move gain $\Delta_{i_1,\ldots,i_k}$ for simultaneously flipping these k variables can be computed as follows:

$$\begin{aligned} \Delta_{i_1,\ldots,i_k} &= \Delta_{i_1} \\ &+ \Delta_{i_2} + 2q_{i_1 i_2}(1 - 2x_{i_1})(1 - 2x_{i_2}) \\ &+ \Delta_{i_3} + 2q_{i_1 i_3}(1 - 2x_{i_1})(1 - 2x_{i_3}) + 2q_{i_2 i_3}(1 - 2x_{i_2})(1 - 2x_{i_3}) \\ &+ \Delta_{i_4} + 2q_{i_1 i_4}(1 - 2x_{i_1})(1 - 2x_{i_4}) + 2q_{i_2 i_4}(1 - 2x_{i_2})(1 - 2x_{i_4}) \\ &+ 2q_{i_3 i_4}(1 - 2x_{i_3})(1 - 2x_{i4}) \\ &\quad \vdots \qquad\qquad\qquad\qquad \vdots \\ &= \sum_{r=1}^{k} \Delta_{i_r} + 2 \sum_{r=1}^{k-1} \sum_{s=r+1}^{k} q_{i_r i_s}(1 - 2x_{i_r})(1 - 2x_{i_s}) \end{aligned} \tag{9.3}$$

The neighbor solution selection strategy decides the neighbor solution that is used to replace the current solution x during the local search. The first move improvement and the best move improvement are two strategies often used. The first move improvement strategy scans the neighborhood of x and chooses the first neighbor solution that is better than x (i.e., with an improving move gain). The best move improvement strategy exhaustively explores the neighborhood of x and retains the neighbor solution with the highest move gain.

9.2 Fast Solving Heuristics

Boros et al. [6] developed a Devour Digest Tidy-up procedure (DDT). On the basis of the posiform representation Z of QUBO, DDT includes the Devour, Digest and Tidy-up phases. Devour identifies a term T from L (L denotes the set of all the elements of Z) with the largest coefficient and places it into S. Digest draws logical conclusions by assigning the disjunctive equation of all the elements in S equaling to 0 (in terms of minimization). If no logical conclusion can be drawn, then T is simply removed from L to S, and return to Devour. Otherwise, Tidy-up begins to substitute the logical conclusions previously drawn into Z. The above DDT procedure repeats until L becomes an empty set. Experiments indicated that DDT is especially effective on problem instances of low density.

Consider the case that the DDT method simultaneously sets several variables with value 1 or 0 would result in worse results than to give inferred assignment to only one variable, Glover et al. [11] proposed several one-pass heuristics to guarantee that in each pass only one variable gets the implied assignment. The difference among the proposed one-pass heuristics lies in the different strategies of evaluating contributions of variables. Experimental comparisons among the proposed one-pass heuristics showed that some of them perform quite effectively for certain problem instances, but no single method dominates on every problem instance.

Hanafi et al. [12] devised five alternative DDT heuristics based on different representations of the QUBO formulation, where DDT1 to DDT4 methods respectively have standard, posiform, bi-form and negaform representations and DDT5 has a posiform representation along with a one-pass mechanism. An obviously additional difference of their DDT alternatives from [6, 11] concerns the use of a *r-flip* local search procedure to improve solutions obtained by DDT constructions. Extensive tests on small, medium and large benchmark instances showed that (1) DDT3 with the bi-form representation generally produces the best results for medium and large instances; (2) the proposed *r-flip* local search brings significant result improvements with only a slight increase of time consumption.

Merz and Freisleben [20] proposed a greedy construction heuristic to quickly obtain an improved solution. It starts from a solution with all variables assigned to be 0.5 (the so called third state). At each construction step, it searches a variable and a value either 0 or 1 for the variable such that assigning the value to the variable makes the gain function, i.e., objective function increment of the resulting solution compared to the previous solution, is maximized. This operation repeats until each variable of the solution vector changes its initial value from 0.5 to 1 or 0. Since this greedy construction method always obtains the same solution for a given problem instance, a randomized greedy variant was proposed to overcome the deterministic drawback. For one thing, randomly pick a variable and randomly assign a value 0 or 1 to it for the first step of the construction. For another thing, select a variable with a probability proportional to the gain value instead of picking a variable with the maximized gain value.

9.3 Local Search Based Methods

9.3.1 Simulated Annealing

Alkhamis et al. [1] presented a simulated annealing based heuristic (SA-AHA) according to the traditional simulated algorithm framework. It begins with a randomly generated solution and an initial temperature. At each iteration SA-AHA generates a random 1-flip move. If this is an improving move, it is performed; Otherwise, it may still be accepted with a probability $e^{-\Delta/T}$ where Δ indicates the objective function difference between the two solutions and T is the current temperature constant. After the above procedure conducts a certain number of iterations, the temperature is decreased with reference to a cooling function. The above procedure is repeated until no solution has been accepted for 10 consecutive temperatures or when the temperature has fallen below a pre-specified value. Tested on problem instances with up to 100 variables and comparisons with several bounding techniques based algorithms indicated that SA-AHA outperforms these compared methods. Especially, SA-AHA is able to solve hard problem instances very efficiently while bounding algorithms can not solve them in a reasonable computation time. Additional experiments indicated that the efficiency of the SA-AHA algorithm is not affected by matrix density.

Beasley [3] proposed another simulated annealing algorithm (SA-B). The basic iterative procedure of SA-B is the same as SA-AHA. However, in SA-B each iteration applies a different temperature value to determine the probability to accept a worse move. In addition, a local search procedure based on the first improvement strategy is utilized to perform a post-optimization of the solution from the annealing process. Experimental results for 45 instances with up to 500 variables indicated that SA-B converges fast to the best solutions than the reference algorithms but obtains inferior solution quality for several instances. In addition, the author generated 60 instances with up to 2500 variables available from OR-Library. Results on this new set of benchmark instances showed that SA-B is especially effective for 10 largest instances with 2500 variables.

Katayama and Narihisa [14] designed a similar implementation of the simulated annealing methodology as SA-AHA, called SA-KN. An obvious characteristic of SA-KN different from SA-AHA and SA-B lies in the fact that it adopts multiple annealing processes to enhance the search ability. Experimental results for problem instances with variables ranging from 500 to 2500 indicated that SA-KN achieves especially competitive performances for the largest OR-Library instances.

9.3.2 Tabu Search

Glover et al. [9] introduced an adaptive memory tabu search (AMTS) algorithm that uses the 1-flip move and two types of memory structures to record recency and frequency information. Strategic oscillation is employed to alternate between constructive phases (progressively setting variables to 1) and destructive phases (progressively setting variables to 0), which are triggered by critical events, i.e., when the next move causes the objective value to decrease. The amplitude of the oscillation is adaptively controlled by a span parameter. Computational results for instances with up to 500 variables showed that AMTS outperforms the best exact and heuristic methods previously reported in the literature.

Beasley [3] proposed a 1-flip move based tabu search algorithm (TS-B). It begins from an initial solution with each variable assigned to be 0 and marked as non-tabu. During each iteration it conducts a best non-tabu move. This performed move is then marked as tabu for a specified number of following iterations. If the current iteration finds a better solution than the best solution found so far, a local search procedure with first-improvement strategy is launched to further improve this new solution. TS-B repeats the above procedure until the current iteration reaches the maximum allowed iteration. Notice that TS-B does not incorporate the fast evaluation technique and also neglects an aspiration criterion.

Palubeckis [23] examined five multistart tabu search strategies (MSTS) dedicated to the construction of the initial solution. Each multistart tabu search algorithm employs a tabu search procedure (TS-P) to enhance solution quality and a multi-start strategy to produce a new initial solution located in a more promising area. Notice that TS-P is very similar to TS-B except that TS-P employs a tactic to get 1-flip moves fast evaluated. The first restart strategy produces a new initial solution in a random way. The second restart strategy identifies a candidate set of variables that are prone to change their values when moving from the current solution to an optimal one. Then it applies a steepest ascent algorithm that only considers variables in the candidate set and keeps the other variables fixed at specific values. The third one employs a randomized greedy constructive method. The fourth one incorporates a set of elite solutions and calculates the probability of each variable with value 1 in this set. If the probability for a given variable is larger than 0.5, then this variable receives value 1 in the resulting new solution; otherwise it receives value 0. The last restart strategy uses a perturbation scheme of changing the problem instance at hand, followed by a short run of tabu search on the modified instance. Experiments evaluated on 25 largest instances from OR-Library and a set of randomly generated larger and denser instances demonstrated that the algorithm with the second restart strategy (MST2) is the best among the proposed algorithms.

Palubeckis [24] developed an iterated tabu search algorithm (ITS) that combines a tabu search procedure to improve the solution quality and a perturbation mechanism to create a new initial solution. The tabu search procedure is exactly the one used in [23]. The perturbation mechanism is operated as follows. First, it constructs a candidate list of a specified size which consists of variables with the largest 1-flip move gains with regard to a local optimal solution. Then it randomly selects a variable from this set and flips this variable to move toward a new solution. Finally, it updates the corresponding move gains of variables caused by the move. The above procedure is repeated until the number of perturbed variables reaches the specified count. Experimental results indicated that despite its simplicity, ITS is very competitive compared to other state-of-the-art algorithms.

Lü et al. [19] studied neighborhood union and token-ring search methods to combine 1-flip (N1) and 2-flip (N2) moves within a tabu search algorithm. The 2-flip move based tabu search considers a constrained set of 2-flip moves requiring that flipping each involved variable produces the move gain ranked top $3\sqrt{n}$ among all the 1-flip moves. In this way, the computational efforts of exploring the neighborhood N2 can be greatly reduced. The neighborhood union includes the strong neighborhood union ($N1 \bigsqcup N2$) that picks each move from both N1 and N2 and the selective neighborhood union ($N1 \bigcup N2$) that select a move from N1 with probability p and N2 with probability $1 - p$. The token ring search (N1 $\rightarrow$ N2) continuously performs moves within a single neighborhood until no improvement is possible and then switches to the other neighborhood to carry out moves in the same fashion. Experimental results on random large instances indicated that selective union is superior to the other two neighborhood combinations.

Liu et al. [15] proposed a hybrid r-flip/1-flip tabu search algorithm (HLS) which switches among a hybrid local search phase, a destruction phase and a construction phase. First, the hybrid local search phase that hybrids 1-flip and r-flip local search is launched. This phase behaves like a basic variable neighborhood search procedure [13] but excludes useless r-flip moves by several orders according to a theorem. When no improved move is found, the hybrid local search phase terminates. Meantime, the destruction phase is followed to carry out the 1-flip move with the least damage to the current solution. The performed move is marked as tabu and the destruction phase continues until an improving non-tabu move occurs. At this point, a construction phase is triggered to perform the best non-tabu move. If the obtained solution is better than the best solution ever found, the algorithm returns to the hybrid local search phase. If no variable exists that can make further improvement, the algorithm then returns to the destruction phase. Tested results showed the superiority of the proposed hybrid r-flip/1-flip tabu search especially for solving large instances with high density.

Shylo and Shylo [26] developed a global equilibrium search (GES) algorithm that performs multiple temperature cycles. Each temperature cycle alternates between an initial solution generation phase and a tabu search phase. The use of information from the whole search history helps to determine the probability of a variable receiving value 1 in the generated solution. The tabu search procedure performs the best 1-flip move with an additional requirement that this move leads to a solution far enough from a reference set in terms of hamming distance. Experimental results showed that GES performs quite well in terms of solution quality and computing time.

Wang et al. [27] devised GRASP-TS and GRASP-TS/PM algorithms that hybrid GRASP with tabu search. GRASP-TS uses a basic GRASP algorithm with single solution search while GRASP-TS/PM launches each tabu search by introducing a population management strategy based on an elite reference set. Specifically, GRASP-TS uses an adaptive random greedy function to construct an initial solution from scratch. GRASP-TS/PM makes use of a restart/recovery strategy to produce a solution, in which partial solution components inherit corresponding elements of an elite solution fetched from a population and the remaining solution components are rebuilt as in the GRASP-TS procedure. Experiments indicated that GRASP-TS and GRASP-TS/PM are very competitive with state-of-the-art algorithms.

9.4 Population Based Search Methods

Amini et al. [2] presented a scatter search approach (SS) that is mainly composed of a diversification generation method, a solution improvement method, a reference set update method, a subset generation method and a solution combination method. The diversification generation method systematically generates a collection of diverse trial solutions based on a seed solution in a way of setting an incremental parameter that determines which bits of the seed solution should be flipped. The improvement method performs a compound move that sequentially cycles among 1-flip, 2-flip and 3-flip candidate moves until no attractive move can be identified. The reference set update method replaces solutions in the reference set with new candidate solutions according to the quality measurement. In order to build a new solution, a linear combination of selected solutions from the reference set is applied. Since some variables would receive fractional values in the combined solution, a rounding procedure is followed to make this solution feasible. Experiments showed that the proposed scatter search method is very robust, especially for large problem instances.

Lodi et al. [16] introduced an evolutionary heuristic (EH) with the following features. First, EH uses a preprocessing phase to fix certain variables at their optimal values and reduce the problem size. This type of fixation belongs to permanent fixation since for each successive round of local search, these variables are excluded from consideration. Second, a local search procedure based on the alternation between construction and destruction phases is employed to get an improved solution. Finally, EH uses a uniform crossover operator to generate offspring solutions, where variables with common values in parental solutions are temporarily fixed in this round of local search. Experimental results showed that EH can match the best known results for problem instances with up to 500 variables in a very short computation time. A further analysis demonstrated that the preprocessing phase is effective for small problem instances but is impossible to reduce the problem size for large ones.

Merz and Freisleben [21] devised a hybrid genetic algorithm (GLS-MF), in which a simple local search is incorporated into the traditional genetic algorithm. The local search procedure uses the 1-flip move and best move improvement strategy. The crossover operator is a variant of uniform crossover, requiring the generated offspring solution has the same hamming distance from the parents. Once the newly generated offspring solution satisfies the updating criterion, it becomes a member of the population and replaces the worst solution. A diversification component is launched when the average hamming distance of the population drops below a threshold $d = 10$ or the population is not updated for more than 30 consecutive generations. Experimental results showed that the simple evolutionary algorithm alone is able to find the best known results for problem instances with less than 200 variables but for larger instances, it is essential to incorporate local search to attain high quality solutions.

Lü et al. [18] proposed a hybrid metaheuristic approach (HMA) which integrates a basic tabu search procedure into a genetic search framework. First, HMA combines a traditional uniform crossover operator with a diversification guided path relinking operator to guarantee the quality and diversity of an offspring solution. Second, HMA replaces the Hamming distance by a new distance by reference to variable's importance and employs a quality-and-distance criterion to update the population as in GTA. Finally, a tabu search procedure is responsible for intensified examination around the offspring solutions. Computational results showed HMA is among the best performing procedures for solving challenging QUBO problem instances.

9.5 Selected Metaheuristic Approaches for QUBO

9.5.1 Diversification-Driven Tabu Search

Glover et al. [10] presented a diversification-driven tabu search (D^2TS) algorithm that alternates between a basic tabu search procedure and a memory-based perturbation procedure guided by a long-term memory. The general scheme of D^2TS works as follows. Starting from a random initial solution, D^2TS uses tabu search to reach local optimum. Then, the perturbation operator is applied to displace the solution to a new region, whereupon a new round of tabu search is launched. To achieve a more effective diversification, the perturbation operator is guided by information from a special memory structure for obtaining improved results in this context.

The tabu search procedure employs the 1-flip neighborhood. Each time a move is carried out, the reverse move is forbidden for the next $TabuTenure$ iterations. The tabu tenure is set to be $TabuTenure(i) = tt + rand(10)$, where tt is a selected constant and rand(10) takes a random value from 1 to 10. Once a move is performed, a subset of move values affected by the move is updated using a fast incremental evaluation technique. Accompanying this rule, a simple aspiration criterion is applied that permits a move to be selected in spite of being tabu if it leads to a solution better than the best solution found so far. The TS procedure stops when the best solution cannot be improved within a given number of moves.

The perturbation procedure includes assigning a score to each variable, selecting a certain number of highly-scored variables (critical elements), and perturbing the solution using the chosen critical elements. The scoring function depends on the information of several memory structures, including a flipping frequency vector $FlipFreq(i)$, an elite set of solutions *EliteSol* and a consistency vector

$EliteFreq(i)$. $FlipFreq(i)$ records the number of times the a variable x_i has been flipped from the beginning until the current iteration, which is collected in the tabu search phase. *EliteSol* stores a set of locally optimal solutions found by tabu search. Each time a new local optimum is found that has the objective value superior to that of the worst local solution in *EliteSol*, the new solution replaces this worst solution. $EliteFreq(i)$ records the total number of times a variable x_i is assigned value 1 in the elite solutions currently stored in *EliteSol*. This memory is used to favor retaining the value assignments that occur more often in the best solutions found to date. The scoring function ranks each variable by taking into account its flip frequency $FlipFreq(i)$ and its elite value frequency $EliteFreq(i)$, which takes the following form:

$$Score(x_i) = \frac{EliteFreq(i)(r - EliteFreq(i))}{r^2} + \beta(1 - \frac{FlipFreq(i)}{maxFreq}) \quad (9.4)$$

The selection step sorts all the variables in non-increasing order according to their scores and then adaptively selects a specified number of critical variables to be randomly assigned a value 0 or 1. The higher the score a variable has, the greater the probability it will be chosen. The perturbation step flips the values of the selected critical variables. This perturbed solution is then used to initiate a new round of tabu search.

9.5.2 Memetic Search

Merz and Katayama [22] conducted a landscape analysis and observed that (1) local optima of the QUBO instances are contained in a small fraction of the search space; (2) the fitness of local optima and the distance to the optimum are correlated. Based on this, they designed a memetic algorithm (MA-MK) in which an innovative variation as the crossover operator is utilized to generate good starting solutions. The MA-MK algorithm is composed of population initialization, randomized k-opt local search, crossover and variation, as well as selection and diversification strategies.

The population initialization procedure repeats generating individuals in the following way until the population size reaches 40. The method to generate an individual includes two steps. The first step employs a randomized greedy heuristic introduced in [20] to produce a seeding solution. The second step applies a randomized k-opt local search to optimize the solution to local optimum.

The randomized k-opt local search is based on the ideas of Lin and Kernighan for solving the traveling salesman problem (TSP) and the graph partitioning problem that searches a small fraction of the k-opt neighborhood efficiently. The randomized k-opt local search performs the following iterations until performing a k-flip move can not yield an improving solution. For each k-opt iteration, all the bits of a solution are sorted in a random order and only the bits of getting the positive move gains are flipped. The bit with the maximum move gain is subsequently flipped. The above-mentioned procedure are repeated until all the bits have been flipped. The best solution is recorded as the resulting solution in this iteration.

The crossover operator introduced the move gain to determine the variation of the offspring solution in order to prevent rediscovering local optima already visited to the most extent. Specifically, the common and the non-common bits of the parent solutions are identified and the initial offspring solution is set to be any parent solution. The bits in the non-common and common sets are operated alternatively. For the non-common set, all the non-common bits with the positive 1-flip move gains are identified and such a bit is randomly selected. For the common set, the common bit with the maximum associated 1-flip move gain is identified even if the move gain is negative. If a bit is flipped, it is removed from the corresponding set. The above-mentioned procedure is repeated for a number of times equal to the size of the non-common set.

In each generation, a new population needs to be formed after offspring individuals are generated. Among the old individuals in the previous generation and the newly generated offspring individuals, those with the highest fitness are selected to maintain the restricted population size. If no new best individual in the population was found for more than 30 generations, a diversification restart strategy is triggered. All the individuals except for the best one in the population are mutated by flipping randomly chosen $n/3$ bits for each individual of length n. After that, each individual is optimized by the randomized k-opt local search to obtain a renewal set of local optima and the search is started again with the newly diversified population.

9.5.3 Path Relinking

Wang et al. [28] proposed two path relinking algorithms, which is composed of a reference set initialization method, a solution improvement method, a reference set update method, a relinking method and a path solution selection method. The proposed algorithms differ from each other mainly on the way they generate the path, one employing a greedy strategy and the other employing a random construction.

The general scheme of the path relinking algorithm works as follows. It starts with the creation of an initial set of elite solutions RefSet, based on which an index set of pairwise solutions PairSet is generated. For each index pair (i, j), a relinking method is applied to generate two paths connecting the elite solutions x^i and x^j, one of which is from x^i to x^j and the other is from x^j to x^i. Then, one solution on each path is selected according to a path solution selection method and refined by a solution improvement method using the same tabu search procedure as in [18]. The resulting solution is subject to the RefSet updating procedure. The above-mentioned procedure is terminated once all the elements in PairSet are examined. If the given stopping criterion is not satisfied, RefSet and PairSet are rebuilt to continue the search.

The RefSet initialization method is used to construct an elite set of high-quality solutions, where each solution is obtained in two steps. The first step generates a randomized solution, where each variable receives value 0 or 1 with an equal probability of 0.5. The second step employs a tabu search based solution improvement method to refine the quality of this solution. Afterwards, the RefSet updating procedure is invoked. The improved solution is permitted to be added into RefSet if it is distinct from any solution in RefSet and better than the worst solution. Once this condition is satisfied, the worst solution is replaced by the improved solution. When PairSet becomes empty, RefSet is recreated. The best solution previously found becomes a member of the new RefSet and the remaining solutions are generated in the same way as in constructing RefSet in the first round.

The path relinking method builds a path connecting an initiating solution where the path starts with and a guiding solution where the path ends at. The path consists of a sequence of intermediate solutions, each of which is generated by exploring the neighborhood of the initiating and guiding solutions. To be specific, identify the set of non-common variables NC where the initiating solution $x^i = x(0)$ and the guiding solution x^g have different values. Meanwhile, initialize another vector where each entry Δ_t denotes the objective difference resulting after flipping the variable $x_t \in NC$ from the previous solution $x(k-1)$ on the path. The path relinking method performs a total of $|NC| - 1$ iterations to construct a path. At each path construction step k, either use a greedy strategy to select the variable having the maximum $\Delta_{t \in NC}$ value in the algorithm PR1 or use a random strategy to randomly select a non-common variable in the algorithm PR2. The path solution $x(k)$ is determined by assigning the same value as the guiding solution x^g for the newly chosen variable and assigning the same values as $x(k-1)$ for all the other variables.

Since two consecutive solutions on a relinking path differ only by flipping a single variable, it is not productive to apply an improvement method to each solution on the path since many of these solutions would lead to the same local optimum. Hence, the path solution selection method chooses only a single solution on the path explored via path relinking. Specifically, it first constructs a candidate solution list that includes the solutions with a Hamming distance of at least $|NC|/3$ from both the initiating and guiding solutions. Then, the candidate solution with the maximum objective value is picked for further amelioration by the solution improvement method.

9.5.4 Automatic Grammar-Based Design of Heuristic Algorithms

de Souza and Ritt [25] designed an automatic algorithm by combining the problem-specific components of state-of-the-art algorithms from the literature. The designed automatic algorithm employs a grammar in Backus-Naur form to model the space of heuristic strategies and their parameters. The grammar is composed of a set of rules, through which the heuristic algorithms can be instantiated. All heuristic strategies are categorized into construction methods, search methods and recombination methods. The construction methods start from an empty solution and apply the greedy randomized heuristics to iteratively set value 0 or 1 to each variable. The search methods include local search components, tabu search components and iterated local search components, which start with an initial solution and perform 1-flip moves to explore the search space. The differences among local search components lie in the move selection strategies that transfer from the current solution to its neighbor solution, such as the first improvement strategy, a round-robin strategy, a best improvement strategy, etc. Two tabu search components are differentiated, one of which always selects the best move and the other selects a random move with a small probability. In addition to different local search strategies, the iterated local search components include perturbation strategies of random perturbation, least-loss perturbation and frequency based perturbation as well as different strategies to define the perturbation strength. The recombination methods employ the path relinking algorithms that evolve a population of elite solutions.

To perform exploration of the grammar based search space, they used the irace tool described in [17] to rank different candidates of the algorithms found in the literature of QUBO and newly generated algorithms. The irace tool implements an iterated racing procedure that iteratively selects a candidate and an instance from a set of elite candidates and calls the resulting algorithm. After performing each iteration, the average difference between the objective value of the found solution and the best known value is used as a result metric for irace to rank different candidates. The process is repeated until irace has finished the specified number of algorithm runs. Experimental results indicate that the automatic approach can find algorithms that outperform state-of-the-art algorithms.

9.5.5 A Systematic Evaluation of Heuristics

Dunning et al. [7] implemented a total of 37 Max-Cut and QUBO heuristics selected from the literature and performed a systematic evaluation on a library of 3296 instances. Because no single heuristic outperforms all others across all problem instances, a regression tree model is built for each of the 37 heuristics to determine the rank on a given instance. Based on this, key insights into when heuristics perform well or poorly according to the problem instance characteristics are identified. Moreover, a random forest model is built for each heuristic to predict the probability of a heuristic that will perform the best for a given instance. By selecting a set of heuristics with the highest predicted probability, an algorithm portfolio is finally constructed to produce the best solution. Unlike many hyper-heuristics found in the literature, the proposed approach does not construct a new heuristic by selecting from a set of heuristic components. In addition, it does not include and implement several advanced algorithms and thus lacks of comparisons with these algorithms.

Results indicate that the proposed algorithm portfolio dominates each of the 37 Max-Cut and QUBO heuristics it combines. The open-source implementations of the heuristics are publicly available[1]

9.6 Computational Results

9.6.1 Benchmark Instances

Three sets of test problems are often used in the QUBO literature to evaluate the performance of algorithms. The first set is composed of 10 large instances from the OR-Library [3, 4].[2] They all have a density of 0.1 and are named by b2500.1,. . .,b2500.10. The second set of benchmarks consists of 21 randomly generated large problem instances named p3000.1,. . .,p7000.3 with sizes ranging from n=3000 to 7000 and with densities from 0.5 to 1.0 [23, 24].[3] These large instances are particularly challenging QUBO problems, especially in the case of instances with more than 5000 variables. The third set of benchmarks includes 54 instances derived from the MaxCut problem, named G1,. . .,G54, with variable sizes ranging from n=800 to 2000 [5, 8].[4] These instances are created by using a machine-independent graph generator, composed of toroidal, planar and random weighted graphs with weight values 1, 0 or -1. The small test instances from the OR-Library whose sizes range from n = 500 to 1000 can be solved relatively easily by many algorithms.

It is worth noting that a completely fair comparison of algorithms is impossible since the compared algorithms are implemented by different authors and run under different conditions. The presented comparisons on the QUBO instances as well as that on the MaxCut problem are thus presented only for indicative purposes and should be interpreted with caution.

9.6.2 Computational Results on the QUBO Instances

The PR1 and PR2 algorithms were tested on a PC with Pentium 2.83 GHz CPU and 8 GB RAM. The running CPU time for each run of PR1 and PR2 is set to 60 seconds for solving the 10 OR-Library instances and set to 5, 10, 20, 30 and 50 minutes for solving the instances of second set with 3000, 4000, 5000, 6000 and 7000 variables. The time limits are comparable to that used by the Diversification-Driven

[1] https://github.com/MQLib/MQLib.

[2] http://people.brunel.ac.uk/~mastjjb/jeb/orlib/bqpinfo.html.

[3] http://www.soften.ktu.lt/~gintaras/ubqop_its.html.

[4] http://www.stanford.edu/~yyye/yyye/Gset.

Table 9.1 Average results comparison on the instances of the first set

		Average solution gap (i.e., $BKR - f_{avg}$)					
Instance	*BKR*	PR1 [28]	PR2 [28]	ITS [24]	MST2 [23]	D^2TS [10]	MA [22]
b2500.1	1515944	0	0	0	0	0	13
b2500.2	1471392	0	58	9	0	0	645
b2500.3	1414192	13	0	11	11	0	173
b2500.4	1507701	0	0	0	0	0	0
b2500.5	1491816	0	0	0	0	0	55
b2500.6	1469162	0	0	0	0	0	190
b2500.7	1479040	0	0	0	0	0	416
b2500.8	1484199	0	0	0	0	0	3
b2500.9	1482413	0	0	0	0	0	321
b2500.10	1483355	0	0	0	0	0	446
Average		1.3	5.8	2	1.1	0	226

Tabu Search (D^2TS) [10], Iterated Tabu Search (ITS) [24], MultiStart Tabu Search (MST2) [23], Memetic Algorithm (MA) [22] and Automatic Algorithm (AAC_R) [25] after considering the computing performance of different machines.

Table 9.1 shows the results obtained by the 6 reference algorithms for solving the 10 bxxx.y instances. Columns 1 and 2 respectively give the instance name and the best known result BKR reported in the literature. The following columns list the average solution gap to the best known result $BKR - f_{avr}$. Given that all the reference algorithms are capable of finding the best known results, we do not report for each instance the tabulated solution gap 0 between the best solution value found by each algorithm and the best known result. The last row "Average" indicates the summary of each algorithm's average performance over this set of instances.

As shown in Table 9.1, D^2TS is able to reach the best known results during each run for all the 10 instances. PR1 and PR2 perform slightly worse by failing for 1 instance. The average solution gaps to the best known results obtained by PR1, PR2, ITS and MST2 are 1.3, 5.8, 2 and 1.1, respectively, which are quite small compared to the solution values. MA performs the worst among all the algorithms by obtaining an average solution gap of 226 to the best known results.

Tables 9.2 and 9.3 show the best and average solution gaps to the best known results for solving the 21 pxxx.y instances. We replace the algorithm MA by AAC_R since the latter is recently proposed and reports much better results. Table 9.2 indicates that PR1 and PR2 achieve the best known results for all the 21 challenging instances. AAC_R and D^2TS perform slightly worse since they fail to reach the best known results for 1 and 2 instances, respectively. ITS and MST2 obtain the worst gaps of 306.8 and 308.9 on average with respect to the best solution found. Table 9.3 indicates AAC_R performs the best with an average solution gap of 211.7. PR1 and PR2 obtain the average solution gaps of 457.1 and 690.4, respectively, which are slightly worse than AAC_R. D^2TS obtains the worst average solution gap of 2082.9 among all the algorithms.

Table 9.2 Best results comparison on the instances of the second set

		Best solution gap (i.e., $BKR - f_{best}$)					
Instance	*BKR*	PR1 [28]	PR2 [28]	ITS [24]	MST2 [23]	D^{2}TS [10]	AAC$_R$ [25]
p3000.1	3931583	0	0	0	0	0	0
p3000.2	5193073	0	0	0	0	0	0
p3000.3	5111533	0	0	0	0	0	0
p3000.4	5761822	0	0	0	0	0	0
p3000.5	5675625	0	0	0	0	0	0
p4000.1	6181830	0	0	0	0	0	0
p4000.2	7801355	0	0	0	0	0	0
p4000.3	7741685	0	0	0	0	0	0
p4000.4	8711822	0	0	0	0	0	0
p4000.5	8908979	0	0	0	0	0	0
p5000.1	8559680	0	0	700	325	325	0
p5000.2	10836019	0	0	0	582	0	0
p5000.3	10489137	0	0	0	0	0	0
p5000.4	12252318	0	0	934	1643	0	0
p5000.5	12731803	0	0	0	0	0	0
p6000.1	11384976	0	0	0	0	0	0
p6000.2	14333855	0	0	88	0	0	0
p6000.3	16132915	0	0	2729	0	0	0
p7000.1	14478676	0	0	340	1607	0	0
p7000.2	18249948	0	0	1651	2330	104	8
p7000.3	20446407	0	0	0	0	0	0
Average		0	0	306.8	308.9	20.4	0.4

9.6.3 Computational Results on the MaxCut Instances

The maximum cut problem can be naturally transformed into the QUBO model. Given an undirected graph $G = (V, E)$ with vertex set $V = \{1, \ldots, n\}$ and edge set $E \subseteq V \times V$, each edge $e(i, j)$ is associated with a weight w_{ij}, the maximum cut problem (MaxCut) asks for a partition of V into two disjoint subsets such that the total weight of the cut (edges crossing the two subsets) is maximized. Formally, the objective function of MaxCut is:

$$\text{Maximize:} \quad f(x) = \sum_{i=1}^{n}\sum_{j=1}^{n} w_{ij} x_i (1 - x_j), \tag{9.5}$$

$$\text{subject to:} \quad x_i \in \{0, 1\}, \quad i = 1, \ldots, n.$$

Table 9.3 Average results comparison on the instances of the second set

Instance	f_{prev}	Average solution gap (i.e., $BKR - f_{avg}$)					
		PR1 [28]	PR2 [28]	ITS [24]	MST2 [23]	D^2TS [10]	AAC$_R$ [25]
p3000.1	3931583	0	80	0	0	0	0
p3000.2	5193073	0	0	97	97	0	0
p3000.3	5111533	36	72	344	287	0	108
p3000.4	5761822	0	0	154	77	0	0
p3000.5	5675625	90	279	501	382	0	49
p4000.1	6181830	0	0	0	0	0	0
p4000.2	7801355	71	314	1285	804	0	323
p4000.3	7741685	0	64	471	1284	0	3
p4000.4	8711822	0	0	438	667	0	0
p4000.5	8908979	491	385	572	717	0	0
p5000.1	8559680	612	918	971	581	656	387
p5000.2	10836019	620	499	1068	978	12533	339
p5000.3	10489137	995	318	1266	1874	12876	77
p5000.4	12252318	1258	1168	1952	2570	1962	554
p5000.5	12731803	51	166	835	1233	239	37
p6000.1	11384976	201	822	57	34	0	18
p6000.2	14333855	221	577	1709	1269	1286	148
p6000.3	16132915	1744	2017	3064	2673	787	672
p7000.1	14478676	935	1523	1139	2515	2138	903
p7000.2	18249948	1942	2986	4301	3814	8712	828
p7000.3	20446407	332	2311	3078	7868	2551	0
Average		457.1	690.4	1109.6	1415.4	2082.9	211.7

The following corresponding relation is apparent in comparison with the formulation of QUBO:

$$q_{ii} = \sum_{j=1, j \neq i}^{n} w_{ij}, \quad q_{ij} = -w_{ij}, \quad (i \neq j) \tag{9.6}$$

According to Eq. (9.6), a MaxCut problem instance can be reformulated into a QUBO instance. Thus, the algorithms designed for solving the QUBO problem are directly applicable to the MaxCut problem.

Table 9.4 reports the computational results on the 54 MaxCut instances. For each execution of the compared algorithms, the time limit for solving each instance is set to be 30 minutes. Columns 1 gives the instance name. Columns 2 to 4 list the best solution values found by PR1, PR2 and D^2TS. Columns 5 to 8 list the average solution values found by PR1, PR2, AAC$_M$ and AAC$_R$. Both AAC$_M$ and AAC$_R$ are proposed in [25], where AAC$_R$ uses the pxxx.y instances of as training inputs to irace and AAC$_M$ is trained on the MaxCut instances. The last row "Matched" indicates the number of instances where each algorithm obtains the best results

Table 9.4 Computational results comparison on the MaxCut instances

	f_{best}			f_{avg}			
Instance	PR1 [28]	PR2 [28]	D^{2}TS [10]	PR1 [28]	PR2 [28]	AAC$_M$ [25]	AAC$_R$ [25]
G1	11624	11624	11624	11624	11624	11624	11607.8
G2	11620	11620	11620	11620	11620	11620	11606.2
G3	11620	11620	11620	11620	11620	11622	11611.2
G4	11646	11646	11646	11646	11646	11646	11633.2
G5	11631	11631	11631	11631	11631	11631	11620.1
G6	2178	2178	2178	2178	2178	2178	2172.1
G7	2006	2006	2006	2006	2006	2006	1995.8
G8	2005	2005	2005	2005	2005	2005	1998.5
G9	2054	2054	2054	2054	2054	2054	2044.4
G10	2000	2000	2000	2000	1999.8	2000	1989.8
G11	564	564	564	564	564	564	556.4
G12	556	556	556	556	556	556	546
G13	582	582	580	582	582	582	568
G14	3063	3064	3061	3062.1	3062.6	3062.6	3055.9
G15	3050	3050	3050	3049.3	3049.3	3049.9	3035.8
G16	3052	3052	3052	3051.3	3051.4	3051.9	3038.4
G17	3047	3047	3046	3045.5	3046.4	3046.7	3034.3
G18	992	992	991	992	992	992	974.3
G19	906	906	904	906	906	906	886.9
G20	941	941	941	941	941	941	915.1
G21	931	931	931	931	931	931	903.5
G22	13359	13359	13359	13353.5	13354.5	13349.2	13338.4
G23	13342	13342	13342	13333	13331.6	13332.1	13330
G24	13337	13333	13337	13327.3	13325.3	13321.9	13321.1
G25	13338	13339	13332	13328	13328.2	13329.2	13324.9
G26	13324	13326	13328	13313.7	13312.3	13314.8	13311.7
G27	3337	3336	3336	3327.3	3326.9	3328.2	3316.2
G28	3296	3296	3295	3286	3288.9	3287.7	3285.2
G29	3404	3405	3391	3395.2	3391.9	3390.2	3383.4
G30	3412	3411	3403	3404.6	3404.8	3405.5	3397.8
G31	3306	3306	3288	3299.7	3299.5	3300.7	3293.8
G32	1408	1410	1406	1400.9	1404.6	1401.3	1375.4
G33	1382	1382	1378	1373.9	1376.1	1373.3	1352.2
G34	1382	1384	1378	1375.4	1378.2	1376	1352.3
G35	7674	7679	7678	7663.3	7670.8	7668.8	7642.5
G36	7666	7671	7670	7653.1	7658.7	7660	7633.5
G37	7673	7682	7682	7663.3	7667.9	7670	7645
G38	7674	7682	7683	7663.4	7670.4	7671.7	7642.5
G39	2402	2407	2397	2391.3	2391.1	2395.2	2341.5
G40	2394	2399	2390	2381.2	2383.3	2387.8	2324.9

(continued)

Table 9.4 (continued)

	f_{best}			f_{avg}			
Instance	PR1 [28]	PR2 [28]	D^{2}TS [10]	PR1 [28]	PR2 [28]	AAC$_M$ [25]	AAC$_R$ [25]
G41	2402	2404	2400	2380	2388.9	2393	2329.2
G42	2475	2478	2469	2462.3	2466.2	2467.8	2404.9
G43	6660	6660	6660	6660	6659.9	6660	6649.3
G44	6650	6650	6639	6649.9	6649.9	6650	6641.4
G45	6654	6654	6652	6653.9	6653.9	6654	6647.4
G46	6649	6649	6649	6648.2	6648.8	6649	6641.5
G47	6657	6657	6665	6656.6	6656.8	6656.9	6648
G48	6000	6000	6000	6000	6000	6000	6000
G49	6000	6000	6000	6000	6000	6000	6000
G50	5880	5880	5880	5880	5880	5880	5875
G51	3848	3848	3847	3844.6	3846.4	3846.6	3832.4
G52	3851	3851	3849	3847.6	3848.4	3849.9	3836.1
G53	3849	3850	3848	3846.9	3847.7	3848	3835.2
G54	3852	3851	3851	3848.6	3847.8	3850.1	3836.3
Matched	38	47	29	25	27	44	2

among the compared algorithms. In terms of the best solution values, PR2 matches the best results for 47 out of 54 instances, performing better than PR1 and D^{2}TS that match 38 and 29 best results, respectively. In terms of the average solution values, AAC$_M$ performs the best by matching the best results for 44 instances, much better than AAC$_R$ that only matches 2 best results. This indicates that the behavior of the automatic approach is closely correlated to the problem structure.

Acknowledgments This work was partially supported by the National Natural Science Foundation of China (No. 71971172), the Natural Science Basic Research Program of Shaanxi (No. 2020JM-089), the Social Science Fund of Shaanxi (No. 2019S051), the Fundamental Research Funds for the Central Universities (No. D5000210834).

References

1. T.M. Alkhamis, M. Hasan, M.A. Ahmed, Simulated annealing for the unconstrained quadratic pseudo-boolean function. Eur. J. Oper. Res. **108**, 641–652 (1998)
2. M.M. Amini, B. Alidaee, G. Kochenberger, A scatter search approach to unconstrained quadratic binary programs, in *New Ideas in Optimization*, ed. by D. Corne, M. Dorigo, F. Glover (McGraw-Hill, New York, 1999), pp. 317–329
3. J.E. Beasley, Heuristic algorithms for the unconstrained binary quadratic programming problem. Ph.D. Thesis, The Management School Imperial College, London (1998)
4. J.E. Beasley, Obtaining test problems via Internet. J. Global Optim. **8**, 429–433 (1996)
5. S.J. Benson, Y. Ye, X. Zhang, Mixed linear and semidefinite programming for combinatorial and quadratic Optimization. Optim. Methods Softw. **11**, 515–544 (1999)
6. E. Boros, P.L. Hammer, X. Sun, The ddt method for quadratic 0-1 minimization, RRR 39-89, RUTCOR Research Center (1989)

7. I. Dunning, S. Gupta, J. Silberholz, What works best when? A systematic evaluation of heuristics for max-cut and qubo. INFORMS J. Comput. **30**, 608–624 (2018)
8. C. Helmberg, F. Rendl, A spectral bundle method for semidefinite programming. SIAM J. Optim. **10**(3), 673–696 (2000)
9. F. Glover, G.A. Kochenberger, B. Alidaee, Adaptive memory tabu search for binary quadratic programs. Manag. Sci. **44**, 336–345 (1998)
10. F. Glover, Z. Lü, J.K. Hao, Diversification-driven tabu search for unconstrained binary quadratic problems. 4OR: A Quart. J. Oper. Res. **8**, 239–253 (2010)
11. F. Glover, C. Rego, B. Alidaee, G. Kochenberger, One-pass heuristic for large-scale unconstrained binary quadratic problems. Eur. J. Oper. Res. **137**, 272–287 (2002)
12. S. Hanafi, A.R. Rebai, M. Vasquez, Several versions of the devour digest tidy-up heuristic for unconstrained binary quadratic problems. J. Heuristics **19**, 645–677 (2013)
13. P. Hansen, N. Mladenović, Variable neighborhood search, in *Handbook of Metaheuristics*, ed. by F. Glover, G. Kochenberger. International Series in Operations Research Management Science, vol. 57 (2003), pp. 145–184
14. K. Katayama, H. Narihisa, Performance of simulated annealing-based heuristic for the unconstrained binary quadratic programming problem. Eur. J. Oper. Res. **134**, 103–119 (2001)
15. W. Liu, D. Wilkins, B. Alidaee, A hybrid multi-exchange local search for unconstrained binary quadratic program. Working Papers Series, Hearin Center For Enterprise Science (2006)
16. A. Lodi, K. Allemand, T.M. Liebling, An evolutionary heuristic for quadratic 0-1 programming. Eur. J. Oper. Res. **119**, 662–670 (1999)
17. M. López-Ibáñez, J. Dubois-Lacoste, L.P. Cáceres, M. Birattari, T. Stützle, The irace package: iterated racing for automatic algorithm configuration. Oper. Res. Perspect. **3**, 43–58 (2016)
18. Z. Lü, F. Glover, J.K. Hao, A hybrid metaheuristic approach to solving the ubqp problem. Eur. J. Oper. Res. **207**, 1254–1262 (2010)
19. Z. Lü, F. Glover, J.K. Hao, Neighborhood combination for unconstrained binary quadratic problems, in *MIC-2009 Post-Conference Book*, ed. by M. Caserta, S. Voss (Springer, Berlin, 2012), pp. 49–61
20. P. Merz, B. Freislebeng, Greedy and local search heuristics for unconstrained binary quadratic programming. J. Heuristics **8**, 197–213 (2002)
21. P. Merz, B. Freisleben, Genetic algorithms for binary quadratic programming, in *Proceedings of the 1999 Genetic and Evolutionary Computation Conference*, vol. 1 (1999), pp. 417–424
22. P. Merz, K. Katayama, Memetic algorithms for the unconstrained binary quadratic programming problem. Biosystems **78**, 99–118 (2004)
23. G. Palubeckis, Multistart tabu search strategies for the unconstrained binary quadratic optimization problem. Ann. Oper. Res. **131**, 259–282 (2004)
24. G. Palubeckis, Iterated tabu search for the unconstrained binary quadratic optimization problem. Informatica **17**, 279–296 (2006)
25. M. de Souza, M. Ritt, Automatic grammar-based design of heuristic algorithms for unconstrained binary quadratic programming, in *EvoCOP 2018*, ed. by A. Liefooghe, M. Lopez-Ibanez. Lecture Notes in Computer Science, vol. 10782 (2018), pp. 67–84
26. V.P. Shylo, O.V. Shylo, Solving unconstrained binary quadratic programming problem by global equilibrium search. Cybern. Syst. Anal. **47**, 889–897 (2011)
27. Y. Wang, Z. Lü, F. Glover, J.K. Hao, Probabilistic GRASP-Tabu search algorithms for the UBQP problem. Comput. Oper. Res. **40**, 3100–3107 (2013)
28. Y. Wang, Z. Lü, F. Glover, J.K. Hao, Path relinking for unconstrained binary quadratic programming. Eur. J. Oper. Res. **223**, 595–604 (2012)

Chapter 10
The Bipartite QUBO

Abraham P. Punnen

Abstract This chapter deals with the bipartite quadratic unconstrained binary optimization problem (BQUBO) which is closely related QUBO, both as a generalization and as a particular case. In this sense many of the results discussed for QUBO in the previous chapters extend to BQUBO. Here we focus primarily on results that exploit the special structure of BQUBO. In particular, we consider computational complexity, polynomially solvable special cases, approximation algorithms, MILP formulations, and exact and heuristic algorithms.

10.1 Introduction

The *bipartite quadratic unconstrained binary optimization problem* (BQUBO) is a variation of the quadratic unconstrained binary optimization problem (QUBO) with special structural properties. Let $\mathbf{Q}$ be an $m \times n$ real matrix with its (i, j)th element as q_{ij}, $\mathbf{c}^T = (c_1, c_2, \ldots, c_m) \in \mathbb{R}^m$, $\mathbf{d}^T = (d_1, d_2, \ldots, d_n) \in \mathbb{R}^n$, $\mathbf{y}^T = (y_1, y_2, \ldots, y_n) \in \{0, 1\}^n$ and $\mathbf{x}^T = (x_1, x_2, \ldots, x_m) \in \{0, 1\}^m$. Then, the BQUBO can be stated as the mathematical programming problem

$$\begin{aligned} \text{Maximize} \quad & f(\mathbf{x}, \mathbf{y}) = \mathbf{x}^{\mathrm{T}}\mathbf{Q}\mathbf{y} + \mathbf{c}^T\mathbf{x} + \mathbf{d}^T\mathbf{y} \\ \text{Subject to} \quad & \\ & \mathbf{x} \in \{0, 1\}^m, \mathbf{y} \in \{0, 1\}^n \end{aligned}$$

An instance of the BQUBO is uniquely defined when $\mathbf{Q}, \mathbf{c}$ and $\mathbf{d}$ are specified. Therefore, we sometimes denote a BQUBO instance by the triplet $(\mathbf{Q}, \mathbf{c}, \mathbf{d})$. The BQUBO is equivalent to the *unconstrained bilinear program* (UBLP) [43, 44] obtained by replacing the binary restrictions on the variables $\mathbf{x}$ and $\mathbf{y}$ with $\mathbf{x} \in [0, 1]^m$ and $\mathbf{y} \in [0, 1]^n$. There is extensive literature on bilinear programs and some

A. P. Punnen (✉)
Department of Mathematics, Simon Fraser University, Surrey, BC, Canada
e-mail: apunnen@sfu.ca

A. P. Punnen (ed.), *The Quadratic Unconstrained Binary Optimization Problem*,
https://doi.org/10.1007/978-3-031-04520-2_10

of these are relevant to the special case of UBLP. Despite the equivalence between the BQUBO and the UBLP, we represent a BQUBO using binary variables and focus primarily on the combinatorial properties, unless otherwise stated.

Before getting into further details on the BQUBO, let us now briefly discuss some of the notations used in this chapter. We follow the same notational convention as discussed in Chap. 1. That is, All matrices are represented using bold capital letters and elements of a matrix are represented by the corresponding small letters along with accents, if any, and the location coordinates. For example the (i, j)th element of the matrix $\mathbf{A}$ is a_{ij}, of the matrix $\bar{\mathbf{B}}$ is $\bar{b}_{ij}$, and of the matrix $\mathbf{D}^k$ is d^k_{ij}. Similarly, vectors are represented by boldface small letters along with appropriate accents, as applicable, and elements of a vector is represented using the same letter (without boldface) along with its location coordinate and accents, if any. For example, the ith element of the vector $\mathbf{c}$ is c_i, of the vector $\mathbf{x}^k$ is x^k_i, and of the vector $\tilde{\mathbf{v}}$ is $\tilde{v}_i$. Exceptions to this rule will be stated explicitly and operators such as transpose etc. are not considered as accents in the case of vectors. The zero vector in any dimension is represented by $\mathbf{0}$. Additional notations will be introduced as need arises. Also we denote $\mathcal{M} = \{1, 2, \ldots, m\}$ and $\mathcal{N} = \{1, 2, \ldots, n\}$. Throughout this chapter, we assume that $m \leq n$, without loss of generality.

Many well-studied optimization problems can be formulated as a BQUBO. In Chapter 1, we have seen that when $m = n$ and $\mathbf{Q}$ is symmetric and positive semidefinite, a QUBO can be solved as a BQUBO and vice versa [42, 43]. Further, any instance $(\mathbf{Q}, \mathbf{c}, \mathbf{d})$ of a BQUBO can be formulated as the QUBO $(\mathbf{Q}', \mathbf{c}')$ or the QUBO $(\mathbf{Q}'', \mathbf{c}')$, where

$$\mathbf{Q}' = \left[\begin{array}{c|c} \mathbf{O}_{m\times m} & \frac{1}{2}\mathbf{Q} \\ \hline \frac{1}{2}\mathbf{Q}^T & \mathbf{O}_{n\times n} \end{array}\right], \ \mathbf{Q}'' = \left[\begin{array}{c|c} \mathbf{O}_{m\times m} & \mathbf{Q} \\ \hline \mathbf{O}_{n\times m} & \mathbf{O}_{n\times n} \end{array}\right], \ \mathbf{c}' = \left[\begin{array}{c|c} \mathbf{c} & \mathbf{d} \end{array}\right], \text{ and}$$

$\mathbf{O}_{n\times n}$, $\mathbf{O}_{m\times m}$, and $\mathbf{O}_{n\times m}$ are zero matrices. In this sense, BQUBO is a special case of QUBO and hence the results derived for the QUBO are applicable for the BQUBO as well.

A graph theoretic interpretation of the BQUBO can be given as follows. Let $G = (V_1, V_2, E)$ be a bipartite graph with $V_1 = \{1, 2, \ldots, m\}$ and $V_2 = \{1, 2, \ldots, n\}$. The edge set $E = \{(i, j) : q_{ij} \neq 0\}$ and let q_{ij} be the weight of the edge (i, j) in G. Also, the weight of the node $i \in V_1$ is c_i and the weight of the node $j \in V_2$ is d_j. Then, the BQUBO is to find an $S \subseteq V_1 \cup V_2$ such that the subgraph of G induced by S have the largest sum of the weights of its edges and nodes [63–65].

In the preceding chapters, we have discussed two primary forms of QUBO, one with 0-1 variables and the other with variables taking values -1 or 1. The later was called the Ising QUBO. The corresponding version of the BQUBO, called the Ising BQUBO, can be stated as the mathematical programming problem

$$\begin{aligned} &\text{Maximize } \phi(\mathbf{x}, \mathbf{y}) = \mathbf{x}^\mathbf{T}\mathbf{A}\mathbf{y} + \mathbf{b}^T\mathbf{x} + \mathbf{p}^T\mathbf{y} \\ &\text{Subject to} \\ &\qquad \mathbf{x} \in \{-1, 1\}^m, \mathbf{y} \in \{-1, 1\}^n. \end{aligned}$$

where $\mathbf{b}^T = (b_1, b_2, \ldots, b_m)$ and $\mathbf{p}^T = (p_1, p_2, \ldots, p_n)$. An instance of the Ising BQUBO can be represented by the 3-tuple $(\mathbf{A}, \mathbf{b}, \mathbf{p})$. The BQUBO and the Ising BQUBO without the linear terms are said to be in *homogeneous form*.

Theorem 10.1 *For any BQUBO there exists an equivalent BQUBO which is in the homogeneous form. Likewise, for any Ising BQUBO there exists an equivalent Ising BQUBO which is in the homogeneous form.*

Proof We prove the result for the Ising BQUBO and the proof for the case of BQUBO is similar. Consider the instance $(\mathbf{A}, \mathbf{b}, \mathbf{p})$ of the Ising BQUBO. Introduce two new variables, say x_{m+1} and y_{n+1}. Consider the $(m+1) \times (n+1)$ matrix

$$\hat{\mathbf{A}} = \left[\begin{array}{c|c} \mathbf{A} & \mathbf{b} \\ \hline \mathbf{p}^T & M \end{array}\right], \tag{10.1}$$

where M is a large number. Since $\phi(\mathbf{x}, \mathbf{y}) = \phi(-\mathbf{x}, -\mathbf{y})$ for any homogeneous Ising BQUBO and the entry in $a_{m+1,n+1}$ is M, without loss of generality we can assume that $x_{m+1} = y_{n+1} = 1$ in an optimal solution to the Ising BQUBO $(\hat{\mathbf{A}}, \mathbf{0}_{m+1}, \mathbf{0}_{n+1})$. This accurately counts the linear terms of the Ising QUBO $(\mathbf{A}, \mathbf{b}, \mathbf{p})$ for an optimal solution. However, this adds a constant M to the objective function value of every optimal solution of $(\hat{\mathbf{A}}, \mathbf{0}_{m+1}, \mathbf{0}_{n+1})$ compared to that of the Ising QUBO $(\mathbf{A}, \mathbf{b}, \mathbf{p})$, and this does not affect the optimal solution set for either problem. Thus, the Ising BQUBO $(\mathbf{A}, \mathbf{b}, \mathbf{p})$ is equivalent to the Ising BQUBO $(\hat{\mathbf{A}}, \mathbf{0}_{m+1}, \mathbf{0}_{n+1})$.

The choice of $a_{m+1,n+1} = M$ was crucial in the above proof which guarantees that x_{m+1} and y_{n+1} will have the same sign in an optimal solution. Just the property that $\phi(\mathbf{x}, \mathbf{y}) = \phi(-\mathbf{x}, -\mathbf{y})$ is not sufficient to establish this since if $x_{m+1} = 1, y_{n+1} = -1$, its negation also have different signs and hence the linear terms may not get counted accurately. For a corresponding result in the case of Ising QUBO, instead of M we could select 0, as we have seen in Chap. 1. Using M in our reduction although preserve optimality, it need not preserve ϵ-optimality and this is a significant difference from the case of the Ising QUBO.

We can also establish equivalence between the BQUBO and the Ising BQUBO [64]. Consider the linear transformation

$$\mathbf{x} = 2\mathbf{w} - \mathbf{e}_m \text{ and } \mathbf{y} = 2\mathbf{z} - \mathbf{e}_n, \tag{10.2}$$

where $\mathbf{e}_m$ and $\mathbf{e}_n$ are all-one vectors in $\mathbb{R}^m$ and $\mathbb{R}^n$, respectively. Under this transformation, $x_i = 1$ is mapped to $w_i = 1$ and $x_i = -1$ mapped to $w_i = 0$. Similarly, $y_j = 1$ is mapped to $z_j = 1$ and $y_j = -1$ is mapped to $z_j = 0$. Thus, using (10.2), the Ising BQUBO can be reduced to the BQUBO

$$\text{Maximize } f(\mathbf{w}, \mathbf{z}) = \mathbf{w}^T\mathbf{Q}^a\mathbf{z} + (\mathbf{c}^a)^T\mathbf{w} + (\mathbf{d}^a)^T\mathbf{z}$$

Subject to

$$\mathbf{w} \in \{0, 1\}^m, \mathbf{z} \in \{0, 1\}^n,$$

where $\mathbf{Q}^a = 4\mathbf{A}$, $\mathbf{c}^a = 2(\mathbf{b} - \mathbf{A}\mathbf{e}_n)$, $\mathbf{d}^a = 2(\mathbf{p} - \mathbf{A}^T\mathbf{e}_m)$ and the constant term $\mathbf{e}_m^T\mathbf{A}\mathbf{e}_n - \mathbf{b}^T\mathbf{e}_m - \mathbf{p}^T\mathbf{e}_n$ is ignored from the objective function. Similarly, using the linear transformation

$$\mathbf{x} = \frac{1}{2}(\mathbf{w} + \mathbf{e}_m) \text{ and } \mathbf{y} = \frac{1}{2}(\mathbf{z} + \mathbf{e}_n) \tag{10.3}$$

a BQUBO can be reduced to the Ising BQUBO

$$\text{Maximize } \phi(\mathbf{w}, \mathbf{z}) = \mathbf{w}^T\mathbf{A}^q\mathbf{z} + (\mathbf{b}^q)^T\mathbf{w} + (\mathbf{p}^a)^T\mathbf{z}$$

$$\text{Subject to}$$

$$\mathbf{w} \in \{-1, 1\}^m, \mathbf{z} \in \{-1, 1\}^n$$

where $\mathbf{A}^q = \frac{1}{4}\mathbf{Q}$, $\mathbf{b}^q = \frac{1}{4}(\mathbf{Q}\mathbf{e}_n) + \frac{1}{2}\mathbf{c}$, $\mathbf{p}^q = \frac{1}{4}\mathbf{Q}^T\mathbf{e}_m + \frac{1}{2}\mathbf{d}$ and the constant term $\frac{1}{4}\mathbf{e}_m^T\mathbf{Q}\mathbf{e}_n + \frac{1}{2}\mathbf{c}^T\mathbf{e}_m + \frac{1}{2}\mathbf{d}^T\mathbf{e}_n$ is ignored from the objective function.

Theorem 10.2 *The BQUBO and the Ising BQUBO are strongly NP-hard.*

Proof We reduce a QUBO to the BQUBO. Consider the instance $(\mathbf{Q}, \mathbf{c})$ of a QUBO. From this, construct the instance $(\bar{Q}, \bar{\mathbf{c}}, \bar{\mathbf{d}})$ of BQUBO, where

$$\bar{\mathbf{Q}} = \mathbf{Q} - 2M\mathbf{I}, \ \bar{\mathbf{c}} = \frac{1}{2}\mathbf{c} + M\mathbf{e} \text{ and } \bar{\mathbf{d}} = \frac{1}{2}\mathbf{c} + M\mathbf{e}, \tag{10.4}$$

$\mathbf{I}$ is the $n \times n$ identity matrix, $\mathbf{e} \in \mathbb{R}^n$ is the all one vector, and M is a large number. Now, it can be verified that an optimal solution $(\mathbf{x}, \mathbf{y})$ to $(\bar{Q}, \bar{\mathbf{c}}, \bar{\mathbf{d}})$ satisfies $x_i = y_i$ for all $i = 1, 2, \ldots, n$ and hence $\mathbf{x}$ is an optimal solution to the QUBO $(\mathbf{Q}, \mathbf{c})$. Since the QUBO is strongly NP-hard, BQUBO is also strongly NP-hard. Further, since any BQUBO can be reduced to the Ising BQUBO using the linear transformation (10.3), it follows that the Ising BQUBO is also strongly NP-hard.

It is well known that the maximum weight cut problem (MWCP) on a general graph G is equivalent to QUBO (see Chaps. 1, 3, and the references therein). In MWCP, if we restrict the graph G to be bipartite, we get an instance of the *bipartite maximum weight cut problem* which we denote by B-MaxCut. Indeed, viewing B-MaxCut as a maximum weight cut problem on a general graph yields an equivalent QUBO. But the support graph of the resulting QUBO need not be bipartite and hence the resulting instance need not be that of a BQUBO. When all edge weights are non-negative, B-MaxCut is a trivial problem. We now show the equivalence between the BQUBO and the B-MaxCut.

Let $G = (V_1, V_2, E)$ be a bipartite graph with $|V_1| = m$ and $|V_2| = n$. Two vectors $\mathbf{x} \in \{-1, 1\}^m$ and $\mathbf{y} \in \{-1, 1\}^n$ defines a cut $(S_1 \cup S_2, T_1 \cup T_2)$ in G if $S_1 = \{i \in V_1 : x_i = 1\}$, $T_1 = \{i \in V_1 : x_i = -1\}$, $S_2 = \{j \in V_1 : y_j = 1\}$, and $T_2 = \{j \in V_2 : y_j = -1\}$. We call $(\mathbf{x}, \mathbf{y})$ the *incidence vector* of the cut $(S_1 \cup S_2, T_1 \cup T_2)$. Let q_{ij} be the weight of the edge (i, j) in G. Then, the value of the cut $(S_1 \cup S_2, T_1 \cup T_2)$ is given by

$$\sum_{i \in S_1, j \in T_2} q_{ij} + \sum_{i \in T_1, j \in S_2} q_{ij} = \sum_{x_i = -y_j} q_{ij} \tag{10.5}$$

Theorem 10.3 ([64]) *The Ising BQUBO* $(\mathbf{A}, \mathbf{0}, \mathbf{0})$ *and B-MaxCut are equivalent in the following sense:*

1. *For any instance of the Ising BQUBO, it is possible to construct a complete bipartite graph G such that an optimal solution to the B-MaxCut problem on G gives an optimal solution to the Ising BQUBO.*
2. *For any instance of B-MaxCut on a bipartite graph* $G = (V_1, V_2, E)$ *with* $|V_1| = m$ *and* $|V_2| = n$*, it is possible to construct an instance of the homogeneous Ising BQUBO with an* $m \times n$ *cost matrix* $\mathbf{A}$ *such that an optimal solution to the Ising BQUBO* $(\mathbf{A}, \mathbf{0}, \mathbf{0})$ *gives an optimal solution to the B-MaxCut problem on G.*

Proof The objective function $\phi(\mathbf{x}, \mathbf{y})$ of the Ising BQUBO $(\mathbf{A}, \mathbf{0}, \mathbf{0})$ can be written as

$$\phi(\mathbf{x}, \mathbf{y}) = \sum_{ij} a_{ij} x_i y_j = \sum_{x_i = y_j} a_{ij} - \sum_{x_i = -y_j} a_{ij} = \sum_{i=1}^{m} \sum_{j=1}^{n} a_{ij} - 2 \sum_{x_i = -y_j} a_{ij}$$

Since $\sum_{i=1}^{m} \sum_{j=1}^{n} a_{ij}$ is a constant, maximizing $\phi(\mathbf{x}, \mathbf{y})$ is equivalent to maximizing $-\sum_{\bar{\mathbf{x}}_i = -\bar{\mathbf{y}}_j} a_{ij}$. Thus, by solving the B-MaxCut problem on a complete bipartite graph $K_{m,n}$ with the weight of the edge (i, j) as $-a_{ij}$ solves the Ising BQUB $(\mathbf{A}, \mathbf{0}, \mathbf{0})$.

To establish the second part of the theorem, we show that the B-MaxCut problem on the bipartite graph $G = (V_1, V_2, E$ with edge weights c_{ij} for $(i, j) \in E$ can be solved as an Ising BQUBO $(\mathbf{A}, \mathbf{0}, \mathbf{0})$. Let $(S_1 \cup S_2, T_1 \cup T_2)$ be a cut in G and $(\mathbf{x}, \mathbf{y})$ be the corresponding incidence vector. Let $\delta(S, T) = \sum_{x_i = -y_j} c_{ij}$ be the value of the cut $(S_1 \cup S_2, T_1 \cup T_2)$. Then it can be verified that $g(S, T) = \frac{1}{2} \sum_{(i,j) \in E} c_{ij} - \frac{1}{2} \sum_{(i,j) \in E} c_{ij} x_i y_j$ and, hence, maximizing $\delta(S, T)$ is equivalent to solving the Ising BQUBO $(\mathbf{A}, \mathbf{0}, \mathbf{0})$ where $\mathbf{A}$ is an $m \times n$ matrix with $|V_1| = m$, $|V_2| = n$ and $a_{ij} = -\frac{1}{2} c_{ij}$ if $(i, j) \in E$ and $a_{ij} = 0$, otherwise.

Given any $m \times n$ matrix $\mathbf{Q}$ we define the *support bipartite graph* as the bipartite graph $G = (V_1, V_2, E)$ where $V_1 = \{u_1, u_2, \ldots, u_m\}$, $V_2 = \{v_1, v_2, \ldots, v_n\}$ and $E = \{(u_i, v_j) : u_i \in V_1, v_j \in V_2, q_{ij} \neq 0\}$. From Theorem 10.3, it follows that,

Theorem 10.4 *Let* $\mathbf{A}$ *be an* $m \times n$ *matrix and* $G = (V_1, V_2, E)$ *be the support bipartite graph of* $\mathbf{A}$ *with the weight of an edge* (u_i, v_j) *is* a_{ij}*. Then G has a cut* $(S_1 \cup S_2, T_1 \cup T_2)$ *with capacity K if and only if there exists* $\mathbf{x} \in \{-1, 1\}^m$, $\mathbf{y} \in \{-1, 1\}^n$ *such that* $\mathbf{x}^T \mathbf{A} \mathbf{y} = \sum_{i=1}^{m} \sum_{j=1}^{n} a_{ij} - 2K$.

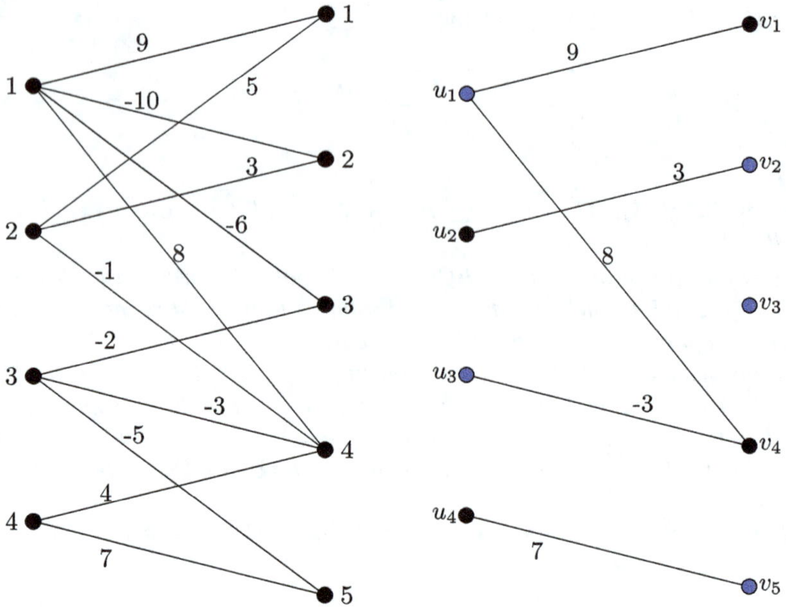

Fig. 10.1 The support bipartite graph of A and the edges of the given cut

To illustrate the result of Theorem 10.4, consider the matrix

$$\mathbf{A} = \begin{pmatrix} 9 & -10 & -6 & 8 & 0 \\ 5 & 3 & 0 & -1 & 0 \\ 0 & 0 & -2 & -3 & -5 \\ 0 & 0 & 0 & 4 & 7 \end{pmatrix}$$

The support bipartite graph $G = (V_1, V_2, E)$, where $V_1 = \{u_1, u_2, u_3, u_4\}$, $V_2 = \{v_1, v_2, v_3, v_4, v_5\}$, of $\mathbf{A}$ with the weight of the edge (u_i, v_j) is assigned as a_{ij} is given in Fig. 10.1 along with edges of the cut $(S_1 \cup S_2, T_1 \cup T_2)$ where $S_1 = \{u_1, u_3\}$, $T_1 = \{u_2, u_4\}$, $S_2 = \{v_2, v_3, v_5\}$ and $T_2 = \{v_1, v_4\}$.

The incidence vector corresponding to the cut is $(\mathbf{x}, \mathbf{y})$ where $\mathbf{x} = (1, -1, 1, -1)$ and $\mathbf{y} = (-1, 1, 1, -1, 1)$. Now $\mathbf{x}^T\mathbf{A}\mathbf{y} = -39$, $\sum_{i=1}^{m}\sum_{j=1}^{n} a_{ij} = 9$ and $K = 24$. Note that $\mathbf{x}^T\mathbf{A}\mathbf{y} = \sum_{i=1}^{m}\sum_{j=1}^{n} a_{ij} - 2K$ is satisfied.

By Theorem 10.4, the Ising BQUBO $(\mathbf{A}, \mathbf{0}, \mathbf{0})$ can be solved by solving the maximum weight cut problem on the support bipartite graph of $\mathbf{A}$ with the weight of the edge (u_i, v_j) as $-a_{ij}$. Further, if $(S_1^* \cup S_2^*, T_1^* \cup T_2^*)$ is an optimal solution to this maximum cut problem with value K^*, then the corresponding incidence vectors $\mathbf{x}^*, \mathbf{y}^*$ will be an optimal solution to the BQUBO $(\mathbf{A}, \mathbf{0}, \mathbf{0})$ with the objective function value $\sum_{i=1}^{m}\sum_{j=1}^{n} a_{ij} + 2K^*$.

10.2 Applications

In Chaps. 1 and 2, we have seen that many combinatorial optimization problems can be modelled as a QUBO. Since QUBO is a special case of BQUBO as established in Theorem 10.2, each of these applications can be modelled as a BQUBO as well. Let us now look at some specific problems where the BQUBO model fits better.

Consider the bipartite graph $G = (V_1, V_2, E)$ with a weight w_{ij} is defined for each edge $(i, j) \in E$. Then, the *maximum weight biclique problem* (MWBP) on the bipartite graph G [6, 70] is to find a biclique in G such that the sum of the weights of its edges is maximized. MWBP is a strongly NP-hard [58] even if $w_{ij} = 1$ for all $(i, j) \in E$. Define

$$q_{ij} = \begin{cases} w_{ij} & \text{if } (i, j) \in E \\ -M & \text{otherwise ,} \end{cases}$$

where M is a large positive number. Then, it can be verified that the BQUBO $(\mathbf{Q}, \mathbf{0}_m, \mathbf{0}_n)$ solves the MWBP. The MWBP has applications in various areas. This include data mining, clustering and bioinformatics [13, 15, 71, 72], anomaly detection in E-commerce, social recommendation, and advertising [50], and formal concept analysis [16]. Consequently, these optimization problems can be solved as a BQUBO. Further, the vertex separator problem on a graph can be formulated as a BQUBO with cardinality constraints [35].

The BQUBO model also arises in approximating a matrix by a rank-one binary matrix [24, 45, 46, 49, 67]. For example, let $\mathbf{D} = (d_{ij})$ be a given $m \times n$ matrix and we want to find an $m \times n$ matrix $\mathbf{B} = (b_{ij})$, where $b_{ij} = u_i v_j$, $u_i \in \{0, 1\}, i = 1, 2, \ldots, m$, and $v_j \in \{0, 1\}, i = 1, 2, \ldots, n$ such that

$$\sum_{i=1}^{m} \sum_{j=1}^{n} (d_{ij} - u_i v_j)^2$$

is minimized. The matrix $\mathbf{B}$ is called a *rank one approximation* of $\mathbf{D}$. Since u_i, v_j are binary variables,

$$\begin{aligned} \sum_{i=1}^{m} \sum_{j=1}^{n} (d_{ij} - u_i v_j)^2 &= \sum_{i=1}^{m} \sum_{j=1}^{n} (d_{ij}^2 - 2d_{ij} u_i v_j + u_i^2 v_j^2) \\ &= \sum_{i=1}^{m} \sum_{j=1}^{n} (d_{ij}^2 - 2d_{ij} u_i v_j + u_i v_j) \\ &= \sum_{i=1}^{m} \sum_{j=1}^{n} d_{ij}^2 + \sum_{i=1}^{m} \sum_{j=1}^{n} (1 - 2d_{ij}) u_i v_j \end{aligned}$$

Therefore, minimizing $\sum_{i=1}^{m}\sum_{j=1}^{n}(d_{ij} - u_i v_j)^2$ is achieved by minimizing $\sum_{i=1}^{m}\sum_{j=1}^{n}(1-2d_{ij})u_i v_j$ which is equivalent to maximizing $\sum_{i=1}^{m}\sum_{j=1}^{n}(2d_{ij} - 1)u_i v_j$. Thus, given the matrix **D**, the matrix **B** can be identified by solving the BQUBO $(\mathbf{Q}, \mathbf{0}, \mathbf{0})$ where $q_{ij} = 2d_{ij} - 1$, for all $i \in \{1, 2, \ldots, m\}$ and $j \in \{1, 2, \ldots, n\}$. Binary matrix factorization arises in mining discrete patterns in binary data [49, 67]. If u_i and v_j are required to be in $\{-1, 1\}$ then the resulting rank one approximation problem can be formulated as a homogeneous Ising BQUBO.

The *maximal sum submatrix problem* (MSSP) studied by Derval [18] and Branders [10] is essentially equivalent to the homogeneous BQUBO. Given an $m \times n$ matrix **Q** with row and column index sets $\{1, 2, \ldots, m\}$ and $\{1, 2, \ldots, n\}$ respectively, the MSSP seeks a subset I of $\{1, 2, \ldots, m\}$ and a subset J of $\{1, 2, \ldots, n\}$ such that

$$\sum_{i \in I}\sum_{j \in J} q_{ij} \tag{10.6}$$

is maximized. Derval [18] considered different variations and generalizations of MSSP with applications in data mining, biclustering , frequent itemset mining, and tiling. Branders [10] considered a penalty parameter in Eq. (10.6) where q_{ij} is replaced by $q_{ij} - \theta$. In this case q_{ij} values above the threshold θ is favored and those below θ is penalized. This is again a homogeneous BQUBO and have applications in the analysis of gene expression data [10].

Another application of BQUBO is to find the *cut-norm* of a matrix [3, 4]. The cut-norm of a matrix **Q**, denoted by $||\mathbf{Q}||_c$, is defined as

$$||\mathbf{Q}||_c = \max_{I \subseteq \mathcal{M}, J \subseteq \mathcal{N}} \left| \sum_{i \in I}\sum_{j \in J} q_{ij} \right|.$$

Thus, the cut-norm $||\mathbf{Q}||_c$ is the optimal objective function value of the binary quadratic program

$$\begin{aligned} \text{CN:} \quad & \text{Maximize} \quad \left|\mathbf{x}^T\mathbf{Q}\mathbf{y}\right| \\ & \text{Subject to:} \quad \mathbf{x} \in \{0, 1\}^m, \mathbf{y} \in \{0, 1\}^n. \end{aligned}$$

From this definition, it follows that the cut norm of **Q** is the largest of the objective function values of the BQUBO problems $(\mathbf{Q}, \mathbf{0}_m, \mathbf{0}_n)$ and $(-\mathbf{Q}, \mathbf{0}_m, \mathbf{0}_n)$. So, the cut norm can be identified by solving two BQUBO problems. The cut-norm of a matrix can be used in developing approximation algorithms for various problems on matrices and dense graphs [21], solving biclustering problems in computational molecular biology [71, 72], and finding regular partition of graphs [4, 5]. An Ising BQUBO formulation for computing the cut norm is given in [4]. This formulation also leads to computing an efficient approximation of the cut norm of a matrix. Let us discus this formulation now.

Let $\text{OPT}(\mathbf{A}, \mathbf{0}, \mathbf{0})$ be the optimal objective function value of the instance $(\mathbf{A}, \mathbf{0}, \mathbf{0})$ of the Ising BQUBO. It may be noted that $\text{OPT}(\mathbf{A}, \mathbf{0}, \mathbf{0}) \geq 0$. The lemma below and its validity proof is from [4].

Lemma 10.1 ([4]) $||\mathbf{A}||_c \leq \text{OPT}(\mathbf{A}, \mathbf{0}, \mathbf{0}) \leq 4||A||_c$.

Proof Let $(\mathbf{x}^*, \mathbf{y}^*)$ be an optimal solution to the Ising BQUBO $(\mathbf{A}, \mathbf{0}, \mathbf{0})$. Since $\text{OPT}(\mathbf{A}, \mathbf{0}, \mathbf{0}) \geq 0$, we have $\sum_{i=1}^m \sum_{j=1}^n a_{ij} x_i^* y_j^* = \left|\sum_{i=1}^m \sum_{j=1}^n a_{ij} x_i^* y_j^*\right|$. Let $X^1 = \{i : x_i^* = 1\}$, $X^2 = \{i : x_i^* = -1\}$, $Y^1 = \{i : y_i^* = 1\}$, and $Y^2 = \{i : y_i^* = -1\}$. Then,

$$
\begin{aligned}
\sum_{i=1}^m \sum_{j=1}^n a_{ij} x_i^* x_j^* &= \left|\sum_{i=1}^m \sum_{j=1}^n a_{ij} x_i^* y_j^*\right| \\
&= \left|\sum_{i \in X^1} \sum_{j \in Y^1} a_{ij} + \sum_{i \in X^2} \sum_{j \in Y^2} a_{ij} + \sum_{i \in X^2} \sum_{j \in Y^1} -a_{ij} + \sum_{i \in X^1} \sum_{j \in Y^2} -a_{ij}\right| \\
&\leq \left|\sum_{i \in X^1} \sum_{j \in Y^1} a_{ij}\right| + \left|\sum_{i \in X^2} \sum_{j \in Y^2} a_{ij}\right| + \left|\sum_{i \in X^2} \sum_{j \in Y^1} -a_{ij}\right| + \left|\sum_{i \in X^1} \sum_{j \in Y^2} -a_{ij}\right| \\
&\leq 4||\mathbf{A}||_c,
\end{aligned}
$$

since each of the absolute values in the sum on the right hand side is at most $||\mathbf{A}||_c$. Thus,

$$
\sum_{i=1}^m \sum_{j=1}^n a_{ij} x_i^* x_j^* \leq 4||\mathbf{A}||_c \tag{10.7}
$$

Also note that $||\mathbf{A}||_c = \sum_{i \in I} \sum_{j \in J} a_{ij}$ or $||\mathbf{A}||_c = -\sum_{i \in I} \sum_{j \in J} a_{ij}$ for some $I \subseteq \mathcal{M}$, $J \subseteq \mathcal{N}$. Now, suppose that $||\mathbf{A}||_c = \sum_{i \in I} \sum_{j \in J} a_{ij}$. Define $x_i = 1$ if $i \in I$ and $x_i = -1$ if $i \notin I$. Similarly, define $y_i = 1$ if $i \in J$ and $y_i = -1$ if $i \notin J$. Then,

$$
\begin{aligned}
||\mathbf{A}||_c = \sum_{i \in I} \sum_{j \in J} a_{ij} &= \sum_{i=1}^m \sum_{j=1}^n a_{ij} \left(\frac{1 + x_i}{2}\right)\left(\frac{1 + y_j}{2}\right) \\
&= \frac{1}{4}\left(\sum_{i=1}^m \sum_{j=1}^n a_{ij} + \sum_{i=1}^m r_i x_i + \sum_{j=1}^n s_j y_j + \sum_{i=1}^m \sum_{j=1}^n a_{ij} x_i y_j\right)
\end{aligned} \tag{10.8}
$$

where $r_i = \sum_{j=1}^n a_{ij}$ and $s_j = \sum_{i=1}^m a_{ij}$. The absolute value of each of the sums inside the bracket on the right hand side is at most $\text{OPT}(\mathbf{A}, \mathbf{0}, \mathbf{0})$ and hence $||\mathbf{A}||_c \geq \text{OPT}(\mathbf{A}, \mathbf{0}, \mathbf{0})$. The case when $||\mathbf{A}||_c = -\sum_{i \in I} \sum_{j \in J} a_{ij}$ can be analyzed in a similar way and this completes the proof.

Lemma 10.2 *If the row and column sums of* $\mathbf{A}$ *are zeros, then* $\text{OPT}(\mathbf{A}, \mathbf{0}, \mathbf{0}) = 4||\mathbf{A}||_c$.

Proof When each row and column sums of $\mathbf{A}$ is equal to zero, in equation (10.8), $r_i = 0$ for $i = 1, 2, \ldots, m$, $s_i = 0$ for $i = 1, 2, \ldots, n$ and $\sum_{i=1}^{m}\sum_{j=1}^{n} a_{ij} = 0$. Thus, $||\mathbf{A}||_c \leq \frac{1}{4}\text{OPT}(\mathbf{A}, \mathbf{0}, \mathbf{0})$. The result now follows from Lemma 10.1.

Thus, if the row and column sums of a matrix $\mathbf{A}$ are zeros, its cut norm can be identified by solving a BQUBO.

Theorem 10.5 ([22, 41]) *For any matrix* $\mathbf{B}$*, there exists an associated matrix* $\mathbf{A}$ *with row and column sums equal to zero such that* $||\mathbf{A}||_c = ||\mathbf{B}||_c$.

Proof Let $\mathbf{B}$ be an $m \times n$ matrix. Now, construct the $(m + 1) \times (n + 1)$ matrix $\mathbf{A}$ such that

$$\mathbf{A} = \left[\begin{array}{c|c} \mathbf{B} & -\mathbf{r} \\ \hline -\mathbf{s}^T & b \end{array}\right] \tag{10.9}$$

where the ith element of the vector $\mathbf{r}$ is $r_i = \sum_{j=1}^{n} b_{ij}$, the jth element of the vector $\mathbf{s}$ is $s_j = \sum_{i=1}^{m} b_{ij}$ and the scalar $b = \sum_{i=1}^{m}\sum_{j=1}^{n} b_{ij}$. Let $I \subseteq \{1, 2, \ldots, m\}$ and $J \subseteq \{1, 2, \ldots, n\}$. Now define I' and J' where

$$I' = \begin{cases} I \text{ if } m+1 \notin I \\ \{1, 2, \ldots m\} \setminus I \text{ if } m+1 \in I \end{cases} \quad \text{and} \quad J' = \begin{cases} J \text{ if } n+1 \notin J \\ \{1, 2, \ldots n\} \setminus J \text{ if } n+1 \in J \end{cases}$$

Then, $\left|\sum_{i \in I, j \in J} a_{ij}\right| = \left|\sum_{i \in I', j \in J'} b_{ij}\right|$ and hence it follows that $||\mathbf{A}||_c = ||\mathbf{B}||_c$.

In view of Lemma 10.2 and Theorem 10.5, the cut norm of a matrix $\mathbf{B}$ can be identified by solving an Ising BQUBO $(\mathbf{A}, \mathbf{0}, \mathbf{0})$, where $\mathbf{A}$ is defined as in the proof of Theorem 10.5.

10.3 MILP Formulations

In Chap. 6, we have seen various MILP formulations of the QUBO. The reader is referred to the discussions there and the references provided for details on some of the terminologies used here. Further, those models can easily be adapted to the case of BQUBO. Because of this, we will not consider a detailed discussion on integer programming or semidefinite programming models for BQUBO in this chapter. Instead, we simply highlight two basic models that are used in other parts of this chapter and indicate the simplifications achieved by exploiting the special structure of BQUBO. For each $i \in \mathcal{M}$, let $R_i = \{j \in \mathcal{N} : q_{ij} \neq 0\}$.

We use the variable w_{ij} to represent the product x_iy_j. Then, the linearization theorem from Chap. 6 can be used to introduce additional constraints so that w_{ij} is a 0-1 variable with $w_{ij} = 1$ if and only if $x_i = y_j = 1$. For QUBO, we used the variable y_{ij} to represent the product x_ix_j. Thus, $y_{ij} = y_{ji}$ and this fact was exploited in some linearizations of QUBO [59]. However, for BQUBO, w_{ij} need not be equal to w_{ji} since $w_{ji} = x_jy_i \neq x_iy_j$. Keeping this in mind, the BQUBO counterpart of the Glover-Woolsey linearization of QUBO can be stated as

$$\text{GWB:} \quad \text{Maximize} \quad \sum_{i=1}^{m}\sum_{j\in R_i} q_{ij}w_{ij} + \sum_{i=1}^{m} c_ix_i + \sum_{j=1}^{n} d_jy_j$$

$$\text{Subject to:} \quad x_i + y_j - w_{ij} \leq 1 \text{ for all } j \in R_i, i = 1, 2, \ldots, m, \tag{10.10}$$

$$w_{ij} - x_i \leq 0 \text{ for all } j \in R_i, i = 1, 2, \ldots, m \tag{10.11}$$

$$w_{ij} - y_j \leq 0 \text{ for all } j \in R_i, i = 1, 2, \ldots, m \tag{10.12}$$

$$x_i \in \{0, 1\} \text{ for all } i = 1, 2 \ldots, m, \tag{10.13}$$

$$y_j \in \{0, 1\} \text{ for all } j = 1, 2 \ldots, n. \tag{10.14}$$

$$w_{ij} \geq 0 \text{ for all } j \in R_i, i = 1, 2, \ldots, m. \tag{10.15}$$

The model GWB was considered earlier by many authors [10, 68, 69]. Exploiting the structure of BQUBO, we now show that the binary restrictions in GWB are required only for the x-variables or y-variables or w-variables. Let GWB_x be the version of GWB with constraint (10.14) is relaxed to $0 \leq y_j \leq 1$ and hence we have binary restrictions only on the x-variables. Similarly, let GWB_y be the version of GWB with constraint (10.13) is relaxed to $0 \leq x_i \leq 1$ and GWB_w be the version of GWB with constraints (10.13) and (10.14) are relaxed respectively to $0 \leq x_i \leq 1$ and $0 \leq y_j \leq 1$ but constraint (10.17) is replaced with $w_{ij} \in \{0, 1\}$ for all $j \in R_i, i = 1, 2, \ldots, m$.

Theorem 10.6 ([69]) *GWB_x, GWB_y and GWB_w are valid MILP models for the BQUBO.*

Proof Let us first show that GWB_x is a valid MILP model for the BQUBO. When $\mathbf{x}$ is fixed at some vector in $\{0, 1\}^m$, the coefficient matrix corresponding to the resulting constraints (10.10) and (10.11) is totally unimodualr since it can be viewed as the transpose of the node arc incidence matrix of some directed graph. This implies that when $\mathbf{x}$ is fixed as a specific vector in $\{0, 1\}^m$, the coefficient matrix of GWB_x is totally unimodualr and the RHS is also a binary vector. Thus a 0-1 optimal solution exists for GWB_x. The case of GWB_y can be handled analogously. Now, let us consider GWB_w. Here both constraints (10.14) and (10.13) are relaxed to respectively to $0 \leq y_j \leq 1$ and $0 \leq x_i \leq 1$ and replace the constraint (10.15) with $w_{ij} \in \{0, 1\}$ in GWB. Thus, if w_{ij} is fixed at 0 or 1 for all i and j, the coefficient matrix corresponding to constraints (10.10) is totally unimodualr since it can be viewed as the transpose of the incidence matrix of an undirected bipartite graph which is known to be totally unimodualr. Thus, it follows that when w_{ij} is fixed at binary values, the coefficient matrix of the remaining constraints of GWB_w is totally unimodular. Further, in this case the righthand side coefficients are also integer establishing that GWB_w has a $0-1$ optimal solution.

The formulation GWB_x shows that the BQUBO can be solved as a sequence of $O(2^m)$ linear programs. Thus, when $m = O(\log n)$ the problem can be solved in polynomial time. A more efficient algorithm for this case will be discussed later. The formulation GWB can be strengthened by adding additional valid inequalities. See [69] for various valid inequalities for the *bipartite Boolean quadric polytope*.

Let $R_i^+ = \{j \in N : q_{ij} > 0\}$ and $R_i^- = \{j \in N : q_{ij} < 0\}$. Then the optimality restricted version (see Chap. 6) of GWB is

$$\text{GWBO:} \quad \text{Maximize} \quad \sum_{i=1}^{m}\sum_{j\in R_i} q_{ij}w_{ij} + \sum_{i=1}^{m} c_i x_i + \sum_{j=1}^{n} d_j y_j$$

$$\text{Subject to:} \quad x_i + y_j - w_{ij} \le 1 \text{ for all } j \in R_i^-, i = 1, 2, \ldots, m, \tag{10.16}$$

$$w_{ij} - x_i \le 0 \text{ for all } j \in R_i^+, i = 1, 2, \ldots, m \tag{10.17}$$

$$w_{ij} - y_j \le 0 \text{ for all } j \in R_i^+, i = 1, 2, \ldots, m \tag{10.18}$$

$$x_i \in \{0, 1\} \text{ for all } i = 1, 2 \ldots, m, \tag{10.19}$$

$$y_j \in \{0, 1\} \text{ for all } j = 1, 2 \ldots, n. \tag{10.20}$$

$$w_{ij} \ge 0 \text{ for all } j \in R_i^-, i = 1, 2, \ldots, m. \tag{10.21}$$

As discussed in the case of QUBO in Chap. 6, we can aggregate (weighted or unweighted) blocks of constraints from GWB to obtain valid MILP models for BQUBO with reduced number of constraints. For example, the constraint block (10.10) in GWB can be replaced by

$$\sum_{j\in R_i} (y_j - w_{ij}) \le |R_i|(1 - x_i), \text{ for } i = 1, 2, \ldots, m$$

to obtain a valid MILP model for BQUBO. A weighted version of the above constraint is also valid. i.e.

$$\sum_{j\in R_i} \alpha_{ij}(y_j - w_{ij}) \le \left(\sum_{j\in R_i} \alpha_{ij}\right)(1 - x_i), \text{ for } i = 1, 2, \ldots, m$$

were $\alpha_{ij} > 0$ can be used to replace the constraint block (10.10) to obtain a valid MILP formulation of BQUBO. Aggregation however, can weaken the LP relaxation. This can be mitigated by choosing the weights α_{ij} appropriately. See Chap. 6 and [59, 60] for further details.

Let $S_j = \{i \in \mathcal{M} : q_{ij} \neq 0\}$. Weighted aggregation of the constraints in blocks (10.11) and (10.12) yields

$$\sum_{j\in R_i} \beta_{ij} w_{ij} \leq \left(\sum_{j\in R_i} \beta_{ij}\right) x_i, i = 1, 2, \ldots, m \tag{10.22}$$

$$\sum_{i\in S_j} \gamma_{ij} w_{ij} \leq \left(\sum_{i\in S_j} \gamma_{ij}\right) y_j, j = 1, 2, \ldots, n. \tag{10.23}$$

where $\beta_{ij} > 0, \gamma_{ij} > 0$. In GWB, we can replace the constraint blocks (10.11) and (10.12) with (10.22) and (10.23) to obtain yet another valid MILP formulation for BQUBO. In this case, it is important to add the upper bound constraints $w_{ij} \leq 1$ for all i and j. Choosing the weights $\beta_{ij} > 0$ and $\gamma_{ij} > 0$ appropriately, the LP relaxation of this aggregated model can be made competitive with that of GWB [60] (see also Chap. 6). Combining all the aggregations discussed above, we get the following MILP model for BQUBO.

$$\text{GWBA:} \quad \text{Maximize} \quad \sum_{i=1}^{m}\sum_{j\in R_i} q_{ij} w_{ij} + \sum_{i=1}^{m} c_i x_i + \sum_{j=1}^{n} d_j y_j$$

$$\text{Subject to:} \quad \sum_{j\in R_i} \alpha_{ij}(y_j - w_{ij}) \leq \left(\sum_{j\in R_i} \alpha_{ij}\right)(1 - x_i), i \in \mathcal{M} \tag{10.24}$$

$$\sum_{j\in R_i} \beta_{ij} w_{ij} \leq \left(\sum_{j\in R_i} \beta_{ij}\right) x_i, i \in \mathcal{M} \tag{10.25}$$

$$\sum_{i\in S_j} \gamma_{ij} w_{ij} \leq \left(\sum_{i\in S_j} \gamma_{ij}\right) y_j, j \in \mathcal{N} \tag{10.26}$$

$$x_i \in \{0, 1\}, i \in \mathcal{M}, y_j \in \{0, 1\}, j \in \mathcal{N} \tag{10.27}$$

$$0 \leq w_{ij} \leq 1, j \in R_i, i \in \mathcal{M}. \tag{10.28}$$

10.3.1 Compact Formulations

The compact MILP formulations discussed in Chap. 6 for QUBO can also be adapted for BQUBO in the natural way. Let $h_i(y) = \sum_{j\in R_i} q_{ij} y_j$. Then, the objective function $\psi(\mathbf{x}, \mathbf{y})$ of BQUBO can be written as

$$\psi(\mathbf{x}, \mathbf{y}) = \sum_{i=1}^{n} x_i h_i(\mathbf{y}) + \sum_{i=1}^{m} c_i x_i + \sum_{j=1}^{n} d_j y_j \tag{10.29}$$

For any $\mathbf{y} \in \{0, 1\}^n$, $h_i(\mathbf{y})$ is a real number satisfying

$$\ell_i = \sum_{j \in R_i} \min\{0, q_{ij}\} \leq h_i(\mathbf{y}) \leq u_i = \sum_{j \in R_i} \max\{0, q_{ij}\}. \tag{10.30}$$

Applying Glover's linearization theorem (see Chap. 6) on the product $x_i h_i(\mathbf{y})$ in (10.29), we get the compact formulation [26]

$$\text{GLB:}\quad \text{Maximize} \quad \sum_{i=1}^{n} w_i + \sum_{i=1}^{m} c_i x_i + \sum_{j=1}^{n} d_j y_j$$

$$\text{Subject to:}\quad w_i \leq u_i x_i, \ \text{ for } i = 1, 2, \ldots, m \tag{10.31}$$

$$w_i \leq \ell_i x_i + \sum_{j \in R_i} q_{ij} y_j - \ell_i, \ \text{ for } i = 1, 2, \ldots, m \tag{10.32}$$

$$w_i \geq \ell_i x_i, \ \text{ for } i = 1, 2, \ldots, m \tag{10.33}$$

$$w_i \geq u_i x_i + \sum_{j \in R_i} q_{ij} y_j - u_i, \ \text{ for } i = 1, 2, \ldots, m \tag{10.34}$$

$$x_i, y_j \in \{0, 1\} \text{ for } i \in \mathcal{M}, j \in \mathcal{N}. \tag{10.35}$$

An optimality restricted version of GLB can also be obtained. Note that the general constraints of GLB can be written as

$$\max\{u_i x_i + \sum_{j \in R_i} q_{ij} y_j - u_i, \ell_i x_i\} \leq w_i \leq \min\{u_i x_i, \ell_i x_i + \sum_{j \in R_i} q_{ij} y_j - \ell_i\}, i = 1, 2, \ldots, m. \tag{10.36}$$

Since we have a maximization objective with the coefficient of w_i in the objective function is one for all i, the inequalities (10.33) and (10.34) can be discarded from GL without affecting optimality [1]. This leads to the optimality restricted version, GLBO, of GLB:

$$\text{GLBO:}\quad \text{Maximize} \quad \sum_{i=1}^{m} w_i + \sum_{i=1}^{m} c_i x_i + \sum_{j=1}^{n} d_j y_j$$

$$\text{Subject to:}\quad w_i \leq u_i x_i, \ \text{ for } i = 1, 2, \ldots, m \tag{10.37}$$

$$w_i \leq \ell_i x_i + \sum_{j \in R_i} q_{ij} y_j - \ell_i, \text{ for } i = 1, 2, \ldots, m \tag{10.38}$$

$$x_i, y_j \in \{0, 1\} \text{ for } i \in \mathcal{M}, j \in \mathcal{N}. \tag{10.39}$$

The model GLBO was also given in [18] and [9]. Let z_i be the slack variable in constraint (10.37). Eliminating the unrestricted variable w_i using (10.37), we get the MILP model

$$\text{GLOB1:} \quad \text{Maximize} \quad \sum_{i=1}^{m}(u_i x_i - z_i) + \sum_{i=1}^{m} c_i x_i + \sum_{j=1}^{n} d_j y_j$$

$$\text{Subject to:} \quad u_i x_i - z_i \leq \ell_i x_i + \sum_{j \in R_i} q_{ij} y_j - \ell_i, \text{ for } i = 1, 2, \ldots, m \tag{10.40}$$

$$x_i, y_j \in \{0, 1\} \text{ for } i \in \mathcal{M}, j \in \mathcal{N} \tag{10.41}$$

$$z_i \geq 0, i \in \mathcal{M}. \tag{10.42}$$

for the BQUBO [1].

The LP relaxation of these compact formulations are relatively weak. However, they can be strengthened by adapting methods discussed in Chap. 6. Since these are relatively straightforward modifications from the corresponding models for QUBO, the details are omitted.

10.4 Polynomially Solvable Special Cases

In Chap. 3, we have seen various polynomially solvable special cases of the QUBO. Since the BQUBO can be solved as a QUBO, whenever the transformed QUBO satisfies conditions of polynomial time solvability of the QUBO, we can solve the BQUBO in polynomial time. Let us now consider some additional polynomially solvable cases that exploit the special properties of the BQUBO. Several of the results in this section are from [65].

Theorem 10.7 ([65]) *The BQUBO can be solved in* $O(mn2^m)$ *time. When* $m = O(\log n)$ *it is solvable in* $O(n^2 \log n)$ *time but remains Strongly NP hard if* $m = O(\sqrt[k]{n})$ *for any fixed constant k.*

Proof Choose any $\mathbf{x}' \in \{0, 1\}^m$ and compute a $\mathbf{y}' \in \{0, 1\}^n$ with

$$y'_j = \begin{cases} 1 & \text{if } \sum_{i=1}^{m} q_{ij} x'_i + d_j > 0 \\ 0 & \text{otherwise} . \end{cases} \tag{10.43}$$

Then, the solution $(\mathbf{x}', \mathbf{y}')$ is optimal for the restricted BQUBO where $\mathbf{x} = \mathbf{x}'$ is enforced. Now, repeat this process by restricting $\mathbf{x} = \mathbf{x}'$ to each of the vectors $\mathbf{x}'$ in $\{0, 1\}^m$ and compute the best corresponding solution $(\mathbf{x}', \mathbf{y}')$ as discussed in Eq. (10.43). The overall best solution generated by this process will be an optimal solution to the BQUBO. For any given $\mathbf{x}' \in \{0, 1\}^m$, the corresponding best solution $(\mathbf{x}', \mathbf{y}')$ along with its objective function value can be identified in $O(mn)$ time. Thus, an optimal solution to BQUBO can be obtained in $O(mn\sum_{k=0}^{m}\binom{n}{k}) = O(mn2^m)$. When $m = O(\log n)$, this complexity becomes $O(n^2 \log n)$.

We now use a self-reduction technique to establish the complexity of the BQUBO when $m = O(\sqrt[k]{n})$ for a fixed k. From an instance $(\mathbf{Q}, \mathbf{c}, \mathbf{d})$ of the BQUBO construct another instance $(\mathbf{Q}', \mathbf{c}', \mathbf{d}')$ of BQUBO where

$$\mathbf{Q}' = \left[\begin{array}{c|c} \mathbf{Q} & \mathbf{O}^1 \\ \hline \mathbf{O}^2 & \mathbf{O}^3 \end{array}\right], \mathbf{c}' = \left[\begin{array}{c|c} \mathbf{c} & \tilde{\mathbf{0}} \end{array}\right], \text{ and } \mathbf{d}' = \left[\begin{array}{c|c} \mathbf{d} & \hat{\mathbf{0}} \end{array}\right].$$

Here, $\mathbf{O}^1$, $\mathbf{O}^2$ and $\mathbf{O}^3$ are zero matrices with dimensions $m \times n^k$, $(n-m) \times (n-m)$, and $(n-m) \times n^k$ respectively, $\tilde{\mathbf{0}}$ is a zero vector of dimension $n-m$ and $\hat{\mathbf{0}}$ is a zero vector of dimension n^k. It can be verified that the instance $(\mathbf{Q}', \mathbf{c}', \mathbf{d}')$ of BQUBO constructed above satisfies the hypothesis of the theorem. Further, for every solution to BQUBO $(\mathbf{Q}, \mathbf{c}, \mathbf{d})$, there exists a solution to the BQUBO $(\mathbf{Q}', \mathbf{c}', \mathbf{d}')$, such that the corresponding objective function values coincide and viceversa. Since the BQUBO is strongly NP-hard, the result follows.

An analogous result for the maximum cut problem was proved in [52]. A variation of the construction discussed in the proof of the first part of Theorem 10.7 has been discussed in Chap. 6 in the context of decomposition upper bounds [14] and in Chap. 3 for a corresponding polynomially solvable case of the QUBO.

A matrix $\mathbf{Q}$ is said to be a *sum matrix*, if there exist constants $a_1, a_2, \ldots, a_m$ and $b_1, b_2, \ldots, b_n$ such that $q_{ij} = a_i + b_j$ for all i, j. A *sum* BQUBO is the special case of the BQUBO when $\mathbf{Q}$ is a sum matrix. Let us now discuss an algorithm to solve the sum BQUBO in polynomial time, which was proposed originally by Punnen, Karapetyan, and Sriprathak [65]. When $\mathbf{Q}$ is a sum matrix,

$$\begin{aligned} f(\mathbf{x}, \mathbf{y}) &= \mathbf{x}^T\mathbf{Q}\mathbf{y} + \mathbf{c}^T\mathbf{x} + \mathbf{d}^T\mathbf{y} \\ &= \sum_{i=1}^{m} a_i x_i \sum_{j=1}^{n} y_j + \sum_{j=1}^{n} b_j y_j \sum_{i=1}^{m} x_i + \sum_{i=1}^{m} c_i x_i + \sum_{j=1}^{n} d_j y_j. \\ &= \sum_{i=1}^{m} (\lambda a_i + c_i) x_i + \sum_{j=1}^{n} (\mu b_j + d_j) y_j \end{aligned}$$

where $\sum_{j=1}^{n} y_j = \lambda$ and $\sum_{i=1}^{m} x_i = \mu$. Note that $\lambda \in \{0, 1, 2, \ldots, n\}$ and $\mu \in \{0, 1, 2, \ldots, m\}$. Thus, our BQUBO can be written as

$$\text{SUM:} \qquad \text{Maximize} \sum_{i=1}^{m}(\lambda a_i + c_i)x_i + \sum_{j=1}^{n}(\mu b_j + d_j)y_j$$

$$\text{Subject to} \sum_{j=1}^{n} y_j = \lambda, \sum_{i=1}^{m} x_i = \mu$$

$$x_i \in \{0, 1\}, y_j \in \{0, 1\}, i = 1, 2 \ldots, m;\ j = 1, 2, \ldots, n$$

$$\lambda \in \{0, 1, 2, \ldots, n\}, \mu \in \{0, 1, 2, \ldots, m\}.$$

For a fixed value of λ and μ, the optimization problem SUM decomposes into two problems,

$$\text{P1}(\lambda, \mu)\text{: Maximize} \sum_{i=1}^{m}(\lambda a_i + c_i)x_i \qquad \text{Subject to} \sum_{i=1}^{m} x_i = \mu \qquad x_i \in \{0, 1\}, i = 1, 2 \ldots, m.$$

and

$$\text{P2}(\lambda, \mu)\text{: Maximize} \sum_{i=1}^{n}(\mu b_j + d_j)y_j \qquad \text{Subject to} \sum_{j=1}^{n} y_j = \lambda \qquad y_j \in \{0, 1\}, j = 1, 2 \ldots, n.$$

For a fixed λ and μ, let $\mathbf{x}(\lambda, \mu)$ be an optimal solution to P1(λ, μ) with optimal objective function value $f_1(\lambda, \mu)$ and $f_2(\lambda, \mu)$ be the optimal objective function value of P2(λ, μ) for an optimal solution $\mathbf{y}(\lambda, \mu)$. Choose λ^*, μ^* such that

$$f_1(\lambda^*, \mu^*) + f_2(\lambda^*, \mu^*) = \max_{\lambda \in M^0} \max_{\mu \in N^0} \left(f_1(\lambda, \mu) + f_2(\lambda, \mu)\right)$$

where $M^0 = \{0, 1, \ldots, m\}$ and $N^0 = \{0, 1, 2, \ldots, n\}$. Then, $\mathbf{x}(\lambda^*, \mu^*), y(\lambda^*, \mu^*)$ will be an optimal solution to the sum BQUBO.

Theorem 10.8 ([65]) *The sum BQUBO can be solved in $O(mn \log n)$ time.*

Proof Based on the discussion above, we need an efficient way to generate the $O(mn)$ values $f_1(\lambda, \mu)$ and $f_2(\lambda, \mu)$ for $\lambda \in \{0, 1, \ldots, n\}$, $\mu \in \{0, 1, \ldots, m\}$ and their associated solutions $\mathbf{x}(\lambda, \mu)$ and $\mathbf{y}(\lambda, \mu)$. Let π be a permutation of $1, 2, \ldots, m$ that generates the descending arrangement

$$\lambda a_{pi(1)} + c_{\pi(1)} \geq \lambda a_{pi(2)} + c_{\pi(2)} \geq \cdots \geq \lambda a_{pi(m)} + c_{\pi(m)}$$

for a fixed λ. Then $\mathbf{x} = \mathbf{0}$ is an optimal solution to P1$(\lambda, 0)$ with $f_1(\lambda, 0) = 0$. For a fixed λ and $\mu = 1, 2, \ldots, m$, an optimal solution to P1(λ, μ) is given by

$$\mathbf{x}_{\pi}(i)(\lambda, \mu) = \begin{cases} 1, & \text{for } i \leq \mu \\ 0, & \text{otherwise} \end{cases}$$

and $f_1(\lambda, \mu) = f_1(\lambda, \mu - 1) + (\lambda a_{\pi(\mu)} + c_{\pi(\mu)})$. Thus, for a given λ, we can compute all of $f_1(\lambda, \mu)$, $\mu = 0, 1, 2, \ldots, m$ and the associated solutions in $O(m)$ time and hence $f_1(\lambda, \mu)$ for all values of λ and μ can be generated in $O(mn)$ time. For a given λ, the ordering π can be identified in $O(m \log m)$ time and this is repeated n times. Thus, the overall complexity of generating $f_1(\lambda, \mu)$ for all values of λ and μ is $O(mn \log m)$. Likewise, $f_2(\lambda, \mu)$ and the associated solution for all values of λ and μ can be generated in $O(mn \log n)$ time. Since $m \leq n$, the result follows.

Theorem 10.9 *BQUBO can be solved in polynomial time, if $q_{ij} \geq 0$ for all i, j or $q_{ij} \leq 0$ for all i, j.*

Proof Consider the MILP formulation GWBO of the BQUBO. If $q_{ij} \geq 0$ for all i, j, then the set R_i^- used in GWBO is empty. Thus, the constraint coefficient matrix of GWBO is totally unimodular and hence GWBO can be solved as a linear program. When $q_{ij} \leq 0$ for i, j, the set $R^+ = \emptyset$. In this case, the coefficient matrix of the LP relaxation of GWBO (after adding the redundant constraint $w_{ij} \leq 1$) will be of the type

$$A = \begin{pmatrix} \mathbf{B} & \mathbf{I}_1 \\ \mathbf{I}_2 & \mathbf{O} \end{pmatrix}$$

where $\mathbf{I}_1$ is the negative of an identity matrix of size $mn \times mn$, $\mathbf{I}_2$ is an identity matrix of size $(m + n) \times (m + n)$, $\mathbf{O}$ is a zero matrix and $\mathbf{B}^T$ is the incidence matrix of an undirected bipartite graph. Note that $\mathbf{B}^T$ is totally unimodular and hence $\mathbf{A}$ is totally unimodular. Thus GWBO can be solved as a linear program and the result follows.

It may be noted that the condition $q_{ij} \geq 0$ yields a polynomial time algorithm for the general QUBO, as shown in Chap. 3. However, the general QUBO is NP-hard even if $q_{ij} \in \{-1, 0\}$ for all i, j (see Chap. 3) but Theorem 10.9 shows that the BQUBO is solvable in polynomial time for all instances where $q_{ij} \leq 0$ for all i, j. This part of Theorem 10.9 was proved in [8].

The polynomial solvability established in Theorem 10.9 can be extended to a more general class of matrices where the number of negative elements or the number of positive elements in $\mathbf{Q}$ are small. Let $U \subseteq \mathcal{M}$ and $V \subseteq \mathcal{N}$. We call the pair (U, V) a *non-negative cover* (*non-positive cover*) of $\mathbf{Q}$ if the matrix $\mathbf{Q}'$ obtained from $\mathbf{Q}$ by deleting rows corresponding to U and columns corresponding to V has only non-negative (non-positive) entries.

Lemma 10.3 *For any matrix* $\mathbf{Q}$*, a non-negative cover (non-positive cover)* (U, V) *such that* $|U \cup V|$ *is the smallest can be identified in polynomial time.*

Proof Construct the bipartite graph $G = (\mathcal{M}, \mathcal{N}, E)$ where $(i, j) \in E$ if and only if $q_{ij} < 0$. Then a non-negative cover of $\mathbf{Q}$ with smallest cardinality is precisely a minimum vertex cover of G. Since the vertex cover problem on a bipartite graph can be solved in polynomial time, a non-negative cover of $\mathbf{Q}$ with smallest cardinality can be identified in polynomial time. The proof for the case of non-positive cover is similar.

Theorem 10.10 ([65]) *Let* (U, V) *be a minimum cardinality non-negative (non-positive) cover of* $\mathbf{Q}$*. Then, the BQUBO can be solved in polynomial time if* $|U \cup V|$ *is* $O(\log n)$ *and the problem is strongly NP-hard if* $|U \cup V|$ *is* $O(\sqrt[k]{n})$ *for some fixed* k*.*

Proof Consider an instance $(\mathbf{Q}, \mathbf{c}, \mathbf{d})$ of the BQUBO. Suppose (U, V) be a non-negative cover of $\mathbf{Q}$. Now, fix the variables x_i for $i \in U$ and y_j for $j \in V$ at 0-1 values results in a BQUBO with the associated cost matrix have non-negative elements. By Theorem 10.9 such a BQUBO can be solved in polynomial time. Since there are at most $2^{|U \cup V|}$ ways to fix the variables associated with U and V, the BQUBO can be solved in polynomial time if $|U \cup V| = O(\log n)$. The case of non-positive cover can be proved in the same way.

The second part of the theorem can be proved by reducing a general BQUBO to a BQUBO satisfying the hypothesis of the theorem that preserves objective function values of the corresponding solutions. This can be achieved by increasing the number of columns (or rows) of $\mathbf{Q}$ to a sufficiently large number, yet polynomial for fixed k and filling these columns (rows) with entries 0.

It may be noted that Theorem 10.10 allows arbitrary $\mathbf{c}$ and $\mathbf{d}$. Now let us look at the case when the rank of $\mathbf{Q}$ is one and we refer to this as *rank-one* BQUBO. Note that the matrix $\mathbf{Q}$ is of rank one if and only if there exist vectors $\mathbf{a}^T = (a_1, a_2, \ldots, a_m)$ and $\mathbf{b}^T = (b_1, b_2, \ldots, b_n)$ such that $q_{ij} = a_i b_j$ for all i, j. In Chap. 3, it is proved that QUBO is NP-hard if $\mathbf{Q}$ is of rank one and $\mathbf{c}$ is arbitrary [12]. Unlike QUBO, we now show that BQUBO can be solved in polynomial time when $\mathbf{Q}$ is of rank one and $\mathbf{c}$ and $\mathbf{d}$ are arbitrary. To establish this, we use parametric linear programming. This method is successfully applied in solving low rank bilinear programs [44] and some combinatorial optimization problems with multiplicative objective functions [62]. Our algorithm for the rank-one BQUBO is closely related to both of these works. The specific details presented below are from [65].

The rank-one BQUBO can be written as

$$\text{Maximize } (\mathbf{a}^T\mathbf{x})(\mathbf{b}^T\mathbf{y}) + \mathbf{c}^T\mathbf{x} + \mathbf{d}^T\mathbf{y}$$

$$\text{Subject to: } \mathbf{x} \in \{0, 1\}^m, \mathbf{y} \in \{0, 1\}^n,$$

As indicated in Sect. 10.1, we can relax the binary restrictions on the variables and write the rank-one BQUBO as the bilinear program

$$\text{BLP:} \quad \text{Maximize } (\mathbf{a}^T\mathbf{x})(\mathbf{b}^T\mathbf{y}) + \mathbf{c}^T\mathbf{x} + \mathbf{d}^T\mathbf{y}$$
$$\text{Subject to: } \mathbf{x} \in [0, 1]^m, \mathbf{y} \in [0, 1]^n.$$

Now, consider the parametric linear program,

$$\text{BLP}(\lambda): \quad \text{Maximize } \lambda\mathbf{b}^T\mathbf{y} + \mathbf{c}^T\mathbf{x} + \mathbf{d}^T\mathbf{y}$$
$$\text{Subject to: } \mathbf{a}^T\mathbf{x} = \lambda$$
$$\mathbf{x} \in [0, 1]^m, \mathbf{y} \in [0, 1]^n,$$

where $\lambda \in [\underline{\lambda}, \bar{\lambda}]$ and $\underline{\lambda} = \sum_{i=1}^{m} \min\{a_i, 0\}$ and $\bar{\lambda} = \sum_{i=1}^{n} \max\{a_i, 0\}$. Solving BLP($\lambda$) for all values of λ and choosing the overall best solution will solve BLP (i.e. the rank-one BQUBO). For a fixed value of λ, BLP(λ) decomposes into two linear programs

$$\text{L1}(\lambda): \quad \text{Maximize } \mathbf{c}^T\mathbf{x} \qquad \text{Subject to: } \mathbf{ax} = \lambda, \mathbf{x} \in [0, 1]^m$$

and

$$\text{L2}(\lambda): \quad \text{Maximize } \lambda\mathbf{b}^T\mathbf{y} + \mathbf{d}^T\mathbf{y} \qquad \text{Subject to } \mathbf{y} \in [0, 1]^n.$$

Thus, for fixed λ, let $\mathbf{x}(\lambda)$ be an optimal solution to L1(λ) and $\mathbf{y}(\lambda)$ be an optimal solution to L2(λ). It can be verified that

$$y(\lambda)_j = \begin{cases} 1 & \text{if } \lambda b_j + d_j \geq 0 \\ 0 & \text{otherwise .} \end{cases}$$

Let us now show that only $O(m)$ values of $\lambda \in [\underline{\lambda}, \bar{\lambda}]$ are relevant in computing an optimal solution for BLP. Let $f_1(\lambda)$ and $f_2(\lambda)$ respectively be the optimal objective function value of L1(λ) and L2(λ), as a function of λ. It is well known that $f_1(\lambda)$ is a piece-wise linear concave function and $f_2(\lambda)$ is a piece-wise linear convex function [53]. Thus, $f_1(\lambda) + f_2(\lambda)$ need not be convex or concave but it is piece-wise linear. Solving BLP amounts to finding the global maximum of $f_1(\lambda) + f_2(\lambda)$ for $\lambda \in [\underline{\lambda}, \bar{\lambda}]$. Let $\underline{\lambda} = \lambda_1, \lambda_2, \ldots, \lambda_p = \bar{\lambda}$ be the consecutive breakpoints of $f_1(\lambda)$. Then, $f_1(\lambda)$ is linear in $[\lambda_i, \lambda_{i+1}]$ and $f_2(\lambda)$ is convex in $[\lambda_i, \lambda_{i+1}]$, $i = 1, 2, \ldots, p-1$. Therefore, $f_1(\lambda) + f_2(\lambda)$ is a convex function in $[\lambda_i, \lambda_{i+1}]$, for $i = 1, \ldots, p-1$ and hence the maximum of $f_1(\lambda) + f_2(\lambda)$ over $\lambda \in [\lambda_i, \lambda_{i+1}]$ is attained either at λ_i or at λ_{i+1}. Thus,

$$\max_{\lambda \in [\underline{\lambda}, \bar{\lambda}]} (f_1(\lambda) + f_2(\lambda)) = \max_{i \in \{1,2,\ldots,p\}} (\{f_1(\lambda_i) + f_2(\lambda_i))$$

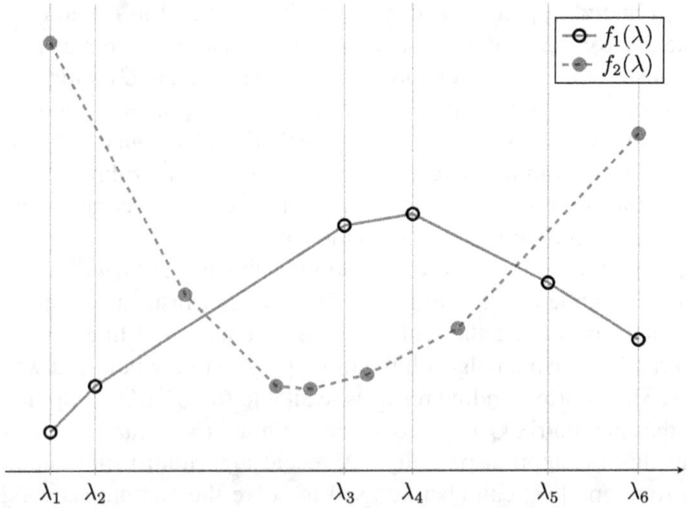

Fig. 10.2 Rank one QUBO example

Now, we need an efficient way to generate the points $\lambda_1, \lambda_2, \ldots, \lambda_p$ and compute an optimal basic feasible solution $\mathbf{x}(\lambda_i)$ of $L1(\lambda)$ at each λ_i, $i = 1, 2, \ldots, p$. This can be achieved relatively easily as shown in [65], which is discussed below (Fig. 10.2).

Let $B = \left\{\frac{c_i}{a_i} : i = 1, 2, \ldots, m,\ a_i \neq 0\right\}$ and $\frac{c_{\pi 1}}{a_{\pi_1}} > \frac{c_{\pi 2}}{a_{\pi 2}} > \cdots > \frac{c_{\pi r}}{a_{\pi r}}$ be a descending arrangement of all distinct elements of B. Let $B(k) = \left\{i : \frac{c_{\pi_k}}{a_{\pi_k}} = \frac{c_i}{a_i}\right\}$. Then the breakpoints of $f_1(\lambda)$ are given by

$$\lambda_1 = \underline{\lambda} \text{ and } \lambda_k = \lambda_{k-1} + \sum_{i \in B(k)} |a_i| \text{ for } k = 2, 3, \ldots, r.$$

An optimal basic feasible solution to L1(λ) at $\lambda = \lambda_k$ for $k = 1, 2, \ldots, r$ can be identified recursively as

$$x_i^1 = \begin{cases} 1 & \text{if } a_i = 0 \text{ and } c_i > 0 \text{ or } a_i < 0 \\ 0 & \text{otherwise,} \end{cases}$$

and

$$x_i^k = \begin{cases} x_i^{k-1} & \text{if } i \notin B(k) \\ 1 & \text{if } i \in B(k) \text{ and } a_i > 0 \\ 0 & \text{otherwise,} \end{cases}$$

Given $f_1(\lambda_k)$ and $\mathbf{x}^k$, $f_1(\lambda_{k+1})$ and $\mathbf{x}^{k+1}$ can be identified in $O(|B(k+1)|)$ time. The complexity for generating these solutions and the breakpoints are dominated by that of constructing the descending arrangement which is $O(m \log m)$. Note that $f_1(\lambda)$ has at most $m+1$ breakpoints. Given $\mathbf{x}^k$ a corresponding solution $\mathbf{y}^k$ can be computed in $O(n)$ time. This leads to a complexity of $O(mn)$. When m is large, say not $O(\log n)$, we can improve the complexity of the algorithm to $O(n \log n)$ by generating all the breakpoints of $f_2(\lambda)$ as well and recursively compute the solutions $\mathbf{y}(\lambda)$. For details of this approach we refer to [65].

The basic idea discussed above for solving the BQUBO with a rank-one $\mathbf{Q}$ matrix can be extended to any BQUBO. In this case, instead of using the single parametric analysis, we use the multi-parametric analysis of linear programs with bounded variables. Such an algorithm will be polynomially bounded when rank of $\mathbf{Q}$ is fixed [65]. A corresponding result is available for QUBO, with the additional restriction that the matrix $\mathbf{Q}$ is positive semidefinite. (See Chap. 3 for discussions on relaxing this assumption partially.) A recent algorithm using enumeration of faces of a zonotope [38] can also be used to solve the continuous version of the QUBO in polynomial time when the rank of $\mathbf{Q}$ is fixed. This algorithm can also be used to solve the BQUBO $(\mathbf{Q}, \mathbf{c}, \mathbf{d})$ when the rank of $\mathbf{Q}$ is fixed. We first solve the continuous version and if the resulting solution is not binary, a rounding procedure can be applied to obtain an optimal solution to the BQUBO. The next section presents a valid rounding procedure that can be used for this purpose.

10.4.1 Polynomially Solvable Biclique Problems

In Sect. 10.2, we have seen that the maximum weight biclique problem (MWBCP) can be formulated as a BQUBO. In fact any BQUBO can also be viewed as a maximum weight biclique problem on a complete bipartite graph. This follows from the fact that any BQUBO is equivalent to a BQUBO in homogeneous form and we can allow edges of weight zero in the MWBCP. Thus polynomially solvable special cases of the MWBCP yield corresponding polynomially solvable special cases of the BQUBO. Let us now review some results in this direction.

A biclique is said to be *maximal* in a graph G if it is not a proper subgraph of any other biclique in G. The number of maximal bicliques in a graph could be of exponential size in the number of vertices. But there are interesting graph classes where the number of maximal bicliques is polynomially bounded. This include chordal bipartite graphs and convex bipartite graphs [54]. Further, in such graphs the maximal bicliques can be enumerated in polynomial time. Thus, if the cost matrix $\mathbf{Q}$ of the homogeneous BQUBO $(\mathbf{Q}, \mathbf{0}, \mathbf{0})$ is such that there exists an optimal solution to the BQUBO $(\mathbf{Q}, \mathbf{0}, \mathbf{0})$ that corresponds to a maximal bliclique in an associated bipartite graph $G = (V_1, V_2, E)$ then the BQUBO can be solved by enumerating the maximal bicliques in G. For example, suppose that $q_{ij} \in \{0, 1, -nm\}$. Consider the complete bipartite graph $K_{m,n}$ corresponding to this BQUBO. An optimal solution to the BQUBO cannot correspond to an induced subgraph of $K_{m,n}$ containing an edge of weight $-nm$. So drop all edges of negative weights from the $K_{m,n}$. If the resulting graph contains a spanning subgraph which is convex and include all the edges of weight one and possibly some edges of weight zero as well, then we can solve the BQUBO by enumerating the polynomial number of maximal bicliques in this spanning subgraph. Recall that BQUBO is NP-hard even if q_{ij} has only two distinct entries and hence the example given above is non-trivial. Pandey, Sharma, and Jain [57] showed that when the edge weights are positive, the MWBCP can be solved in polynomial time on bipartite permutation graphs. Thus, if the missing edges have weights are 0 or have a value $q_{ij} \leq -mn$ then the corresponding BQUBO can also be solved polynomial time.

10.5 Approximation Algorithms

Theoretical analysis of approximation algorithms are carried out using different performance measures. This include relative performance ratio [75], differential ratio [17], domination ratio [31, 33, 34, 61], domination number [31, 79], comparison to average value of solutions [56, 61, 73] etc. These measures are defined in Chap. 8 and hence we use them without formally defining here.

We first prove a non-approximability result for BQUBO in terms of the relative performance ratio.

Theorem 10.11 *Unless NP = ZPP, there does not exist any polynomial time $n^{1-\epsilon}$-approximation algorithm for BQUBO for any $\epsilon > 0$, even if* $\mathbf{c}$ *and* $\mathbf{d}$ *are zero vectors and* $\mathbf{Q}$ *contains only two distinct non-zero entries.*

Proof We reduce the maximum clique problem to BQUBO in such a way that ϵ-optimality is preserved for $\epsilon > 0$. Let $G = (V, E)$ be a graph on n nodes. Now construct the $n \times n$ matrix $\mathbf{Q}$ where $q_{ii} = 1, q_{ij} = q_{ji} = 0$ if $(i, j) \in E$ and $-(n+1)$ if $(i, j) \notin E$. The optimal objective function value of the BQUBO constructed here is clearly positive. Let $(\mathbf{x}, \mathbf{y})$ be any $n^{1-\epsilon}$-optimal solution to BQUBO. Note that if $x_i = y_j = 1$ then (i, j) must be an edge in G. Let $S = \{i : x_i = y_i = 1\}$. Then the subgraph of G induced by S in G must be a clique in G and the BQUBO objective function value $f(\mathbf{x}, \mathbf{y})$ of $(\mathbf{x}, \mathbf{y})$ is $|S|$. Further, if $(\mathbf{x}^0, \mathbf{y}^0)$ is an optimal solution to the BQUBO then $f(\mathbf{x}^0, \mathbf{y}^0) = |S^0|$ where $S^0 = \{i : x_i^0 = y_i^0 = 1\}$. Thus, S must be an $n^{1-\epsilon}$-optimal solution for the maximum clique problem on G. The result follows from the fact that unless NP = ZPP, there does not exist any polynomial time $n^{1-\epsilon}$-approximation algorithm for the maximum clique problem [20].

For various non-approximability results on the maximum (weight) biclique problem we refer to [6, 51, 70] and these results extend in a natural way to BQUBO.

10.5.1 Domination Analysis

The discussion in this subsection is based on [63]. Let us now consider algorithms for BQUBO that are guaranteed to produce solutions with values no worse than the average value of all solutions. We will also examine the domination ratio of such algorithms. For more details on domination analysis of algorithms we refer to Chap. 8 and the papers [31, 34]. Let $\mathcal{F}$ be the family of all feasible solutions of the BQUBO $(\mathbf{Q}, \mathbf{c}, \mathbf{d})$. i.e. $\mathcal{F} = \{(\mathbf{x}, \mathbf{y}) : \mathbf{x} \in \{0, 1\}^m, \mathbf{y} \in \{0, 1\}^n\}$. For any $(\mathbf{x}, \mathbf{y}) \in \mathcal{F}$, there are 2^m choices for $\mathbf{x}$ and 2^n choices for $\mathbf{y}$. Thus $|\mathcal{F}| = 2^{m+n}$. The average value of all solutions of the BQUBO $(\mathbf{Q}, \mathbf{c}, \mathbf{d})$ is denoted by $\mathcal{A}(Q, c, d)$. For any two solutions $(\mathbf{x}, \mathbf{y})$ and $(\mathbf{x}', \mathbf{y}')$ in $\mathcal{F}$, we say that $(\mathbf{x}, \mathbf{y})$ *dominates* $(\mathbf{x}', \mathbf{y}')$ if $f(\mathbf{x}, \mathbf{y}) \geq f(\mathbf{x}', \mathbf{y}')$. In this case, we also say that $(\mathbf{x}', \mathbf{y}')$ is *dominated by* $(\mathbf{x}, \mathbf{y})$. A solution to the BQUBO is said to be *no worse than average* if its objective function value is at least $\mathcal{A}(Q, c, d)$.

Theorem 10.12 ([63]) $\mathcal{A}(Q, c, d) = \frac{1}{4}\sum_{i=1}^{m}\sum_{j=1}^{n} q_{ij} + \frac{1}{2}\sum_{i=1}^{m} c_i + \frac{1}{2}\sum_{j=1}^{n} d_j.$

Proof Consider $m + n$ independent random variables $x_1, x_2, \ldots, x_m$ and $y_1, y_2, \ldots, y_n$ which take values 0 or 1 with probability $\frac{1}{2}$ each. Then, the expected value $\mathbf{E}(x_i)$ and $\mathbf{E}(y_i)$ of the variables x_i and y_i is $\frac{1}{2}$. Since x_i and y_j are independent, the expected value $\mathbf{E}(x_i y_j)$ of $x_i y_j$ is $\mathbf{E}(x_i)\mathbf{E}(y_j) = \frac{1}{4}$. Now, consider the random variable

$$X = \sum_{i=1}^{m}\sum_{j=1}^{n} q_{ij}x_i x_j + \sum_{i=1}^{m} c_i x_i + \sum_{j=1}^{n} d_j y_j.$$

By the linearity property of expectation, the expected value $\mathbf{E}(X)$ of X is

$$\frac{1}{4}\sum_{i=1}^{m}\sum_{j=1}^{n} q_{ij} + \frac{1}{2}\sum_{i=1}^{m} c_i + \frac{1}{2}\sum_{j=1}^{n} d_j$$

and the result follows.

Although we have a closed form formula to calculate the average of the objective function values of all solutions of the BQUBO, computing the median of the objective function values of these solutions is hard [63].

Let us now define an equivalence relation among the solutions in $\mathcal{F}$. Two solutions $(\mathbf{x}, \mathbf{y})$ and $(\mathbf{x}', \mathbf{y}')$ are equivalent, denoted by $(\mathbf{x}, \mathbf{y}) \sim (\mathbf{x}', \mathbf{y}'$, if either $\mathbf{x}' = \mathbf{e}^m - \mathbf{x}$ or $\mathbf{y}' = \mathbf{e}^n - \mathbf{y}$, where $\mathbf{e}^k$ is the vector in $\mathbb{R}^k$ with all components equal to 1. It can be verified that the relation $\sim$ is reflexive, symmetric, and transitive and hence its is an equivalence relation on $\mathcal{F}$ which partitions $\mathcal{F}$ into disjoint equivalence classes. The equivalence class corresponding to $(\mathbf{x}, \mathbf{y})$ is

$$\mathcal{E}(\mathbf{x}, \mathbf{y}) = \{(\mathbf{x}, \mathbf{y}), (\mathbf{e}^m - \mathbf{x}, \mathbf{y}), (\mathbf{x}, \mathbf{e}^n - \mathbf{y}), (\mathbf{e}^m - \mathbf{x}, \mathbf{e}^n - \mathbf{y})\}.$$

Also, it is easy to verify that

$$\sum_{(\mathbf{x},\mathbf{y})\in\mathcal{E}(\mathbf{x},\mathbf{y})} f(\mathbf{x},\mathbf{y}) = 4\mathcal{A}(Q,c,d). \tag{10.44}$$

The collection of all solutions in $\mathcal{F}$ with objective function value no better than $\mathcal{A}(Q,c,d)$ is denoted by $\mathcal{D}$. i.e. $\mathcal{D} = \{(\mathbf{x},\mathbf{y}) \in \mathcal{F}: f(\mathbf{x},\mathbf{y}) \le \mathcal{A}(Q,c,d)\}$.

Theorem 10.13 ([63]) *For any instance* $(\mathbf{Q},\mathbf{c},\mathbf{d})$ *of BQUBO,* $|\mathcal{D}| \ge 2^{m+n-2}$.

Proof Let $\mathcal{E}(\mathbf{x}^k,\mathbf{y}^k)$, $k = 1,2,\ldots,t$ be the family of all equivalence classes on $\mathcal{F}$ defined by the equivalence relation $\sim$. Note that $t = 2^{m+n-2}$. By Eq. (10.44)

$$\min_{(\mathbf{x},\mathbf{y})\in\mathcal{E}(\mathbf{x}^k,\mathbf{y}^k)} f(\mathbf{x},\mathbf{y}) \le \mathcal{A}(Q,c,d) \text{ for each } k = 1,2,\ldots,t.$$

Since $t = 2^{m+n-2}$, the result follows.

The lower bound 2^{m+n-2} established on $|\mathcal{D}|$ in Theorem 10.13 is tight and it is possible to construct examples of BQUBO where $|\mathcal{D}| = 2^{m+n-2}$. Consider the matrix $\mathbf{Q}$ where $q_{nn} = -\alpha$ for $\alpha > 0$ and all other entries are zeros. Also, $\mathbf{c}$ and $\mathbf{d}$ are zero vectors. Then, $\mathcal{A}(Q,c,d) = \frac{-\alpha}{4}$ and the set of solutions $(\mathbf{x},\mathbf{y})$ with $f(\mathbf{x},\mathbf{y}) \le \mathcal{A}(Q,c,d)$ is precisely $\mathcal{D} = \{(\mathbf{x},\mathbf{y})|x_m = y_n = 1\}$ and $|\mathcal{D}| = 2^{m+n-2}$.

Computing a solution to the BQUBO with objective function value no worse than $\mathcal{A}(Q,c,d)$ is very easy. First choose a random solution $(\mathbf{x},\mathbf{y})$. Then, by equation (10.44), the solution with largest objective function value in $\mathcal{E}(\mathbf{x},\mathbf{y})$ is no worse than $\mathcal{A}(Q,c,d)$. In particular, the best solution amongst $\{(\mathbf{e}^m,\mathbf{e}^n), (\mathbf{e}^m,\mathbf{0}^n), (\mathbf{0}^m,\mathbf{e}^n), (\mathbf{0}^m,\mathbf{0}^n)\}$ have its objective function value no worse than $\mathcal{A}(Q,c,d)$ [63]. Despite this simplicity in computing a solution with the objective function value no worse than $\mathcal{A}(Q,c,d)$, reasonable looking heuristic algorithms may not guarantee a solution that is no worse than average.

For example, let us examine the *alternating algorithm* [24, 39, 40, 49] which can be stated as follows. Given a partial solution $\mathbf{x}^0 \in \{0,1\}^m$, find a best $\mathbf{y}^0 \in \{0,1\}^n$ using the formula

$$y_j^0 = \begin{cases} 1 & \text{if} \sum_{i=1}^{m} q_{ij}x_i^0 + d_j > 0, \\ 0 & \text{otherwise,} \end{cases} \tag{10.45}$$

This results in the solution $(\mathbf{x}^0,\mathbf{y}^0)$. Now fix $\mathbf{y}^0$ and find the best $\mathbf{x}^1 \in \{0,1\}^m$ using the following formula

$$x_i^1 = \begin{cases} 1 & \text{if} \sum_{j=1}^{n} q_{ij}y_j^0 + c_i > 0, \\ 0 & \text{otherwise,} \end{cases} \tag{10.46}$$

This produces the solution $(\mathbf{x}^1, \mathbf{y}^0)$.

This process of alternatively fixing $\mathbf{x}$ and $\mathbf{y}$ can be continued until no improvement is possible by fixing either the $\mathbf{x}$ or the $\mathbf{y}$ variables. Experimental analysis shows that this algorithm produces reasonable quality solutions [39, 40]. However, the worst-case behaviour of the alternating algorithm in comparison to $\mathcal{A}(Q, c, d)$ is poor. We illustrate this using the following example from [63].

Consider the instance of BQUBO where $\mathbf{Q}$ is an $n \times n$ matrix with $q_{11} = 1$ and $q_{nn} = \alpha > 3$. All other entries of $\mathbf{Q}$ are zeros and $\mathbf{c}$ and $\mathbf{d}$ are zero vectors. The starting partial solution $\mathbf{x}^0$ is given by $x_1^0 = 1$ and $x_i^0 = 0$ for $i \neq 1$. The algorithm will find the solution $\mathbf{y}^0$ where $y_1^0 = 1$ and $y_j^0 = 0$ for $j \neq 1$. The solution $(\mathbf{x}^0, \mathbf{y}^0)$ is locally optimal and the algorithm terminates with its objective function value 1. The optimal objective function value however is $\alpha + 1$ and $\mathcal{A}(Q, c, d) = \frac{\alpha+1}{4}$. Thus, the objective function value of the solution produced by the alternating algorithm for this instance of the BQUBO with the given selection of the starting partial solution, is worse than $\mathcal{A}(Q, c, d)$. Also, as α increases, the quality of the solution produced by the algorithm deteriorates and becomes an arbitrarily bad solution.

By choosing the starting partial solution appropriately, and using repeated runs with different starting solutions, the worst case performance of the alternating algorithm can be improved. More specifically, run the alternating algorithm twice, once with starting partial solution $\mathbf{x}^0 = \mathbf{e}^m$ and then repeat the algorithm with starting solution $\mathbf{x}^0 = \mathbf{0}^m$. The best solution produced from these two runs will have an objective function value no worse than $\mathcal{A}(Q, c, d)$. In fact, a better domination number can be established for such an algorithm [63] than the general purpose domination number established for algorithms producing no worse than average solutions.

Let us now look at another algorithm, called the rounding algorithm, which rounds a fractional solution to obtain a 0-1 solution. This is similar to the [0, 1]-rounding algorithm for QUBO discussed in Chap. 8 with some minor differences. Let $(\mathbf{x}, \mathbf{y})$ be a fractional solution with $\mathbf{x} \in [0, 1]^m$ and $\mathbf{y} \in [0, 1]^n$ and $(\mathbf{x}^*, \mathbf{y}^*)$ be the 0-1 solution obtained from the fractional solution $(\mathbf{x}, \mathbf{y})$ using the rounding scheme:

$$y_j^* = \begin{cases} 1 & \text{if } d_j + \sum_{i=1}^{m} q_{ij} x_i > 0, \\ 0 & \text{otherwise,} \end{cases} \tag{10.47}$$

$$x_i^* = \begin{cases} 1 & \text{if } c_i + \sum_{j=1}^{n} q_{ij} y_j^* > 0, \\ 0 & \text{otherwise.} \end{cases} \tag{10.48}$$

Theorem 10.14 ([63]) *$f(\mathbf{x}^*, \mathbf{y}^*) \geq f(\mathbf{x}, \mathbf{y})$. Further, when $\mathbf{x}$ and $\mathbf{y}$ are all-half vectors, $f(\mathbf{x}^*, \mathbf{y}^*) \geq \mathcal{A}(Q, c, d)$.*

Proof Recall that the objective function $f(\mathbf{x}, \mathbf{y})$ of the BQUBO is given by

$$\sum_{i=1}^{m}\sum_{j=1}^{n} q_{ij}x_iy_j + \sum_{i=1}^{m} c_ix_i + \sum_{j=1}^{n} d_jy_j = \sum_{j=1}^{n}\left(\sum_{i=1}^{m} q_{ij}x_i + d_j\right)y_j + \sum_{i=1}^{m} c_ix_i$$

$$\leq \sum_{j=1}^{n}\left(\sum_{i=1}^{m} q_{ij}x_i + d_j\right)y_j^* + \sum_{i=1}^{m} c_ix_i = \sum_{j=1}^{n} d_jy_j^* + \sum_{i=1}^{m}\left(\sum_{j=1}^{n} q_{ij}y_j^* + c_i\right)x_i$$

$$\leq \sum_{j=1}^{n} d_jy_j^* + \sum_{i=1}^{m}\left(\sum_{j=1}^{n} q_{ij}y_j^* + c_i\right)x_i^* = f(\mathbf{x}^*, \mathbf{y}^*).$$

Corollary 10.1 *The rounding algorithm produces a solution to the BQUBO that is no worse than average in $O(mn)$ time when the starting solution is $(\mathbf{x}, \mathbf{y})$ have all of its components equal to half. Further, in this case, the domination ratio of the rounding algorithm is $\frac{1}{4}$*

Proof Note that the objective function value of a solution with all components equal to half is $\mathcal{A}(Q, c, d)$. The result now follows from Theorems 10.13 and 10.14.

It is possible to establish stronger bounds for the BQUBO by using (modifying) other related algorithms discussed in Chapter 8. The details are omitted.

Let us now establish a non-approximability result for the BQUBO related to the domination ratio (domination number) of heuristic algorithms. The theorem below and its validity proof is from [63].

Theorem 10.15 ([63]) *Let $\alpha > \beta$ be relatively prime natural numbers bounded above by a polynomial function of the input length of BQUBO and $\delta = \frac{\alpha}{\beta}$. Unless P=NP, no polynomial time heuristic algorithm for the BQUBO can have its domination number more than $2^{m+n} - 2^{\lceil \frac{m+n}{\delta} \rceil}$.*

Proof Suppose that a polynomial time algorithm $\mathcal{A}$ for the BQUBO exists with domination number at least $2^{m+n} - 2^{\lceil \frac{m+n}{\delta} \rceil} + 1$. Using this algorithm, we show that an optimal solution to BQUBO can be identified. Consider any instance $(\mathbf{Q}, \mathbf{c}, \mathbf{d})$ of the BQUBO. Now, construct the instance $(\mathbf{Q}', \mathbf{c}', \mathbf{d}')$ where $\mathbf{Q}' = (q'_{ij})$ is an $\alpha\beta m \times \alpha\beta n$ matrix defined by

$$q'_{ij} = \begin{cases} q_{ij} & \text{if } 1 \leq i \leq m \text{ and } 1 \leq j \leq n, \\ 0 & \text{otherwise} \end{cases}$$

and $\mathbf{c}'$ and $\mathbf{d}'$ are respectively vectors in $\mathbb{R}^{\alpha\beta m}$ and $\mathbb{R}^{\alpha\beta n}$ such that

$$c'_i = \begin{cases} c_i & \text{if } 1 \leq i \leq m, \\ 0 & \text{otherwise} \end{cases} \text{ and } d'_j = \begin{cases} d_j & \text{if } 1 \leq j \leq n, \\ 0 & \text{otherwise.} \end{cases}$$

By our assumption on α and β, the instance $(\mathbf{Q}', \mathbf{c}', \mathbf{d}')$ of the BQUBO constructed above have polynomial size in relation to $(\mathbf{Q}, \mathbf{c}, \mathbf{d})$. Also, if $(\mathbf{x}', \mathbf{y}')$ is an optimal solution to the BQUBO $(\mathbf{Q}', \mathbf{c}', \mathbf{d}')$ then $(\mathbf{x}^0, \mathbf{y}^0)$ will be an optimal solution to the BQUBO $(\mathbf{Q}, \mathbf{c}, \mathbf{d})$, where $\mathbf{x}^0$ consists of the first m components of $\mathbf{x}'$ and $\mathbf{y}^0$ consists of the first n components of $\mathbf{y}'$. There are $2^{\alpha\beta(m+n)}$ feasible solutions for $(\mathbf{Q}', \mathbf{c}', \mathbf{d}')$ of which at least $2^{\alpha\beta(m+n)-(m+n)}$ are optimal. So the maximum number of non-optimal solutions is $2^{\alpha\beta(m+n)} - 2^{\alpha\beta(m+n)-(m+n)}$. Solve the BQUBO instance $(\mathbf{Q}', \mathbf{c}', \mathbf{d}')$ using $\mathcal{A}$ and let $(\mathbf{x}', \mathbf{y}')$ be the solution produced. The objective function value of $(\mathbf{x}', \mathbf{y}')$ is no worse than that of at least $2^{\alpha\beta(m+n)} - 2^{\lceil \frac{\alpha\beta(m+n)}{\alpha/b} \rceil} + 1 = 2^{\alpha\beta(m+n)} - 2^{\beta^2(m+n)} + 1$ solutions. Since $\alpha > \beta$, we have $2^{\alpha\beta(m+n)} - 2^{\beta^2(m+n)} + 1 > 2^{\alpha\beta(m+n)} - 2^{\alpha\beta(m+n)-(m+n)}$. Thus, $(\mathbf{x}', \mathbf{y}')$ must be an optimal solution to $(\mathbf{Q}', \mathbf{c}', \mathbf{d}')$. As indicated earlier, from $(\mathbf{x}', \mathbf{y}')$, an optimal solution to $(\mathbf{Q}, \mathbf{c}, \mathbf{d})$ can be recovered. The result now follows from NP-completeness of the BQUBO.

From Theorem 10.15 we can see that no polynomial time approximation algorithm exists for the BBQP with domination ratio more than $1-2^{\frac{1-\delta}{\delta}(m+n)}$ for any fixed rational number $\delta > 1$. A non-approximability result similar to Theorem 10.15 is mentioned in Chap. 8 for the case of QUBO and proved in [61] for the traveling salesman problem.

10.5.2 Approximation Algorithms for the Ising BQUBO

Recall that a BQUBO can be mapped to an equivalent Ising BQUBO and vice versa. This transformation adds a constant to the objective function value, which is ignored for optimization purposes. This constant however affects the relative performance ratio of an approximation algorithm. That is, the image of an ϵ-optimal solution for the BQUBO under the linear transformation (10.3) need not be an ϵ-optimal solution for the resulting Ising BQUBO. However, domination ratio and the domination number of an algorithm are not affected by the transformations that adds a constant to the objective function value. Thus, the approximation algorithms and their domination ratio discussed for the BQUBO can be used to obtain corresponding results for the Ising BQUBO and vice versa. In particular, the non-approximability result in terms of the domination ratio established for the BQUBO is valid for the Ising BQUBO as well.

Let us now look at some approximation algorithms for the Ising BQUBO from the point of view of relative performance ratio. As noted in the introduction, any Ising BQUBO $(\mathbf{A}, \mathbf{0}^m, \mathbf{0}^n)$ can be formulated as the Ising QUBO $(\mathbf{A}', \mathbf{0})$ where

$$\mathbf{A}' = \left[\begin{array}{c|c} \mathbf{O}_{m\times m} & \frac{1}{2}\mathbf{A} \\ \hline \frac{1}{2}\mathbf{A}^T & \mathbf{O}_{n\times n} \end{array}\right]$$

and $\mathbf{0}$ is the zero vector in $\mathbb{R}^{m+n}$. Note that $\mathbf{A}'$ is symmetric and the diagonal elements of $\mathbf{A}'$ are zeros. Thus, we can apply the SDP rounding algorithm discussed in Chap. 8 to obtain an approximation algorithm for BQUBO with guaranteed relative performance ratio of $O(\log(m+n))$ in expectation. However, exploiting the bipartite structure, for the similar SDP based randomized rounding idea and leads to a constant performance ratio, in expectation [4]. Let us look at this result briefly.

Let $\mathbf{U} \in \mathbb{R}^{d\times m}$ and $\mathbf{V} \in \mathbb{R}^{d\times n}$ be matrices with respective columns $\mathbf{u}^1, \mathbf{u}^2, \ldots, \mathbf{u}^m$, and $\mathbf{v}^1, \mathbf{v}^2, \ldots, \mathbf{v}^n$, such that $||\mathbf{u}^i|| = 1, i = 1, 2, \ldots, m$ and $||\mathbf{v}^j||, j = 1, 2, \ldots, n$. Then the standard semidefinite programming relaxation of BQUBO can be written as:

$$\text{SDPB:} \qquad \text{Maximize } \eta(\mathbf{U},\mathbf{V}) = \sum_{i=1}^{m}\sum_{j=1}^{n} a_{ij}\langle \mathbf{u}^i \cdot \mathbf{v}^j\rangle$$

Subject to

$$||\mathbf{u}^i|| = 1, ||\mathbf{v}^j|| = 1, i = 1, 2, \ldots, m; j = 1, 2, \ldots, n.$$

where $\langle \mathbf{u}^i \cdot \mathbf{v}^j\rangle$ denotes the dot product of the vectors $\mathbf{u}^i$ and $\mathbf{v}^j$. Let $\mathbf{U}^*, \mathbf{V}^*$ be an optimal solution to SDPB and $\mathbf{x}^*, \mathbf{y}^*$ be an optimal solution to the BQUBO $(\mathbf{A}, \mathbf{0}, \mathbf{0})$. These solutions are related by the *Grothendieck Inequality* which states that

$$\eta(\mathbf{U}^*, \mathbf{V}^*) \leq K f(\mathbf{x}^*, \mathbf{y}^*)$$

where K is the Grothendieck constant [11, 47]. The precise value of K is not known but many authors developed upper and lower bounds on K. For example, it is known that $\frac{\pi}{2} \leq K < \frac{\pi}{2\log(1+\sqrt{2})}$ [11, 47]. This raises the prospect that if we can identify $\mathbf{x} \in [-1, 1]^m, \mathbf{y} \in [-1, 1]^n$ from $\mathbf{U}^*, \mathbf{V}^*$ such that $\eta(\mathbf{U}^*, \mathbf{V}^*)$ is not too far from $f(\mathbf{x}, \mathbf{y})$ we can obtain an ϵ-optimal solution for QUBO by applying our rounding algorithm on $(\mathbf{x}, \mathbf{y})$. Alon and Naor [4] established precisely this using randomized rounding of the SDP solution.

Theorem 10.16 *Let* $\mathbf{U}^*, \mathbf{V}^*$ *be an optimal solution to the SDPB with* $d = m + n$ *and* $\mathbf{x}^*, \mathbf{y}^*$ *be an optimal solution to the BQUBO* $(\mathbf{A}, \mathbf{0}, \mathbf{0})$. *Then it is possible to identify* $\mathbf{x} \in [-1, 1]^m, \mathbf{y} \in [-1, 1]^n$ *using* $\mathbf{U}^*, \mathbf{V}^*$ *and randomized rounding such that*

$$\mathbf{E}\left(\sum_{i=1}^{m}\sum_{j=1}^{n} a_{ij}x_i x_j\right) = \frac{1}{K}\eta(\mathbf{U}^*, \mathbf{V}^*) \geq 0.56 f(\mathbf{x}^*, \mathbf{y}^*),$$

where $\mathbf{E}(.)$ *is the expected value.*

The proof of Theorem 10.16 can be obtained by following analysis similar to that used in Chap. 8 for a corresponding result for QUBO with appropriate modifications. An independent proof of this is given [4]. Other algorithms discussed in Chap. 8 can also be used to obtain guaranteed relative performance ratio for the Ising BQUBO $(\mathbf{A}, \mathbf{0}, \mathbf{0})$ with some additional simplifications.

10.6 Local Search and Metaheuristics

Local search is a powerful and popular heuristic strategy employed in solving a variety of combinatorial optimization problems. Local search and metaheuristic algorithms are discussed in Chap. 9 focussing the QUBO model. Let us now discuss some local search and metaheuristic algorithms for the BQUBO, exploiting the bipartite (bilinear) structure of the problem. A local search algorithm maintains a solution $(\mathbf{x}, \mathbf{y})$ and an associated neighborhood $\mathbb{N}(\mathbf{x}, \mathbf{y})$. If the structure of the neighborhood is simple with small neighborhood size, complete enumeration can be employed to find an improving solution. For more complex neighborhoods, as in the case of very large scale neighborhood search (VLSN search) [2], complete enumeration is not viable and carefully designed neighborhood search algorithms are employed to find an improving solution.

Starting with a solution $(\mathbf{x}, \mathbf{y})$, the local search algorithm tries to find a 'better' solution in $\mathbb{N}(\mathbf{x}, \mathbf{y})$. If no such solution exists in $\mathbb{N}(\mathbf{x}, \mathbf{y})$, the algorithm terminates and output the best solution obtained. If an improving solution is identified, then the algorithm 'move' to this solution replacing the current solution $(\mathbf{x}, \mathbf{y})$ with the improved solution and its neighborhood is explored. There are different criteria used in selecting an improved solution. For example, the first improving criteria selects the first improved solution found in the neighborhood. In the 'best improving' criteria, the entire neighborhood is explored and selects the solution that provides the largest improvement in the objective function value. An intermediate strategy of candidate lists is also used in practice, where a list of solutions of a prescribed size is constructed and a best solution from this list is selected. A local search algorithm is well defined when the following components are specified.

1. Strategy to select a starting solution.
2. Definition of the neighborhood and the neighborhood search algorithm.
3. Selection criteria for an improving solution.

Local search algorithms are often embedded within more complex metaheuristic strategies such as tabu search, variable neighbourhood search etc, to achieve performance enhancements.

10.6.1 Flip Based Neighborhoods

This is perhaps the most natural and well-studied strategy for constructing efficient neighbourhood structures for 0-1 optimization problems. We have seen in Chap. 9 where various metaheuristic algorithms for QUBO make use of flip-based neighborhoods. Let us now discuss such neighborhoods in the context of the BQUBO which has additional flexibility. A *flip* operation changes the value of a variable to its complement. For example, by flipping the i-th component of $\mathbf{x}^T = (x_1, x_2, \ldots, x_m)$, we get the solution vector $(\mathbf{x}')^T = (x_1, \ldots, x_{i-1}, 1 - x_i, x_{i+1}, \ldots, x_m)$. We can apply the flip operation on both $\mathbf{x}$ and $\mathbf{y}$ variables. The flip neighborhood of $(\mathbf{x}, \mathbf{y})$, denoted by $\mathbb{N}^{1,1}(\mathbf{x}, \mathbf{y})$, is the collection of all solutions obtained from $(\mathbf{x}, \mathbf{y})$ by precisely one flip operation.

Let us now discuss an implementation of the local search algorithm of Glover et al. [32] using the $\mathbb{N}^{1,1}(\mathbf{x}, \mathbf{y})$ neighborhood with the best improving solution as the move selection criteria. It is important to compute the objective function value of the candidate solutions in $\mathbb{N}^{1,1}(\mathbf{x}, \mathbf{y})$ efficiently, after each flip operation. To accomplish this, we adapt an evaluation technique, widely used for many binary quadratic optimization problems [27, 28, 48, 76, 77]. Exploiting the structure of the BQUBO, we maintain two arrays, $\Delta(\mathbf{x})$ and $\Delta(\mathbf{y})$, to store the contribution of each flip operation to the objective function value.

Let $\mathbf{x}'$ be the solution obtained from $\mathbf{x}$ by flipping the ith component. Define $\Delta x_i = f(\mathbf{x}', \mathbf{y}) - f(\mathbf{x}, \mathbf{y})$ and $\Delta(\mathbf{x}) = (\Delta x_1, \Delta x_2, \ldots, \Delta x_m)$. Then,

$$\Delta x_i = (x'_i - x_i)\left(c_i + \sum_{j=1}^{n} q_{ij} y_j\right). \tag{10.49}$$

Similarly, let $\mathbf{y}'$ be the solution obtained from $\mathbf{y}$ by flipping the jth component. Define $\Delta y_i = f(\mathbf{x}, \mathbf{y}') - f(\mathbf{x}, \mathbf{y})$ and $\Delta(\mathbf{y}) = (\Delta y_1, \Delta y_2, \ldots, \Delta y_n)$. Then,

$$\Delta y_j = (y'_j - y_j)\left(d_j + \sum_{i=1}^{m} q_{ij} x_i\right). \tag{10.50}$$

The vectors $\Delta(\mathbf{x})$ and $\Delta(\mathbf{y})$ can be initialized in $O(mn)$ time using direct applications of (10.49) and (10.50). For subsequent iterations, these values can be efficiently updated. A detailed description of the algorithm for updating the Δx_i and Δy_j values in $O(m + n)$ time is given in [32]. Using the vectors $\Delta(\mathbf{x})$ and $\Delta(\mathbf{y})$, a best improving solution and the corresponding objective function value can be identified in $O(m + n)$ time.

10.6.1.1 The Flip and Float Neighborhood

The flip neighborhood discussed earlier is standard for many quadratic optimization problems. The flip and float neighborhood introduced by Glover et al. [32] exploits the special structure of the BQUBO. The material presented in this subsection is drawn from [32] with some modifications. Recall that, for a fixed $\mathbf{x} = (x_1, x_2, \ldots, x_m)$, we can find an optimal $\mathbf{y} = \mathbf{y}^*(\mathbf{x})$ yielding the largest improvement in the value of $f(\mathbf{x}, \mathbf{y})$ [39] as

$$y^*(\mathbf{x})_j = \begin{cases} 1 \text{ if } d_j + \sum_{i=1}^{m} q_{ij} x_i > 0; \\ 0 \text{ otherwise.} \end{cases} \tag{10.51}$$

Similarly, the optimal $\mathbf{x} = \mathbf{x}^*(\mathbf{y})$ for a given $\mathbf{y}^T = (y_1, y_2, \ldots, y_n)$ can be identified using the equation

$$x^*(\mathbf{y})_i = \begin{cases} 1 \text{ if } c_i + \sum_{j=1}^{n} q_{ij} y_j > 0; \\ 0 \text{ otherwise.} \end{cases} \tag{10.52}$$

A *Flip*-**x**-*Float*-**y** operation flips a selected component of **x** and the float **y** (i.e. choose the best **y** with respect to the flipped **x**). Similarly, the *Flip*-**y**-*Float*-**x** operation flips one component of **y** and then float **x** with respect to the flipped **y**. There are m possible Flip-**x**-Float-**y** moves and n possible Flip-**y**-Float-**x** moves.

Let $\mathbb{N}^{1,n}$ denotes the Flip **x** -Float **y** neighborhood and $\mathbb{N}^{m,1}$ denotes the Flip **y** - Float **x** neighborhood. While $\mathbb{N}^{1,n}$ contains only m solutions and $\mathbb{N}^{m,1}$ contains only n solutions, their extended neighborhoods [55] respectively contains at least $m2^n$ and $n2^m$ solutions. Note that each flip operation generates an $\mathbf{x}'$ and $\mathbf{y}^*(\mathbf{x}')$ is a better candidate than 2^n possibilities for **y**. Thus, the best solution in $\mathbb{N}^{1,n}$ is no worse than the $m2^n$ solutions in the extended neighborhood of $\mathbb{N}^{1,m}$. Let $\mathbb{N}^1 = \mathbb{N}^{1,n} \cup \mathbb{N}^{m,1}$. Then the best solution in $\mathbb{N}^1$ is no worse than $n2^m + m2^n$ alternative solutions. Yet, we can find the best solution in $\mathbb{N}^1$ in polynomial time.

Let us very briefly discuss some implementation aspects of exploring the neighborhood $\mathbb{N}^1$ for an improved solution. Let $\mathbf{x}^i$ be the solution obtained from **x** by flipping its ith component and $\Delta_{xy}(i)$ be the change in the objective function value caused by flipping the i-th component of **x** and floating **y**. Then,

$$\Delta_{xy}(i) = f(\mathbf{x}^i, \mathbf{y}^*(\mathbf{x}^i)) - f(\mathbf{x}, \mathbf{y}) \tag{10.53}$$

Similarly, let $\mathbf{y}^j$ be the solution obtained from **y** by flipping the jth component of **y** and floating **x** and $\Delta_{yx}(j)$ be the change in the objective function value caused by flipping the j-th component of **y** and floating **x**. Then,

$$\Delta_{yx}(j) = f(\mathbf{x}^*(\mathbf{y}^j), \mathbf{y}) - f(\mathbf{x}, \mathbf{y}) \tag{10.54}$$

For an efficient implementation of a local search algorithm based on the flip and float neighborhood, we need a fast way to compute the value of $\Delta_{xy}(i)$ and $\Delta_{yx}(j)$. Following Glover et al. [32], let us now discuss how to determine $\Delta_{xy}(i)$ efficiently. Let

$$\sigma(\mathbf{x}, j) = d_j + \sum_{i=1}^{m} q_{ij} x_i \tag{10.55}$$

Then,

$$f(\mathbf{x}, \mathbf{y}^*(\mathbf{x})) = \sum_{i=1}^{m} c_i x_i + \sum_{j=1}^{n} \max\,[0, \sigma(\mathbf{x}, j)]. \tag{10.56}$$

For simplicity, let us write Δ_i for $\Delta_{\mathbf{x},\mathbf{y}^*(\mathbf{x})}(i)$. Then,

$$\begin{aligned}\Delta_i &= f(\mathbf{x}^i, \mathbf{y}^*(\mathbf{x}^i)) - f(\mathbf{x}, \mathbf{y}^*(\mathbf{x})) \\ &= \sum_{k=1}^{m} c_k {x^i}_k + \sum_{j=1}^{n} \max\left[0, \sigma(\mathbf{x}^i, j)\right] - \sum_{k=1}^{m} c_i x_k - \sum_{j=1}^{n} \max\left[0, \sigma(\mathbf{x}, j)\right] \\ &= c_i({x^i}_i - x_i) + \sum_{j=1}^{n}\left(\max\left[0, \sigma(\mathbf{x}^i, j)\right] - \max\left[0, \sigma(\mathbf{x}, j)\right]\right) \end{aligned} \tag{10.57}$$

We can compute $\Delta_{yx}(j)$ in an analogous way.

To compute the value of $\Delta_{xy}(i)$ as indicated above we need the values $\sigma(\mathbf{x}, j)$. For this purpose, we maintain a one dimensional array of size n to store $\sigma(\mathbf{x}, j)$ $(j = 1, 2, \ldots, n)$. Initially, we calculate $\sigma(\mathbf{x}, j)$ using Eq. (10.55) and this requires $O(mn)$ time. Also the value $\Delta_{xy}(i)$ can be identified in $O(n)$ time. Subsequent to the *Flip*-**x**-*Float*-$\mathbf{y}(i)$ move, we can update the array $\sigma(\mathbf{x}, j)$ in $O(n)$ time by simply adding $({x'}_i - x_i)q_{ij}$ to each of its elements. For further details we refer to [32].

The flip neighborhood and the flip and float neighborhood belongs to a more general neighborhood structure $\mathbb{N}^{hk}$ considered by Punnen et al. [63]. For any solution $(\mathbf{x}, \mathbf{y})$ of the BQUBO, let $\mathbb{N}^{hk}$ be the set of solutions obtained by flipping at most h components of $\mathbf{x}$ and at most k components of $\mathbf{y}$. Note that $|\mathbb{N}^{h,k}| = \left(\sum_{j=0}^{h}\binom{m}{j}\right)\left(\sum_{i=0}^{k}\binom{n}{i}\right)$. For fixed h and k, the best solution in this neighborhood can be identified in polynomial time. When exactly one h or k is allowed to take values m or n, we a get a more powerful neighborhood. Let $\mathbb{N}^{\alpha} = \mathbb{N}^{m,\alpha} \cup \mathbb{N}^{\alpha,n}$. As indicated in [63], $|\mathbb{N}^{\alpha}| = 2^m \sum_{j=0}^{\alpha}\binom{n}{j} + 2^n \sum_{i=0}^{\alpha}\binom{m}{i} - \sum_{i=0}^{\alpha}\binom{m}{i}\sum_{j=0}^{\alpha}\binom{n}{j}$. However, a best solution in $\mathbb{N}^{\alpha}$ can be identified in polynomial time for a fixed α [39, 63]. It may be noted that a solution produced by the alternating algorithm is locally optimal with respect to the neighborhood $\mathbb{N}^0 = \mathbb{N}^{m,0} \cup \mathbb{N}^{0,n}$. The flip and float neighborhood is precisely $\mathbb{N}^1$, and the flip neighborhood is $\mathbb{N}^{1,1}$. Computational results with local search and tabu search algorithms using the neighborhoods $\mathbb{N}^1$ and $\mathbb{N}^{1,1}$ provided good quality solutions [32]. Nonetheless, a locally optimal solution of such a powerful neighbourhood could be bad and the corresponding objective function value could be less than $\mathcal{A}(Q, c, d)$ even if α is a function of n.

Theorem 10.17 ([63]) *A local search algorithm for the BQUBO* $(\mathbf{Q}, \mathbf{c}, \mathbf{d})$ *with respect to the neighborhood* $\mathbb{N}^{\alpha}$ *could produce a solution with objective function value less than* $\mathcal{A}(Q, c, d)$ *for any* $\alpha \leq \left\lfloor \frac{n}{5} \right\rfloor$.

Corollary 10.2 ([63]) *For any fixed h and k, the objective function value of a locally optimal solution with respect to the neighborhood* $\mathbb{N}^{hk}$ *could be worse than* $\mathcal{A}(Q, c, d)$ *for sufficiently large m and n.*

10.6.2 Metaheuristics

Glover et al. [32] developed a tabu search based metaheuristic algorithm enhancing the flip neighborhood based local search to solve the BQUBO. They also developed another tabu search algorithm using the flip and float neighbourhood. As noted in the previous subsection, the flip and float neighborhood is a very large scale neighborhood (VLSN) and is more powerful but its neighbourhood search algorithm is slower than that of the flip neighbourhood. Thus, Glover et al. [32] combined the two search strategies and proposed a hybrid algorithm integrating tabu search and VLSN search to achieve better overall performance. The algorithm works as follows. Apply the flip neighbourhood based tabu search first and when a local optimum is reached, use the flip and float neighbourhood to escape from the local optimum. This way, the flip and float neighbourhood used as a 'cutting neighbourhood' to get out of the local optimum. If an improvement is obtained using the flip and float neighbourhood, the algorithm switch back to the flip neighborhood based tabu search and the process is continued. Otherwise output the best solution produced. Results of extensive computational experiments reported in [32] suggests that the hybrid algorithm combining VLSN search [2] and simple tabu search could result in heuristics with superior performance for the BQUBO.

Duarte et al. [19] proposed an iterated local search heuristic to solve the BQUBO. The iterated local search first uses a construction heuristic to generate a starting solution. For this purpose, they used two types of heuristics; a pure greedy heuristic and the semi-greedy strategy originally proposed by Hart and Shogan [37]. For the semi-greedy heuristic, they generate a fixed number of solutions and then choose the overall best solution. The improvement part of the algorithm uses simple flip operations to explore better solutions. History based intensification, diversification, and perturbation are also used to enhance the performance of the algorithm. Extensive computational results using Karapetyan and Punnen data set [39] disclosed that the algorithms of Duarte et al. [19] and Glover et al. [32] produced good quality solutions but neither shown superiority over the other.

Karapetyan et al. [40] developed a metaheuristic framework called *Conditional Markov Chain Search* (CMCS) and applied it to solve the BQUBO. They developed different component algorithms such hill-climbers and mutaters etc., and integrated them into a metaheuristic framework. A distinguishing feature of the algorithm is the elimination of experimental fine tuning of the parameters and replaced it with an automated multicomponent fine-tuning procedure. For this purpose the CMCS framework is used. The automated parameter tuning using acquired intelligence produced an algorithm for the BQUBO with superior computational performance.

Wu et al. [78] proposed three tabu search based algorithms to solve the BQUBO. Two of them are based on strategic oscillation while the third incorporates a path-relinking based strategy to solve the BQUBO. All three algorithms worked exceptionally well computationally and the path-relinking based algorithm was able to discover new improved solutions for large benchmark problems. This algorithm also used the flip neighborhood and the flip and float neighborhood originally proposed in [32]. Careful integration of various components, including the use of memory and perturbations, construction and destruction strategies, and path relinking resulted in enhanced performance.

10.7 Exact Algorithms

Before applying any exact (or complex heuristic) algorithms, a preprocessing analysis to fix variables will be useful. Such studies are available for QUBO [29, 36] (see also Chap. 5) and these ideas can be used directly on the BQUBO. We state two such results for the BQUBO and its validity can be established from the corresponding results for QUBO.

Theorem 10.18 *If a variable x_i is zero or one in an optimal solution of the LP relaxation of GWB, then there exists an optimal solution to BQUBO where x_i assumes the same values.*

This is the well known persistency result for QUBO stated in the context of BQUBO. See Chap. 5 and the reference therein for further details. As in the case of QUBO, simple conditions, without using LP relaxation of GWB, can also be derived to fix variables. Let $\ell_i^x = c_i + \sum_{j=1}^n \min\{0, q_{ij}\}$, $u_i^x = c_i + \sum_{j=1}^n \max\{0, q_{ij}\}$, $\ell_j^y = d_j + \sum_{i=1}^m \min\{0, q_{ij}\}$ and $u_j^y = d_j + \sum_{i=1}^m \max\{0, q_{ij}\}$.

Theorem 10.19 *If $\ell_i^x \geq 0$ then $x_i = 1$ in an optimal solution of BQUBO and if $u_i^x \leq 0$ then $x_i = 0$ in an optimal solution. Similarly, if $\ell_j^y \geq 0$ then $y_j = 1$ in an optimal solution of BQUBO and if $u_j^y \leq 0$ then $y_j = 0$ in an optimal solution.*

Proof Without loss of generality assume that $i = m$. Then,

$$f(\mathbf{x}, \mathbf{y}) = \left(\sum_{i=1}^{m-1} c_i x_i + \sum_{j=1}^{n} d_j y_j + \sum_{i=1}^{m-1} \sum_{j=1}^{n} q_{ij} x_i y_j \right) + x_m \Big(c_m + \sum_{j=1}^{n} q_{mj} y_j \Big) \tag{10.58}$$

The first bracket in Eq. (10.58) is independent of x_m. The minimum contribution from the second bracket is ℓ_m^x and if this is positive, then x_m must be at value 1 in every optimal solution and if $\ell_m^x = 0$ then $x_m = 1$ in at least one optimal solution. Likewise, the maximum contribution from the last bracket is u_m^x and if this is negative, then x_m must be zero in every optimal solution and if $u_m^x = 0$ then $x_m = 0$ in at least one optimal solution. The case for the $\mathbf{y}$ variables can be proved analogously.

The BQUBO can be solved using general purpose solvers such as Gurobi, CPLEX etc. using their quadratic 0-1 programming problem solver, or by solving the MILP formulations. Also, any exact algorithm for QUBO can be used to solve the BQUBO. The various upper bounds, MILP formulations, and valid inequalities discussed earlier for BQUBO can be used to develop branch and cut, branch and bound, or other implicit enumeration algorithms but there are not many such exact algorithms published in the literature designed specifically for the BQUBO, exploiting its special structure. There is however one such work published by Duarte et al. [19] and let us discuss this briefly here.

The algorithm is of branch and bound type which maintains a binary tree. Branching is done using the $\mathbf{x}$ variables only since for each assignment of $\mathbf{x}$ variables, the best $\mathbf{y}$ variables can easily be identified. This is a distinguishing feature of BQUBO. Each node of the branch-and-bound tree represents a variable x_i and a branching decision is made assigning $x_i = 1$ for one branch and $x_i = 0$ for the other branch. Also, at each node k of the search tree, the variables corresponding to the nodes on the unique path from k to the root node are already fixed. Let I^k be index set of $\mathbf{x}$ variables that are fixed at value one and F^k be the index set of free $\mathbf{x}$ variables (variables that not yet fixed at zero or one). Then, an upper bound $U(k)$ at node k can be identified using the formula

$$U(k) = \sum_{i \in I^k} c_i + \sum_{i \in F^k} \max(0, c_i) + \sum_{j=1}^{n} \max\left(0, \sum_{i \in I^k} q_{ij} + \sum_{i \in F^k} \max(0, q_{ij}) + d_j\right)$$

Let $\mathbf{x}^k$ be the solution obtained by setting $x_i^k = 1$ for $i \in I^k$ and 0 for $i \notin I^k$. By fixing this $\mathbf{x}^k$ identify the optimal $\mathbf{y}^k$ using formulas discussed earlier. Then $(\mathbf{x}^k, \mathbf{y}^k)$ is a candidate solution. Compare this with the best solution produced so far and update the best solution and its objective function value, if necessary. Let BOBJ be the best objective function value identified sofar. Prune the node k if $U(k) \leq$ BOBJ. The search tree can be explored using a depth-first or breadth-first strategy. A hybrid approach is also used normally, with breadth-first strategy applied first and when the tree size reaches a specified limit and then switch to a depth-first strategy. Duarte et al. [19] reported that this simple branch and bound strategy was competitive with the general purpose Cplex solver.

Branders [9] used constraint programming and MILP to solve the homogeneous BQUBO. The literature on MWBCP also talk about special purpose exact and heuristic algorithms for the problem. Depending on the restrictions involved on wedge weights, special cases of the BQUBO which are still hard, can be solved using these algorithms [7, 23, 66].

10.8 Concluding Remarks

In this chapter, we have discussed various properties, algorithms, and applications of the models BQUBO and the Ising BQUBO. These problems can be viewed as generalizations as well as special cases of QUBO or the Ising QUBO. Both BQUBO and the Ising BQUBO are less studied compared to their QUBO counterparts.

In this chapter we have illustrated how some of the results obtained in the context of QUBO get simplified for the BQUBO. Further research is necessary to understand the linkages between the QUBO and the BQUBO and to make algorithmic advancements for the BQUBO.

References

1. W.P. Adams, R.J. Forrester, A simple recipe for concise mixed 0-1 linearizations. Oper. Res. Lett. **33**, 55–61 (2005)
2. R.K. Ahuja, J.B. Orlin, O. Ergon, A.P. Punnen, A survey of very large-scale neighborhood search techniques. Discr. Appl. Math. **123**, 75–102 (2002)
3. N. Alon, A. Naor, Approximating the cut-norm via Grothendieck's inequality, in *Proceedings of the Thirty-sixth Annual ACM Symposium on Theory of Computing* (2004), pp. 72–80
4. N. Alon, A. Naor, Approximating the cut-norm via Grothendieck's inequality. SIAM J. Comput. **35**, 787–803 (2006)
5. N. Alon, R.A. Duke, H. Lefmann, V. Rödl, R. Yuster, The algorithmic aspects of the Regularity Lemma. J. Algorithms **16**, 80–109 (1994)
6. C. Ambühl, M. Mastrolilli, O. Svensson, Inapproximability results for maximum edge biclique, minimum linear arrangement, and sparsest cut. SIAM J. Comput. **40**, 567–596 (2011)
7. B.P.W. Ames, S.A. Vavasis, Nuclear norm minimization for the planted clique and biclique problems. Math. Program. Ser. B **129**, 69–89 (2011)
8. A. Billionnet, Solving a cut problem in bipartite graphs by linear programming: application to a forest management problem. Appl. Math. Modell. **34**, 1042–1050 (2010)
9. V. Branders, P. Schaus, P. Dupont, Mining a sub-matrix of maximal sum, in *Proceedings of the 6th International Workshop on New Frontiers in Mining Complex Patterns in conjunction with ECML-PKDD*, 2017
10. V. Branders, *Finding submatrices of maximal sum: applications to the analysis of gene expression data*, PhD thesis. UCL - Université Catholique de Louvain, Louvain-la-Neuve, Belgium, 2021
11. M. Braverman, K. Makarychev, Y. Makarychev, A. Naor, The Grothendieck constant is strictly smaller than Krivine's bound. *2011 IEEE 52nd Annual Symposium on Foundations of Computer Science*, 2011, pp. 453-462
12. E. Çela, B. Klinz, C. Meyer, Polynomially solvable cases of the constant rank unconstrained quadratic 0-1 programming problem. J. Combin. Optim. **12**, 187–215 (2006)
13. W. C. Chang, S. Vakati, R. Krause, O. Eulenstein, Exploring biological interaction networks with tailored weighted quasi-bicliques. BMC Bioinform. **13**, 1–9 (2012)
14. P. Chardaire, A. Sutter, A decomposition method for quadratic 0-1 programming. Manage. Sci. **41**, 704–712 (1995)
15. Y. Cheng, G.M. Church, Biclustering of expression data, in *Proceedings of the 8th International Conference on Intelligent Systems for Molecular Biology* (2000), pp. 93–100
16. M. Dawande, P. Keskinocak, J.M. Swaminathan, S. Tayur, On bipartite and multipartite clique problems. J. Algorithms **41**, 388–403 (2001)
17. M. Demange, P. Grisoni, V. Th. Paschos, Differential approximation algorithms for some combinatorial optimization problems. Theor. Comput. Sci. **209**, 107–122 (1998)
18. G. Derval, Finding maximum sum submatrices. Ph.D. Thesis, Louvain School of Engineering, UCL - Université Catholique de Louvain, Louvain-la-Neuve, Belgium, 2021
19. A. Duarte, M. Laguna, R. Martí, J. Sánchez-Oro, Optimization procedures for the bipartite unconstrained 0-1 quadratic programming problem. Comput. Oper. Res. **51**, 123–129 (2014)
20. L. Engebretsen, J. Holmerin, Clique is hard to approximate within $n^{1-o(1)}$, in *Proceedings of International Colloquium on Automata, Languages and Programming*, Geneva (2000), pp. 2–12

21. A. Frieze, R. Kannan, Quick approximation to matrices and applications. Combinatorica **19**, 175–220 (1999).
22. B. Gartner and J. Matoušek, *Approximation Algorithms and Semidefinite Programming* (Springer, New York, 2010)
23. S. Gaspers, D. Kratsch, and M. Liedloff, On independent sets and bicliques in graphs, Algorithmica **62**, 637–658 (2012)
24. N. Gillis, F. Glineur, Low-rank matrix approximation with weights or missing data is NP-Hard. SIAM J. Matrix Anal. Appl. **32**, 1149–1165(2011)
25. N. Gillis and F. Glineur, A continuous characterization of the maximum-edge biclique problem. J. Global Optim. **58**, 439–464 (2014)
26. F. Glover, Improved linear integer programming formulations of nonlinear integer problems. Manage. Sci. **22**, 455–460 (1975)
27. F. Glover, J.K. Hao, Efficient evaluations for solving large 0-1 unconstrained quadratic optimisation problems. Int. J. Metaheurist. **1**(2010), 3–10
28. F. Glover, G.A. Kochenberger, B. Alidaee, Adaptive memory tabu search for binary quadratic programs. Manage. Sci. **44**, 336–345 (1998)
29. F. Glover, M. Lewis, G. Kochenberger, Logical inequality implications for reducing the size and difficulty of quadratic unconstrained binary optimization problems. Eur. J. Oper. Res. **265**, 829–842 (2018)
30. F. Glover, Z. Lü, J.K. Hao, Diversification-driven tabu search for unconstrained binary quadratic problems. 4OR: Quart. J. Oper. Res. **8**, 239–253 (2010)
31. F. Glover, A.P. Punnen, The traveling salesman problem: new solvable cases and linkages with the development of approximation algorithms. J. Oper. Res. Soc. **48**, 502–510 (1997)
32. F. Glover, T. Ye, A.P. Punnen, G. Kochenberger, Integrating tabu search and VLSN search to develop enhanced algorithms: a case study using bipartite Boolean quadratic programs. Eur. J. Oper. Res. **241**, 697–707 (2015)
33. G. Gutin, A. Yeo, A. Zverovitch, Exponential neighborhoods and domination analysis for TSP, in *The Travelling Salesman Problem and Its Variations*, ed. by G. Gutin, A.P. Punne (Kluwer Academic Publishers, Boston, 2002)
34. G. Gutin, A. Yeo, Domination analysis of combinatorial optimization algorithms and problems, in *Graph Theory, Combinatorics and Algorithms*, Operations Research/Computer Science Interfaces Series, ed. by M.C. Golumbic, I.B.A. Hartman, vol. 34 (Springer, Boston, MA, 2005)
35. W.W. Hager, J.T. Hungerford, Continuous quadratic programming formulations of optimization problems on graphs. Eur. J. Oper. Res. **240**, 328–337 (2015)
36. P.L. Hammer, P. Hansen, Logical relations in quadratic 0-1 programming. Revue Roumaine Math. Pures et Appl. **26**, 421–429 (1981)
37. J.P. Hart, A.W. Shogan, Semi-greedy heuristics:an empirical study. Oper. Res. Lett. **6**, 107–114 (1987)
38. M. Hladík, M. Cerný, M. Rada, A new polynomially solvable class of quadratic optimization problems with box constraints. Optim. Lett. **15**, 2331–2341 (2021)
39. D. Karapetyan, A.P. Punnen, Heuristic algorithms for the bipartite unconstrained 0-1 quadratic programming problem (2012). http://arxiv.org/abs/1210.3684
40. D. Karapetyan, A.P. Punnen, A.J. Parkes, Markov chain methods for the bipartite Boolean quadratic programming problem. Eur. J. Oper. Res. **260**, 494–506 (2017)
41. S. Khot, A. Naor, Grothendieck-type inequalities in combinatorial optimization. Commun. Pure Appl. Math. **LXV**, 992–1035 (2011)
42. H. Konno, Maximization of a convex quadratic function under linear constraints. Math. Program. **11**, 117–127 (1976)
43. H. Konno, Maximizing a convex quadratic function over a hypercube. J. Oper. Res. Soc. Jpn. **23**, 171–188 (198)
44. H. Konno, T. Kuno, Y. Yajima, Parametric simplex algorithms for a class of NP-complete problems whose average number of steps is polynomial. Comput. optim. Appl. **1**, 227–239 (1992)

45. M. Koyuturk, A. Grama, N. Ramakrishnan, Compression, clustering, and pattern discovery in very high-dimensional discrete-attribute data sets. IEEE Trans. Knowl. Data Eng. 17, 447–461 (2005)
46. M. Koyuturk, A. Grama, N. Ramakrishnan, Nonorthogonal decomposition of binary matrices for bounded error data compression and analysis. BMC Bioinform. **32**, 1–9 (2006)
47. J.-L. Krivine, Sur la constante de Grothendieck. Comptes Rendus de l'Académie des Sciences Paris Ser. A-B. **284**, 445–446 (1977)
48. Z. Lü, F. Glover, J.K. Hao, A hybrid metaheuristic approach to solving the UBQP problem. Eur. J. Oper. Res. **207**, 1254–1262 (2010)
49. H. Lu, J. Vaidya, V. Atluri, H. Shin, L. Jiang, Weighted rank-one binary matrix factorization, in *Proceedings of the 2011 SIAM International Conference on Data Mining* (SDM), 2011
50. B. Lyu, L. Qin, X. Lin, Y. Zhang, Z. Qian, J. Zhou, Maximum Biclique search at billion scale. **PVLDB 13**, 1359–1372 (2020)
51. P. Manurangsi, Inapproximability of maximum edge Biclique, maximum balanced Biclique and minimum k-cut from the small set expansion hypothesis. Algorithms **11**, 12 (2018)
52. S.T. McCormick, M.R. Rao, G. Rinaldi, Easy and difficult objective functions for max cut. Math. Program. **94**, 459–466 (2003)
53. K.G. Murty, *Linear Programming* (Wiley, New York, 1983)
54. D. Nussbaum, S. Pu, J.R. Sack, T. Uno, H. Zarrabi-Zadeh, Finding maximum edge bicliques in convex bipartite graphs. Algorithmica **64**, 311–325 (2012)
55. J.B. Orlin, D. Sharma, Extended neighborhood: definition and characterization. Math. Program. Ser. A **101**, 537–559 (2004)
56. G. Palubeckis, Heuristics with a worst-case bound for unconstrained quadratic 0-1 programming. Informatica **3**, 225–240 (1992)
57. A. Pandey, G. Sharma, N. Jain, Maximum weighted edge Biclique problem on bipartite graphs, in *Algorithms and Discrete Applied Mathematics*, CALDAM 2020. Lecture Notes in Computer Science, ed. by M. Changat, S. Das, vol. 12016 (Springer, Cham, 2020)
58. R. Peeters, The maximum edge biclique problem is NP-complete. Discr. Appl. Math. **131**, 651–654 (2003)
59. A.P. Punnen, N. Kaur, Revisiting some classical explicit linearizations for the quadratic binary optimization problem. Res. Rep. Department of Mathematics, Simon Fraser Univerity, 2021
60. A.P. Punnen, N. Kaur, Fast heuristics for the quadratic unconstrained binary optimization problem. Res. Rep.. Department of Mathematics, Simon Fraser University, 2021.
61. A.P. Punnen, F.S. Margot, S.N. Kabadi, TSP heuristics: domination analysis and complexity. Algorithmica **35**, 111–127 (2003)
62. A.P. Punnen, K.P.K. Nair, Linear multiplicative programming. Opsearch **34**, 140–154 (1997)
63. A.P. Punnen, P. Sripratak, D. Karapetyan, Average value of solutions for the bipartite Boolean quadratic programs and rounding algorithms. Theor. Comput. Sci. **565**, 77–89 (2015)
64. A.P. Punnen, P. Sripratak, D. Karapetyan, The bipartite unconstrained 0-1 quadratic programming problem: Polynomially solvable cases (2012). arXiv:1212.3736v3
65. A.P. Punnen, P. Sripratak, D. Karapetyan, The bipartite unconstrained 0-1 quadratic programming problem: polynomially solvable cases. Discr. Appl. Math. **193**, 1–10 (2015)
66. S. Shahinpour, S. Shirvani, Z. Ertem, S. Butenko, Scale reduction techniques for computing maximum induced bicliques. Algorithms **10**, 113 (2017)
67. B.H. Shen, S. Ji, J. Ye, Mining discrete patterns via binary matrix factorization, in *Proceedings of the 15th ACM SIGKDD International Conference on Knowledge Discovery and Data Mining*, (2009)
68. P. Sripratak, The bipartite Boolean quadratic programming problem. Ph.D. Thesis, Simon Fraser University (2014)
69. P. Sripratak, A.P. Punnen, T. Stephen, The bipartite Boolean quadric polytope. Discr. Optim. **44**(1), 100657 (2022). https://doi.org/10.1016/j.disopt.2021.100657
70. J. Tan, Inapproximability of maximum weighted edge biclique and its applications. In *Proceedings of the 5th International Conference on Theory and Applications of Models of Computation* (TAMC'08), ed. by M. Agrawal, D. Du, Z. Duan, A. Li (eds.) (Springer-Verlag, Berlin, Heidelberg, 2008), pp. 282–293

71. A. Tanay, R. Sharan, M. Kupiec, R. Shamir, Revealing modularity and organization in the yeast molecular network by integrated analysis of highly heterogeneous genome-wide data. Proc. Natl. Acad. Sci. U.S.A. **101**, 2981–2986 (2004)
72. A. Tanay, R. Sharan, R. Shamir, Discovering statistically significant biclusters in gene expression data. Bioinformatics **18**, 136–144 (2002)
73. G. Tavares, New algorithms for quadratic unconstrained binary optimization problem (QUBO) with applications in engineering and social sciences. Ph.D Thesis, Rutgers University, 2008
74. D. Urošević, Y.I.Y. Alghoul, Z. Amirgaliyeva, N. Mladenović, Less is more: tabu search for bipartite quadratic programming problem, in *Mathematical Optimization Theory and Operations Research*, MOTOR 2019. Lecture Notes in Computer Science, vol. 11548, ed. by M. Khachay, Y. Kochetov, P. Pardalos (Springer, New York, 2019)
75. V.V. Vazirani, *Approximation Algorithms* (Springer, New York, 2010)
76. Y. Wang, Z. Lü, F. Glover, J.K. Hao, Backbone guided tabu search for solving the UBQP problem. J. Heurist. **19**(4), 1–17 (2011)
77. Y. Wang, Z. Lü, F. Glover, J.K. Hao, Path relinking for unconstrained binary quadratic programming. Eur. J. Oper. Res. **223**, 595–604 (2012)
78. Q. Wu, Y. Wang, F. Glover, Advanced tabu search algorithms for bipartite Boolean quadratic programs guided by strategic oscillation and path relinking. INFORMS J. Comput. **32**(1), 74–89 (2019)
79. E. Zemel, Measuring the quality of approximate solutions to zero-one programming problems. Math. Oper. Res. **6**, 319–332 (1981)

Chapter 11
QUBO Software

Brad D. Woods, Gary Kochenberger, and Abraham P. Punnen

Abstract In this chapter, we compile information regarding various QUBO solvers available. This includes general purpose MILP solvers, SDP solvers, codes for heuristics, exact algorithms, and test instances. We include information on solvers and test instances for problems that are equivalent to QUBO and the equivalent formulation can be obtained directly without significant effort. This includes the maximum cut problem, the maximum weight stable set problem, the maximum clique problem, and bilinear programs.

11.1 Introduction

In the preceding chapters we have discussed various algorithms for solving the quadratic unconstrained binary optimization (QUBO) problem and other related problems either by finding a provably optimal solution or by finding a heuristic solution. When evaluating the performance of heuristic algorithms, two metrics are most commonly used: solution quality, and computational time to reach the solution. While these metrics are important, there are other important aspects of software that must be considered in practice. In addition to cost and licensing concerns, a practitioner may also be concerned with the specific language that is used, whether or not a callable library is available, or the complexity of interfacing with large software packages. Other practitioners may be interested in utilizing exotic hardware (Quantum Annealers, Coherent Ising Machines, etc.) either now or

---

B. D. Woods
1Qbit Inc., Vancouver, BC, Canada

G. Kochenberger
Entanglement, Inc., Westminster, CO, USA
e-mail: Gary@entanglement.ei

A. P. Punnen (✉)
Department of Mathematics, Simon Fraser University, Surrey, BC, Canada
e-mail: apunnen@sfu.ca

A. P. Punnen (ed.), *The Quadratic Unconstrained Binary Optimization Problem*,
https://doi.org/10.1007/978-3-031-04520-2_11

in the future to solve QUBO or related problems, so software packages and services with these capabilities may be desired. Other software may be provided under the "Software As A Service" (SaaS) model, where the software is centrally hosted on cloud infrastructure. Here, we do not attempt to provide a comparative study of all available software, rather we attempt to provide the reader with a summary of available codes and present them in an organized manner.

In Chap. 1, we have seen how to write QUBO in other equivalent forms such as the maximum cut problem, the maximum weight stable set problem, the maximum weight clique problem, and bilinear programs. Since the transformation from QUBO to these models are somewhat straightforward, solvers developed for these problems can also be used to solve QUBO. Consequently, we also include a discussion of software for solving these equivalent forms. This is not an exhaustive list of all available software for QUBO, instead we present relevant information on a set of programs that we either have used or attracted our attention while exploring powerful solvers for QUBO.

We classify solvers into the following categories: General purpose solvers, Exact solvers, Heuristic solvers, QUBO Tools, Hardware for QUBO, and Benchmark Instances along with a list of miscellaneous items. There are many interesting codes and solvers related to QUBO available as part of the COIN-OR project, particularly solvers for nonlinear mixed integer programs. We are not listing all of them here but highlight a couple of relevant programs with a note that other similar programs are available.

11.2 General Purpose Solvers

1. **SCIP:** This is a powerful non-commercial suite for solving various mathematical optimization problems. The current version is 7.0.3, released on 12 August 2021. The software is available free of cost for academic work under ZIB academic licence. The licence terms are available at ZIB academic licence. The software requires an external LP solver and the default LP solver used is SoPlex. A variety of optimization models can be solved using SCIP directly or with an appropriate extension. Some specific problem classes that can be handled this way include mixed integer linear programs, mixed integer non-linear programs, constraint integer programs, Pseudoboolean optimization, Satisfiability and variations, multicriteria optimization, and Mixed-integer semidefinite programs. For details on this powerful and flexible software, we refer to
https://www.scipopt.org/
2. **LocalSolver:** This is a general-purpose C++ code for solving mathematical optimization problems. It supports linear, nonlinear, and combinatorial models involving high-level expressions like set variables, arrays, and external functions for black-box optimization. Max-Cut is used, among other combinatorial problems, to illustrate the use of the package.

The current release 10.5 has free 30 day trial and free academic licenses, as well as paid options. The software is available for download at:
https://www.localsolver.com/download.html

3. **Couenne:** This is a mixed integer non-linear programming solver, developed in C++ and part of the COIN-OR project. The solver is available under Eclipse Public License 1.0. For additional details, please visit the COIN-OR website https://www.coin-or.org/projects/ and for program download, please visit
https://github.com/coin-or/Couenne
4. **BONMIN:** This is an open-source C++ code for solving Mixed Integer Non-Linear Programming problems. It is available on COIN-OR under the Common Public License. For additional details, please visit the COIN-OR page of the software https://www.coin-or.org/projects/. BOMIN can also be accessed through NEOS web interface. For additional information on the underlying algorithm, please consult the technical report [4]. For program download, please visit
https://coin-or.github.io/Bonmin/
5. **SYMPHONY:** This is an open-source mixed integer linear program (MILP) solver and a callable library. It can be used to develop customized MILP solvers. SYMPHONY is available under Eclipse Public License 1.0. For additional information, please visit the COIN-OR page of the software https://www.coin-or.org/projects/. For software download, please visit
https://projects.coin-or.org/SYMPHONY
6. **GUROBI:** This is a popular general purpose commercial solver, that can solve linear programs, mixed integer linear programs, quadratic programs, and mixed integer quadratic programs. In Chap. 6 we have seen various mixed integer linear programming formulations of QUBO and a comparative analysis on the effectiveness of Gurobi in solving these and other formulations is given in [26]. Small size QUBO problems can be solved to optimality easily by Gurobi. For larger size problems, Gurobi can be used as heuristic by providing a time limit and collecting the best solution produced. Gurobi also has a binary quadratic program solver which can be used directly to solve QUBO of small size. The pre-solve processor embedded in Gurobi adds cuts automatically (see Chap. 4 for some valid inequalities and cuts.)
The software is free for academic research, through their academic license program. For commercial use, a paid version is required. For further information on the product and services, visit their website:
https://www.gurobi.com/
7. **CPLEX:** This is another popular general purpose commercial solver with somewhat similar capabilities as Gurobi. This can also solve linear programs, mixed integer linear programs, and integer quadratic programs.
The software is free for academic research, through their academic license program. For commercial use, a paid version is required. For further information on the product and services, visit their website:
https://www.ibm.com/analytics/cplex-optimizer

8. **LINDO:** This is a commercial software package capable of solving integer linear, nonlinear, and stochastic programming problems and also global optimization. Academic users can get the software free of cost. The software is available for download at:
https://www.lindo.com/
9. **BARON:** The solver BARON (Branch-And-Reduce Optimization Navigator) is a commercial software marketed by The Optimization Firm. It computes a global optimum of mixed integer non-linear programs. It uses the branch and bound framework and one can use other solves to handle subproblems. It also has many features, including capability to generate McCormick envelop inequalities (see Chap. 1). For additional details, please visit
https://www.theoptimizationfirm.com/home.

11.3 Exact Solvers for QUBO

1. **Some Max Cut:** This is a Python code to solve instances of the QUBO problem. The code finds the optimal solution by belief propagation on the clique intersection graph. The code is free to use and distribute for commercial use under the MIT license. The code repository can be found at:
https://github.com/ongmingyang/some-max-cut
2. **BiqCrunch:** This is a C code (using a FORTRAN library) for solving binary quadratic problems. It uses the branch and bound platform BOB [9] and the nonlinear optimization routine L-BFGS-B [23, 32]. The bounding procedure is based on an improved semidefinite programming algorithm. The code is released as a free and open-source software.
The code uses a format similar to the widely used sparse SDPA format, and contains converters from LP to the BiqCrunch format and from specific problem classes, such as Max-Cut, QUBO and Max-k-Cluster to the BiqCrunch format.
Users are able to submit problems to an online solver at:
https://biqcrunch.lipn.univ-paris13.fr/BiqCrunch/solver/
The current release can be obtained in tar.gz format ("BiqCrunch2.0.tar.gz") from the website:
https://biqcrunch.lipn.univ-paris13.fr/BiqCrunch/repository/solver/
3. **Maxcut:** This source has two codes, maxcut.MaxCutSDP which interfaces external solvers (such as SCS or CVXOPT) to solve the SDP formulation of the Max-Cut problem and maxcut.MaxCutBM which is an implementation of the Burer-Monteiro algorithm to solve a non-convex formulation of the Max-Cut problem [8].
https://github.com/pandrey-fr/maxcut
4. **BiqBin:** This is an exact solver for linearly constrained binary quadratic problems by converting the problem into a Max-Cut and solves the latter by a branch-and-bound algorithm. A parallel version is also available. Users can

upload their problem data to the software web interface. For additional details, we refer to:
http://www.biqbin.eu/

5. **Biq Mac Solver:** This is an implementation of the SDP based branch and bound algorithm solving QUBO [27], where $\mathbf{c}$ is assumed to be the zero vector and there are no restrictions on the diagonal entries of $\mathbf{Q}$. The input data is the $\mathbf{Q}$ matrix. After uploading data, the job will be queued and a time limit of three hours is set for the solver. Output will be communicated by email. By submitting the data, you also agree that the will be stored and could also appear in the Biq Mac Library. Contact information is biqmac@uni-klu.ac.at. For further details, visit the website:
http://biqmac.uni-klu.ac.at/
6. **MADAM:** This is a maximum cut exact solver with parallel capabilities, applying ADMM to solve the underlying SDP relaxations. It is one of the newest exact solvers for max cut (and hence QUBO) introduced by Hrga and Povh [18].

11.4 Heuristics for QUBO

1. **AlphaQUBO:** AlphaQUBO is developed by Fred Glover and Rick Hennig. It is managed and marketed by Meta-Analytics, Inc. AlphaQUBO is actively managed and updated frequently. It is a modern metaheuristic solver designed to solve large-scale QUBO models of the form $\mathbf{x}^T\mathbf{Q}\mathbf{x}$ (minimized or maximized) where $\mathbf{Q}$ is a symmetric matrix of constants. The current capacity is up to one million variables and can handle dense as well as sparse problems. It is a commercial software but a free version with a limit of up to 2500 variables is available on AWS Marketplace. The software is extensively tested and widely used to solve large-scale, real-world problems. Some computational experience with this software is included in Chap. 2. For additional information, please contact admin@meta-analytics.com or visit the web site
http://meta-analytics.com/.
2. **dwave-tabu:** This is a C++ code to compute an approximate solution of QUBO. It implements the multistart tabu search algorithm developed by Palubeckis [24]. The code is released under the Apache License 2.0.
The current release 0.4.2 has a Python interface that wraps the C++ implementation as a sampler for either QUBO or Ising QUBO problems. The code repository can be found at:
https://github.com/dwavesystems/dwave-tabu
3. **Qbsolv:** This is a C++ code to compute an approximate solution of QUBO. The code implements a tabu search heuristic which is used as a subroutine on QUBO subproblems. The algorithm is loosely based off the ideas by Glover [12]. The code is well-documented and released under the open source Apache License 2.0.

There is also a version of the code which has the ability to use a D-Wave Systems Quantum computer to solve the QUBO subproblems, which can be obtained from D-Wave Systems. The open source version is available at:
https://github.com/dwavesystems/qbsolv

4. **QCI qbsolv:** This a commercial C++ implementation of D-Wave Systems' qbsolv heuristic algorithm [5].
   The software is commercially accessible via a cloud service, and information on how to purchase access can be found at the website:
   https://www.quantumcomputinginc.com/qatalyst-core/
5. **OpenJij:** This is a C++ library for solving the Ising QUBO and QUBO through a Python interface. The library is intended to be used to investigate Quantum Annealing machines, and includes several annealing heuristic algorithms which can be run on CPU or GPU.
   The code is licensed under the Apache 2.0 license, and can be found at:
   https://github.com/OpenJij/OpenJij
6. **MQLib:** This is a open source collection of C++ codes implementing a variety of QUBO and Max-Cut heuristic algorithms presented in the academic literature. An experimental analysis of the implemented algorithms is presented by Dunning, Gupta and Silberholz [11]. Included in the package are scripts to test some newly developed heuristics, analysis tools and a set of benchmark instances. The library, released under the MIT license, can be found at:
   https://github.com/MQLib/MQLib
7. **Azure Quantum:** This is a cloud service providing a diverse set of quantum solutions and technologies. The service provides access to a number of heuristic QUBO solvers developed and maintained by 1QBit and Microsoft QIO, which can be run on CPU, GPU or FGPA. Toshiba has partnered to offer the Simulated Bifurcation Machine (SBM) [15] in Azure Quantum, but at the time of writing, this has not yet been made available. The code belongs to an open ecosystem and there is a free 30 day trial without an Azure membership. Details can be found at:
   https://azure.microsoft.com/
8. **BQUBO Markov chain algorithm:** This is heuristic algorithm to solve the bipartite quadratic unconstrained binary optimization problem (BQUBO) developed by Karapetyan, Punnen, and Parks [19]. It is a learning based algorithm. The source code and executables can be downloaded from:
   http://www.cs.nott.ac.uk/ pszdk/bqplib/BbqpCmcs.zip.

11.5 QUBO Tools

1. **dwave-ocean-sdk:** This is a C++ software development kit (SDK) containing tools for QUBO relating to the D-Wave Quantum Annealer [20]. Included are implementations of algorithms for preprocessing [7], finding exact solutions by

enumeration, heuristics [24], and bounding procedures [6] for QUBO and the Ising QUBO.
The code is released under the Apache 2.0 license, and is available at:
https://github.com/dwavesystems/dwave-ocean-sdk
2. **QUBOgen:** This is a Python code that converts several types of combinatorial optimization problems into QUBO problems. It implements the transformations presented by Glover, Kochenberger and Du [14]. The code repository can be found at:
https://github.com/tamuhey/qubogen
3. **pyQUBO:** This is a C++ code that expresses Ising Hamiltonians as QUBO. It is fully integrated with Ocean SDK. The code is released under the Apache 2.0 license and the repository can be found at:
https://github.com/recruit-communications/pyqubo
Also included is a Python Library for embedding various combinatorial optimization problems into QUBO or Ising QUBO form. For details of this library, we refer to the article [31].
4. **dimod:** This is a shared API for samplers written in Python and C++. It can handle discrete variables, linear inequality and equality constraints and samples with higher-order polynomial objectives. It is included in the DWave Ocean SDK.
The code is released under the Apache 2.0 license and can be found at:
https://github.com/dwavesystems/dimod

11.6 Hardware for QUBO

In this section we describe some of the custom built hardware using QUBO that has been brought to market, or is close to being publicly available.

1. **D-Wave Systems Advantage:** This is a quantum annealer which solves Ising problems that can be mapped onto the Pegasus graph structure [10], through the use of quantum entanglement. The Pegasus graph used in the D-Wave Systems quantum annealer contains 5640 qubits and 40484 connections. The quantum annealer can be accessed via the D-Wave Systems cloud service at:
https://www.dwavesys.com/solutions-and-products/cloud-platform/
2. **Fujitsu Digital Annealer:** This is a custom silicon (ASIC) implementation of a parallel-trial simulated annealing algorithm that supports up to 8192-bit, full connectivity, problems [1]. Fujitsu offers access to the device via paid access to their cloud service. A variety of support services are also offered. Details can be found at:
https://www.fujitsu.com/ca/en/services/business-services/digital-annealer/services/index.html
3. **Hitachi CMOS Annealer:** This is an ASIC implementation which implements an annealing machine with CMOS semiconductor circuits [30]. Due to the physical nature of the connections, Ising problems that can be mapped onto the

384 × 384 King's graph can be solved. A cloud service has been developed and released for free. Details can be found at:
https://annealing-cloud.com/en/index.html

4. **NTT LASOLV:** This is a Coherent Ising Machine (CIM) developed by NTT. This is a network of optical parametric oscillators to compute heuristic solutions to Ising problems [17, 29]. Access to the machine at time of this publication has been limited mainly to academic researchers through LCS a cloud computing cluster maintained and developed by NTT. Information on this system can be found at:
https://www.rd.ntt/e/research/JN20191127_h.html

11.7 Miscellaneous

1. **Systems Optimization Laboratory:** Various optimization solvers (linear and non-linear) can be obtained from the Systems Optimization Laboratory, Stanford university. This include among others, MINOS: a Fortran package that can be used for solving linear and non-linear optimization problems, QPOPT: a Fortran package for solving constrained linear least-squares and quadratic programming problems. If the quadratic is convex, the solver provides an optimal solution (for minimization) and when the quadratic is concave (minimization form) it produce a local optimum. Additional details on the available software can be obtained from.
http://stanford.edu/group/SOL/download.html
2. **CSDP:** This is a C Library for solving Semidefinite Programs that implements a predictor corrector variant of the semidefinite programming algorithm of Helmberg, Rendl, Vanderbei, and Wolkowicz. For SDP formulation of QUBO and the Ising QUBO we refer to Chap. 6 for using solutions of SDP relaxation to develop approximation algorithms, we refer to Chap. 8. This is part of the COIN-OR project and is available under the Eclipse Public License. The software is available for download at:
https://github.com/coin-or/csdp
3. **SDPA:** This is an efficient package for solving SDP based on the primal-dual interior-point method. It is written in C++ but python and MATLAB interfaces are also available. The software can be downloaded free of cost from:
http://sdpa.sourceforge.net/

11.8 Benchmark Instances

Experimental analysis of algorithms for QUBO requires representative standard test problems. Many authors contributed to generating test instances with specified properties. In this section, we present information test instance generators and benchmark problems publicly available.

1. **Biq Mac Library** This is a nearly comprehensive source for test problems for QUBO and the maximum cut problem. The library is maintained by Angelika Wiegele who can be contacted at angelika.wiegele@aau.at for additional details or for adding new problem instances. The details on the library and download links are available at:
   http://biqmac.uni-klu.ac.at/biqmaclib.html
   The library contains Beasley instances [2] (available through OR Library), Glover, Kochenberger and Alidaee instances [13] (also available through OR Library), and Billionnet and Elloumi instances [3]. All these instances are for QUBO (either maximization or minimization). Additional problems for maximum cut are also available. This includes Max-Cut instances generated from applications in statistical physics [21, 22].
2. **Verma & Lewis instances:** This dataset corresponds to cubic and quadratic pseudo boolean optimization problems. The name convention used for each file the set is: $data\_d\_n\_m\_k.txt$, wherein d, n, m, and k represents the degree, the number of nodes, the number of edges and the instance identifier respectively. The parameter $n = 200$ nodes and values of $m \in [250, 300, 350, 400, 450, 500, 1000]$ were used and five instances of each type were generated. In each test file, the three or four nodes that belong to each term in the optimization problem are chosen uniformly from the set of available variables. The corresponding coefficient is also chosen uniformly from the interval $[-10, 10]$. Experimental results using this data set is reported in [28].
   https://github.com/amitverma1509/QUBO
3. **OR Library instances:** These data files the Beasley instances. Experimental results using these instances were reported first in the paper [2] and include some large instances. The site is maintained by J. E. Beasley. The cite also contains Glover, Kochenberger and Alidaee instances.
   http://people.brunel.ac.uk/ mastjjb/jeb/orlib/bqpinfo.html
4. **Instance generators:** In [25] Pardalos presented fortran programs for constructing test instances for QUBO. It provides additional information on the difficulty levels and properties of the test problems and includes discussion on test problems of Gulati, Gupta and Mittal [16].
5. **BQUBO Instances:** These are test instances for the bipartite quadratic unconstrained binary optimization problem reported in Karapetyan, Punnen, and Park [19]. There are three classes of instances: small instances (56 mb), moderate instance (11 mb) and large instance (256 mb). The instances can be downloaded respectively from:
   http://www.cs.nott.ac.uk/ pszdk/bqplib/small.rar
   http://www.cs.nott.ac.uk/ pszdk/bqplib/moderate.rar
   http://www.cs.nott.ac.uk/ pszdk/bqplib/large.rar.

References

1. M. Aramon, G. Rosenberg, E. Valiante, T. Miyazawa, H. Tamura, H.G. Katzgraber, Physics-inspired optimization for quadratic unconstrained problems using a digital annealer. Front. Phys. **7**, 48 (2019)
2. J.E. Beasley, Heuristic algorithms for the unconstrained binary quadratic programming problem. Ph.D. Thesis, The Management School Imperial College, London, 1998
3. A. Billionnet, S. Elloumi, Using mixed integer quadratic programming solver for the unconstrained quadratic 0-1 problem. Math. Program. Ser. A **109**, 55–68 (2007)
4. P. Bonami, L.T. Biegler, A.R. Conn, G. Cornuejols, I.E. Grossmann, C.D. Laird, J. Lee, A. Lodi, F. Margot, N. Sawaya, A. Waechter, An algorithmic framework for convex mixed integer nonlinear programs. IBM Res. Rep. RC23771, Oct. 2005
5. M. Booth, J. Berwald, U. Chukwu, J. Dawson, R. Dridi, D. Le, M. Wainger, S.P. Reinhardt, QCI Qbsolv delivers strong classical performance for quantum-ready formulation (2020). Preprint. arXiv:2005.11294
6. E. Boros, P.L. Hammer, Pseudo-Boolean optimization. Discr. Appl. Math. **123**, 155–225 (2002)
7. E. Boros, P.L. Hammer, G. Tavares, Preprocessing of unconstrained quadratic binary optimization. RUTCOR Res. Rep. RRR 10-2006, April 2006, Rutgers University, USA
8. N. Boumal, V. Voroninski, A. Bandeira, The non-convex Burer-Monteiro approach works on smooth semidefinite programs, in *proceedings of NIPS 2016* (2016)
9. B.L. Cun, C. Roucairol, Bob: a unified platform for implementing branch-and-bound like algorithms. Tech. Rep., 95/16, Laboratoire PRiSM, Universite de Versailles-Saint Quentin en Yvelines, 78035 Versailles Cedex, 1995
10. N. Dattani, S. Szalay, N. Chancellor, Pegasus: the second connectivity graph for large-scale quantum annealing hardware (2019). Preprint. arXiv:1901.07636
11. I. Dunning, S. Gupta, J. Silberholz, What works best when? A systematic evaluation of heuristics for max-cut and qubo. INFORMS J. Comput. **30**, 608–624 (2018)
12. F. Glover, Heuristics for integer programming using surrogate constraints. Decis. Sci. **8**, 156–166 (1977)
13. F. Glover, G.A. Kochenberger, B. Alidaee, Adaptive memory tabu search for binary quadratic programs. Manage. Sci. **44**, 336–345 (1998)
14. F. Glover, G. Kochenberger, Y. Du., A tutorial on formulating and using qubo models (2018). Preprint. arXiv:1811.11538
15. H. Goto, K. Tatsumura, A. Dixon, Combinatorial optimization by simulating adiabatic bifurcations in nonlinear Hamiltonian systems. Sci. Adv. **5**(4), eaav2372 (2019)
16. V.P. Gulati, S.K. Gupta, A.K. Mittal, Unconstrained quadratic bivalent programming problem. Eur. J. Oper. Res. **15**, 121–125 (1984)
17. T. Honjo, T. Sonobe, K. Inaba, T. Inagaki, T. Ikuta, Y. Yamada, T. Kazama, K. Enbutsu, T. Umeki, R. Kasahara, K. Kawarabayashi, H. Takesue, 100,000-spin coherent ising machine. Sci. Adv. **7**(40), eabh0952 (2021)
18. T. Hrga, J. Povh, MADAM: a parallel exact solver for for max-cut based on semidefinite programming and ADMM. Comput. Optim. Appl. **80**, 347–375 (2021)
19. D. Karapetyan, A.P. Punnen, A.J. Parkes, Markov chain methods for the bipartite Boolean quadratic programming problem. Eur. J. Oper. Res. **260**, 494–506 (2017)
20. A.D. King, C.C. McGeoch, Algorithm engineering for a quantum annealing platform (2014). Preprint. arXiv:1410.2628
21. F. Liers, M. Jünger, G. Reinelt, G. Rinaldi, Computing exact ground states of hard ising spin glass problems by branch-and-cut, in *New Optimization Algorithms in Physics*, ed. by A. Hartmann, H. Rieger (Wiley, New York, 2004), pp. 47–68
22. F. Liers, Contributions to Determining Exact Ground-States of Ising Spin-Glasses and to their Physics. PhD thesis, Department of Mathematics and Computer Science, Universität zu Köln, 2004

23. J.L. Morales, J. Nocedal, Remark on algorithm 778: L-BFGS-B: fortran subroutines for large-scale bound constrained optimization. ACM Trans. Math. Softw. **38**, 1–4 (2011)
24. G. Palubeckis, Multistart tabu search strategies for the unconstrained binary quadratic optimization problem. Ann. Oper. Res. **131**, 259–282 (2004)
25. P.M. Pardalos, Construction of test problems in quadratic bivalent programming. ACM Trans. Math. Softw. **17**, 74–87 (1991)
26. A.P. Punnen, N. Kaur, Revisiting some classical explicit linearizations for the quadratic binary optimization problem. Res. Rep.. Department of Mathematics, Simon Fraser Univerity, 2021
27. F. Rendl, G. Rinaldi, A. Wiegele, Solving Max-Cut to optimality by intersecting semidefinite and polyhedral relaxations. Math. Program. **121**, 307–335 (2010)
28. A. Verma, M. Lewis, Optimal quadratic reformulations of fourth degree pseudo-Boolean functions. Optim. Lett. **14**, 1557–1569 (2020)
29. Y. Yamamoto, T. Leleu, S. Ganguli, H. Mabuchi, Coherent ising machines quantum optics and neural network perspectives. Appl. Phys. Lett. **117**, 160501 (2020)
30. M. Yamaoka, C. Yoshimura, M. Hayashi, T. Okuyama, H. Aoki, H. Mizuno, A 20k-spin ising chip to solve combinatorial optimization problems with cmos annealing. IEEE J. Solid-State Circ. **51**, 303–309 (2015)
31. M. Zaman, K. Tanahashi, S. Tanaka, PyQUBO: Python library for mapping combinatorial optimization problems to QUBO form (2021). arXiv:2103.01708
32. C. Zhu, R.H. Byrd, P. Lu, J. Nocedal, Algorithm 778: L-BFGS-B: fortran subroutines for large-scale bound-constrained optimization. ACM Trans. Math. softw. **23**, 550–560 (1997)

Index

A. P. Punnen (ed.), *The Quadratic Unconstrained Binary Optimization Problem*,
https://doi.org/10.1007/978-3-031-04520-2

www.ingramcontent.com/pod-product-compliance
Lightning Source LLC
Chambersburg PA
CBHW071947210726
48454CB00039B/68
* 9 7 8 3 0 3 1 0 4 5 1 9 6 *